QUANTIFYING MATTER

QUANTIFYING MATTER

Editor

Vinzent Fertig

scitus
academics

Quantifying Matter

Edited by **Vinzent Fertig**

This Book is Printed on Acid-Free Paper. ♾

ISBN: 9781681178561
Library of Congress Control Number: 2017948883

SCITUS Academics LLC,
616, Corporate Way, Suite 2, 4766,
Valley Cottage, NY 10989

www.scitusacademics.com

Notice

Printed in United States of America.

Preface

Matter is all around us and within us such as anything that we use, touch, eat, etc. is an example of matter. Matter can be as anything that takes up space, and it is composed of miniscule particles called atoms. All stuffs have properties that we can use to categorize them. In a similar way matter has two basic types of properties that we can associate with matter. These properties are known as Physical properties and Chemical properties. All properties of matter are either extensive or intensive and either physical or chemical. Extensive properties, such as mass and volume, depend on the amount of matter that is being measured. Intensive properties, such as density and color, do not depend on the amount of matter. Both extensive and intensive properties are physical properties, which mean they can be measured without changing the substance's chemical identity. Matter can have multiple dimensions or can be invisible to the naked eye. Different tools enable us to observe and record different types of matter and the different properties of the same matter. Familiarizing with the different types of tools for matter and what qualities of matter they gauge can help you to get a better hold on matter.

This book 'Quantifying Matter' illuminates how scientists learned to compute matter and quantify some of its most enthralling and functional properties.

Table of Contents

CHAPTER 1

A Novel Model of the Atom

Hans W. Giertz

Giertz Tech, Gnesta, Sweden

ABSTRACT

In this study all energy in the universe, here called energy quanta, originate from a singularity at the centre of the universe. These energy quanta have different frequencies and at each frequency the energy quanta can have positive or negative spin direction. There is a force of attraction between energy quanta which have exactly the same frequency but opposite spins. This is the dominating force in the universe and accounts for the strong nuclear force, the Coulomb force and the gravitational force. The universe contains one more basic entity; the oscillator quantum which absorbs and re-emits energy quanta at one specific frequency. The oscillator quantum can have positive or negative spin. Thus, there is a force of attraction between oscillator quanta with opposite spins and which amalgamates oscillator quanta into larger structures, i.e. particles (e.g. electron). These particles also have spin at a specific spin frequency and they have positive or negative spin. Thus they absorb and re-emit energy quanta at a frequency specific to the particle and where they can have positive or negative spin. This amalgamates particles into larger structures, e.g. quarks, neutron, proton and atomic nucleus. Using this model enables simple and stringent descriptions of elementary particle physics, electromagnetic theory, gravity, photon and inertial mass. The present model may be a step towards unification of elementary particle physics, general relativity, quantum physics and electromagnetic theory into one comprehensive theory.

Keywords: Electron, Atom, Gravity, Photon, Mass, Strong Force, Standard Model, Theory of Everything

1. INTRODUCTION

Here a novel theoretical model of the atom is proposed. The model builds on a universe which is rather simple. The centre of the universe emits vast amount of energy consisting of energy quanta. An energy quantum has a specific frequency and spin direction which can be positive or negative spin. Forces in the universe originate from these energy quanta. There is a repelling force between energy quanta which have the same frequency and spin direction. There is force of attraction between energy quanta which have the same frequency and opposite spin direction. The centre of the universe emits a broad spectrum of energy quanta with different frequencies and positive as well as negative spin.

Furthermore, the novel model builds on the proposal that the universe consists of only two basic entities; the energy quantum and the oscillator quantum. The oscillator quantum has spin and its spin frequency and spin direction creates resonance with energy quanta which have the same spin frequency and spin direction; thus, it absorbs and re-emits energy quanta with the same frequency and spin direction. Hence, the oscillator quantum is a forced damped oscillator. All oscillator quanta in the universe have the same spin frequency; however positive or negative spin. Thus, there is force of attraction between oscillator quanta with opposite spin because there is force of attraction between re-emitted energy quanta. Clusters of oscillator quanta also have spin at a specific frequency and positive or negative spin direction; thus, they absorb and re-emit energy quanta with the same spin frequency and spin direction. Hence, a cluster of oscillator quanta is also a forced damped oscillator. There is repelling force between clusters where the re-emitted energy quanta have the same spin frequency and spin direction. There is force of attraction between clusters where the re-emitted energy quanta have the same spin frequency and opposite spin direction. It is proposed that the atom consist of different clusters of oscillator quanta. A cluster can constitute an electron, a positron, a quark, a neutron, a proton and the atomic kernel. Each cluster has its specific spin frequency; however, it can have positive or negative spin. This creates force of attraction between electron and positron, between quarks and between neutrons and protons. These forces amalgamate the atom. The exchange of energy quanta which have opposite spin directions are what is called the strong force and the gluon in the Standard Model [1] .

Essential parts of this theory have already been published. However, three major changes are introduced which makes the theory simpler and more accurate. In this novel theory energy quanta have spin and there is a force between energy quanta which have the same frequency and opposite spin (attracting force) or the same spin (repelling force). In previous studies, [2] [3] , it was assumed that energy quanta did not have spin and that there was force of attraction between energy quanta having slightly different frequencies [2] . In this novel model the universe contains only two basic entities; the energy quantum and the oscillator quantum. All oscillator quanta in the universe have the same frequency and they have positive or negative spin; hence, there is only one unique oscillator quantum and it can have positive or negative spin. Previous reports, [2] [3] , assumed that there were a number of oscillator quanta with different frequencies and where each had specific properties. In this novel model different configurations of oscillator quanta provide unique properties like electron, positron, quarks, neutron and proton.

The above theory has been mapped previously on the electron [2] [3] , although with the constraints mentioned above. These reports describe electromagnetic theory in a novel way entirely based on the electron's absorbed and re-emitted energy quanta. The electron charge Q is the re-emitted energy quanta flow density at distance $r = 0$ and the current i is the transport, with velocity v, of this re-emitted energy quanta flow density. The displacement D (electric field) is the re-emitted energy quanta flow density at distance r and the magnetic field H(r), from transported electrons, is the time derivate of observed re-emitted energy quanta flow density at distance r. This previously published electromagnetic theory was compared with traditional electromagnetism [4] [5] . It has been shown how energy quanta originating from the centre of the universe can be measured as well as the direction towards this source [3] . It has been shown that the gravitational force is the result of energy quanta exchanged between bodies [2] . It has been shown that the photon is enforced energy hv super positioned on energy quanta [2] and which was compared with the state of the art model of the photon [6] . It has also been shown that inertial mass is the result of enforced energy hv super positioned on re-emitted energy quanta [2] . However, in [2] gravity, photon and inertial mass were described with the constraints described above. Therefore, an enhanced description of gravity, photon and inertial mass is included in the present theoretical model. This novel description of inertial mass is here compared with today's description based on the Higgs boson and the Higgs field [7] -[11] . The present theoretical model is fundamentally different to the Standard Model. Here all energy originates from the centre of the universe and the atom is complete passive; it absorbs and re-emits energy quanta which results in stored energy, forces and fields. The Standard model, quantum physics [12] , general relativity [13] , electromagnetic theory [4] [5] and string theory [14] describes a different universe where energy, forces and fields are inherent properties of the electron and the atom, detached from external influence. A comparison with these disciplines is made.

2. THEORETICAL MODEL

2.1. General

Today physics is described by different models: the Standard Model [1] , quantum physics [12] , general relativity [13] , electromagnetic theory [4] [5] and string theory [14] . These theories have a common denominator. It is assumed that the characteristics of the electron and atomic nucleus, such as charge, fields, forces and energy, are intrinsic properties. Hence, today's physics claims that they are generated without external influence [1] [4] [5] [12] -[14] . The exception is the Higgs Field and the Higgs Boson [7] -[9] .

This paper proposes that it is quite the opposite. The atom in itself contains no forces or energy, what so ever. The atom is completely passive. All energy in the universe originates from a singularity, positioned at the centre of the universe. This energy is absorbed by the electron and the atomic nucleus and the absorbed and re-emitted energy result in phenomena observed in physics: stored energy, forces, current and fields.

This Section is structured in the following way. It starts with a description of energy quanta. A model of the electron and the positron is presented, including a description of absorbed and re-emitted energy quanta. The same model is then utilized to explain the structure of the nucleus and the strong nuclear force. The strong nuclear force is created by a process similar to the Coulomb force, albeit by energy quanta and oscillator quanta with frequencies specific to the nucleus and the strong nuclear force.

The same model is once again utilized to explain gravity. Gravity is created by a process similar to the Coulomb force, albeit by energy quanta and oscillator quanta with frequency specific to gravity.

The same model is finally utilized to explain the photon and inertial mass.

2.2. Singularity

According to the theoretical model the universe contains a singularity. This singularity has been described previously [2] [3] , although with the constraints described in Section 1. This singularity generates energy quanta at different frequencies and they can have positive or negative spin. The energy quantum has no mass and propagates with the speed of light. An energy quantum is described by its vector δS_ω in the present study, where ω denotes its frequency ω. The energy quantum δS_ω remains unaltered as it travels through space. The singularity emits energy quanta δS_ω uniformly into space creating energy flow density $S_\omega(0)$ at the singularity and energy flow density $S_\omega(R)$ at the distance R. Energy quanta δS_ω are radiated uniformly into space, which results in that $S_\omega(R)$ decreases with the area of the sphere

$$S_\omega(R) = S_\omega(R=0)\frac{1}{4\pi R^2}. \tag{1}$$

The energy quanta can have positive and negative spin and which is denoted δS_ω^+ respectively δS_ω^-.

2.3. Oscillator Quantum

It is proposed that the universe contains only two basic entities: energy quanta and one oscillator quantum. The oscillator quantum is a forced damped oscillator with natural frequency ω_o. It absorbs energy quanta $\delta S_{\omega o}$ and energy flow density $S_{\omega o}(R)$, having the same frequency as the oscillator quantum natural frequency. However, the energy quanta have positive spin $\delta S_{\omega o}^+$ or negative spin $\delta S_{\omega o}^-$ which implies that the oscillator quantum also has positive or negative spin. An oscillator quantum absorbs only energy quanta having the same spin frequency and spin direction. The absorbed energy quanta cannot be stored in the oscillator quantum and are consequently re-emitted uniformly into space as energy flow density $S_{\omega o}^+(r)$ respectively $S_{\omega o}^-(r)$ at the distance r from the oscillator quantum. Hence, the energy flow density re-emitted by the oscillator quantum is

$$S_{\omega o}(r) = \alpha S_{\omega o}(R=0)\frac{1}{4\pi R^2}\cdot\frac{1}{4\pi r^2}, \tag{2}$$

where α is a constant which mirrors the oscillator quantum transfer function. However, at the earth the distance R to the singularity is constant. Equation (2) can, thus, be simplified into

$$S_{\omega o}(r) = S_{\omega o}(r=0)\frac{1}{4\pi r^2}. \tag{3}$$

From now on $S_{\omega o}(r=0)$ is denoted $S_{\omega o}(0)$. Equation (3) describes the energy flow density from one oscillator quantum. There is repelling force between two energy quanta which have the same frequency and spin direction and force of attraction between two energy quanta which have the same frequency but opposite spin directions. Now we picture an energy quantum, $\delta S_{\omega o}$, with frequency ω_o and which passes close to an oscillator quantum with frequency ω_o. The energy quantum imposes an impulse and force $\delta F_{\omega o}$ on the oscillator quantum. This force is directed towards the propagating energy quantum when the energy quantum and the oscillator quantum have opposite spins. This force is directed along the propagating energy quantum when the energy quantum and the oscillator quantum have the same spin. The force $\delta F_{\omega o}$ on the oscillator quantum, as a result of a passing energy quantum $\delta S_{\omega o}$, is

$$\delta \boldsymbol{F}_{\omega o} = \beta \delta \boldsymbol{S}_{\omega o}, \tag{4}$$

where β is a constant. Energy flow density $S_{\omega o}(r)$ re-emitted and propagating from another oscillator quantum, at the distance r, towards this oscillator quantum imposes a force $F_{\omega o}(r)$ on the oscillator quantum

$$\boldsymbol{F}_{\omega o}(r) = \eta \boldsymbol{S}_{\omega o}(r), \tag{5}$$

where η is a constant. According to Equation (3) this can be re-written into

$$\boldsymbol{F}_{\omega o}(r) = \eta S_{\omega o}(0)\frac{\boldsymbol{r}}{4\pi r^3}. \tag{6}$$

The total force acting on n oscillator quanta is

$$\boldsymbol{F}^{n}_{\omega o}(r) = \eta n S_{\omega o}(0)\frac{\boldsymbol{r}}{4\pi r^3}. \tag{7}$$

The total energy flow density $S_{\omega o}(r)$, re-emitted by m oscillator quanta in a cluster, is $mS_{\omega o}(r)$. Consequently, the force $\boldsymbol{F}^{mn}_{\omega o}(r)$ imposed from one cluster of m oscillator quanta, acting on another cluster of n oscillator quanta at distance r, is

$$\boldsymbol{F}^{mn}_{\omega o}(r) = \xi_o mn\frac{\boldsymbol{r}}{4\pi r^3}, \tag{8}$$

where ξ_o is a constant, valid for oscillator quanta with frequency ω_o. Here r is much larger than the distances within each cluster of oscillator quanta. Equations ((5)-(8)) result in repelling force when oscillator quanta in the two clusters have the same spin and force of attraction when they have opposite spins.

2.4. Electron and Positron

It is proposed that the electron consists of two oscillator quanta with opposite spin directions. This creates a strong force between the two oscillator quanta according to Equation (8) and which amalgamates them into a configuration. This configuration of two oscillator quanta also has spin at its spin frequency ω_e, where e denotes electron. Consequently, the configuration of two oscillator quanta is a forced damped oscillator with natural frequency ω_e. The spin is negative, whereby it absorbs energy quanta and re-emits energy flow density $\boldsymbol{S}_{\omega e}^{-}(r)$ in the same way as the oscillator quantum, with the exception that the absorbed and re-emitted energy quanta have the frequency ω_e. The re-emitted energy quanta flow density $S_{\omega e}^{-}(0)$ is the electron charge and the re-emitted energy quanta flow density $\boldsymbol{S}_{\omega e}^{-}(r)$ is the displacement D. The transport of $S_{\omega e}^{-}(0)$ with velocity v is the current and the observed time derivate of $\boldsymbol{S}_{\omega e}^{-}(r)$ at distance r, from travelling electrons, describes the magnetic field H(r). This has been described previously in detail [2] [3] ; however with the constraints described in Section 1. Electrons have the same spin and consequently, they repel each other according to Equation (8). However, where the equation is valid at the electron frequency ω_e and for the electron constant ξ_e;

$$\boldsymbol{F}_{\omega e}^{mn}(r) = \xi_e mn \frac{\boldsymbol{r}}{4\pi r^3}.$$

This expresses the Coulomb force. Hence, there is no principle difference between the

force created between oscillator quanta and the force created between electrons other than that they operate in two different frequency domains.

The positron is identical to the electron; however, it has positive spin and the absorbed energy quanta and the re-emitted energy flow density $\boldsymbol{S}_{\omega e}^{+}(r)$ have positive spin. Consequently, two positrons repel according to Equation (8) because they have the same spin. There is force of attraction between a positron and an electron because they have the same spin frequency and opposite spin directions.

2.5. Atom

It is of course impossible to describe the atom in its exact details. The following is an attempt to describe a common principle which may explain layers of the atom. The description is schematic and simplified and the purpose is only to highlight principles.

It is proposed that the atom contains entities which are smaller than the quark; here they are called electron/positron and micro quark (μq). The electron/positron is a configuration (called ep) which consists of one electron and one positron, amalgamated by the force of attraction according to Equation (8). This configuration has spin at the frequency ω_{ep} and the spin can be positive or negative. Thus, the configuration is a forced damped oscillator with natural frequency ω_{ep}. The configuration absorbs and re-emits energy quanta with the frequency ω_{ep}. Hence, there is force of attraction between electron/positron configurations which have opposite spin directions. A μq can for instance consist of twelve electron/positron configurations oriented in a ring and where adjacent configurations have opposite spins, i.e. the ring consist of ep^+, ep^-, ep^+, ep^-, ep^+ etc. Hence, there is a total force of attraction between adjacent electron/positron configurations in the ring according to Equation (8). This configuration has a spin in itself at its frequency $\omega_{\mu q}$ where μq denotes micro quark. The configuration absorbs energy quanta and re-emits energy flow density $S_{\omega\mu q}(r)$ in the same way as the oscillator quantum, the electron and the electron/positron configuration, with the exception that the absorbed and re-emitted energy quanta have the frequency $\omega_{\mu q}$. The μq can have positive or negative spin. There are three different micro quarks and where the configuration of ep are somewhat different (e.g. different number of electron/positron configurations in the ring) which result in that each type of μq has its specific spin frequency, i.e. $\omega_{\mu q1}$, $\omega_{\mu q2}$ and $\omega_{\mu q3}$ and where each can have positive or negative spin directions.

The configuration may resemble one of the most common structures in chemistry; the biphenyl ring where two rings of each six carbon atoms are connected forming a structure of twelve carbon atoms (and 10 hydrogen atoms) in total. By studying common structures in chemistry we may have an insight into the atom. Physics may be simple and repetitive; guarding those structures which exhibit maximum stability or minimum energy level.

There are three different quarks. The quarks may consist of rings, each consisting of twelve micro quarks (μq) with the same spin frequency. One quark, here called quark type 1 (q_1) may consist of a ring containing the same type of μq, e.g. μq_1. Adjacent μq_1 have opposite spin, i.e. the ring consist of μq_1^+, μq_1^-, μq_1^+, μq_1^-, μq_1^+ and μq_1^- etc. Hence, there is a total force of attraction between adjacent μq in the ring according to Equation (8). This configuration has spin in itself at its natural frequency ω_q, where q denotes quark. The configuration absorbs energy quanta and re-emits energy flow density $S_{\omega q}(r)$ in the same way as the oscillator quantum, the electron, the electron/positron configuration and the μq, with the exception that the absorbed and re-emitted energy quanta have the frequency ω_q. The quark q_1 can have positive and negative spin; hence, they are called q_1^+ and q_1^-. A similar configuration of micro quarks with frequency $\omega_{\mu q2}$constitutes the quark type 2 which can have positive and negative spin called q_2^+ and q_2^-. A third configuration of micro quarks with frequency $\omega_{\mu q3}$ constitutes the quark type 3 called q_3^+ and q_3^-. All three quarks have the same spin frequency ω_q; however, they can have positive or negative spin.

The proton may consist of one q_1^+, one q_2^- and one q_3^+. Thus, there is a total force of attraction within the proton which amalgamates the three quarks according to Equation (8). The proton also contains an embedded positron or one of the μq contains a surplus electron. The proton has spin in itself at its natural frequency ω_p, where p denotes proton. The configuration absorbs energy quanta and re-emits energy flow density $S_{\omega p}(r)$ in the same way as the oscillator quantum, the electron, the micro quarks and the quarks, with the exception that the absorbed and re-emitted energy quanta have the frequency ω_p. The proton can have positive and negative spin.

The neutron may consist of one q_1^-, one q_2^+ and one q_3^-. Thus there is a total force of attraction within the proton which amalgamates the three quarks according to Equation (8). The neutron also contains an embedded positron and an embedded electron or two of the μq contain one surplus electron respectively one surplus positron. The neutron has spin in itself at its frequency ω_n, which is the same as the protons spin frequency. The neutron can have positive or negative spin. The configuration absorbs energy quanta and re-emits energy flow density $S_{\omega n}(r)$ in the same way as the oscillator quantum, the electron, the micro quarks and the quarks, with the exception that the absorbed and re-emitted energy quanta have the frequency ω_n.

The atomic nucleus consists of different configurations of protons and neutrons, depending on type of chemical element. Adjacent protons and neutrons have opposite spin which ensures force of attraction between adjacent protons and neutrons according to Equation (8). The total force of attraction decreases with larger configurations which explains why chemical elements with large atomic weight are less stable. The repelling force between positrons embedded in protons plays a part in the total force; however, smaller than the force created by $S_{\omega np}(r)$.

We now realize one extremely important property of the nucleus. There is a force of attraction within each layer; however, weaker forces between layers. The reason is that each layer operates at a different frequency or in different frequency domains, i.e. each layer consists of forced damped oscillators with natural frequency specific to the layer. Consequently, the interaction between layers is small resulting in a strong total force within each layer. A structure where all layers where operating on the same frequency would result in a total force close to zero. This is the enabler of the strong force or the strong forces.

The atomic nucleus has spin in itself at a frequency which is specific to the chemical element, ω_{ce}, where ce denotes chemical element. The nucleus absorbs energy quanta and re-emits energy flow density $S_{\omega ce}(r)$ in the same way as the oscillator quantum, the electron, micro quarks, quarks, protons and neutrons, with the exception that the absorbed and re-emitted energy quanta have the frequency ω_{ce}. It is proposed that every electron, orbiting the nucleus, constitutes a spin in itself (i.e. the orbital constitutes the spin) and the electron has the same spin frequency, ω_{ce}, as the nucleus; however, opposite spin. Consequently, all orbiting electrons are synchronized to each other. Every electron has spin in itself with frequency ω_e or wavelength λ_e. The only allowed orbitals with radius r are those who satisfy $n\cdot\lambda_e = 2\pi r$. The force of attraction between positrons embedded in protons and orbiting electrons plays a part in the total force between nucleus and orbiting electrons together with the force created by $S_{\omega ce}(r)$ according to Equation (8). Consequently, the electron orbital in e.g. the helium atom is different from the first orbital of e.g. the lead atom. All electron orbitals in a chemical element have the same spin frequency ω_{ce}. From this follows that electrons in the outer orbitals have more kinetic energy than those in the lower orbitals.

2.6. Gravity

This Section explains that the force called gravity is created by a mechanism similar to the mechanism which generates forces between electrons and positrons and within the atomic nucleus; however, gravity encompasses energy quanta with frequency specific to gravity. Gravity based on the present theory has been described previously [2] , although with the constraints described in Section 1. The present theoretical model provides a more stringent and simpler description of gravity. Neutrons, protons, quarks, positrons and electrons contain oscillator quanta with natural frequency ω_0. Neutrons, protons and electrons always contain equal amount of oscillator quanta with positive spin respectively with negative spin, hence there is always the same amount of oscillator quanta with positive spin and with negative spin in bodies. Now we picture two bodies where the distance between the bodies is much larger than the distance between oscillator quanta within the bodies. There is repelling force between oscillator quanta with equal spin and force of attraction between oscillator quanta with opposite spins, according to Equation (8). These forces would normally level out resulting in a total force equal to zero. However, it is proposed that the force of attraction is slightly larger than the repelling force; this results in a small force of attraction, i.e. the gravitational force. Hence, the force F of attraction between two bodies at distance r containing m respectively n oscillator quanta is

$$\boldsymbol{F}_{\text{gravity}}(r) = -\xi_g mn \cdot \frac{\boldsymbol{r}}{4\pi r^3}, \tag{9}$$

where ξ_g is the gravitational constant and the ? sign denotes force of attraction. Equation (9) is derived from Equation (8).

The number of oscillator quanta m in body 1 respectively n in body 2 is proportional to the gravitational mass m_1 of body 1 and the gravitational mass m_2 of body 2. We now recognize the similarity with Newton's metric law of gravity, where G is Newton's gravitational constant [5]

$$F_{\text{gravity}}(r) = \xi_g mn \cdot \frac{1}{4\pi r^2} \approx F_{\text{Newton}} = G m_1 m_2 \frac{1}{r^2}. \tag{10}$$

We also recognize the similarity with Coulomb's law $F_{\text{Coulomb}}(r) = \frac{Q_1 Q_2}{4\pi\varepsilon_0 r^2}$. The reason is that the gravitational force is similar to the force between clusters of electrons, the reason being that both derive from Equation (8).

2.7. Photon

The nature of the photon has been described previously [2] , although with the constraints described in Section 1. The process creating photons is described here in condensed form. Particles, e.g. electron, neutron and proton contain two or many oscillator quanta with natural frequency $\omega_o$as described in Sections 2.4 and 2.5. Each oscillator quantum absorbs energy quanta $\delta S_{\omega o}$ and re-emits energy quanta $\delta S_{\omega o}$. Accelerating or decelerating an electron implies that force and energy is enforced on the electron. This energy can be described as one or many hν. The energy hν cannot be stored in the electron and its oscillator quanta. The energy hν are, as a result, superpositioned on re-emitted energy quanta $\delta S_{\omega o}$. One energy hν is superpositioned (in the oscillator quantum) on one re-emitted energy quantum $\delta S_{\omega o}$ at a time. This new energy quantum $\delta S_{\omega o+\nu}$ has the frequency ω_o and ν. Radio waves, light and gamma rays are energy hν super positioned on energy quanta $\delta S_{\omega o}$. The re-emitted energy quanta can have positive and negative spin, which implies that the photon $\delta S_{\omega o+\nu}$ also can have positive and negative spin. This is confirmed by the classical definition of photon spin [5] [6] .

2.8. Gravitational and Inertial Mass

Gravitational and inertial mass have been described previously [2] , although with the constraints described in Section 1. The process behind gravitational and inertial mass is described here in condensed form.

Gravitational mass is the expression of the mechanism, i.e. atomic oscillator quanta, which absorb and re-emit energy quanta $\delta S_{\omega o}$ and energy flow density $S_o(r)$. This results in gravitational force according to Equations ((9) and (10)). Consequently, the gravitational mass m_1 in Equation (10) is proportional to the number m of oscillator quanta with natural frequency ω_o (creating gravity) in the mass or the body.

Accelerating or decelerating a mass, a particle, an electron or an atom demands enforced force and energy. The mass, particle, electron or atom cannot store this enforced energy. The enforced energy can be described as many hv. The enforced energy hv are superpositioned on re-emitted energy quanta, comprising the energy quantum $\delta S_{\omega o+v}$ as described in Section 2.7. Hence, enforced energy hv are emitted as photons. Thus, inertial mass is the expression of the mechanism, i.e. atomic oscillator quanta, with natural frequency ω_o, which convert enforced energy hv into emitted photons $\delta S_{\omega o+v}$. The inertial mass is proportional to the number m of oscillator quanta in the mass. Consequently, the inertial mass is proportional to the gravitational mass, although inertial and gravitational mass originate from totally different processes. The process creating inertial mass is identical to the process creating photons, as described in Section 2.7. The number of oscillators in a mass or body is normally extremely large. This results in that the emitted photons have very low frequency v. Photons can be detected as TEM waves (Transverse Electromagnetic waves). However, TEM waves with very low frequency cannot be measured with state of the art instruments.

3. DISCUSSION AND CONCLUSIONS

According to the present study the atom is completely passive; forces, fields and energy are not atomic inherent properties. All properties of the atom are the result of external energy quanta which are absorbed and re-emitted by oscillator quanta and different configurations of oscillator quanta. This enables a simple description of atom, electromagnetic theory, gravity, photon and inertial mass. There is a clear distinction between energy, forces and fields on one hand and matter and particles on the other hand. Energy is mediated by energy quanta and sometimes by super positioned energy hv. Forces and fields are mediated by energy quanta. Matter and particles always consist of different configurations of oscillator quanta, amalgamated by absorbed and re-emitted energy quanta.

The Standard Model [1] , quantum physics [12] , general relativity [13] , electromagnetic theory [4] [5] and string theory [14] use a different approach. There is no external influence on the atom in those sciences, except for the Higgs field [7] -[11] . Forces, fields and energy are atomic inherent properties and there is no supply of energy from the outside. For instance the electron radiates displacement (the electric field) [4] [5] which consist of energy; however, there is no supply of energy to the electron. It has been shown that this is theoretically impossible; it is in fact a perpetual mobile machine [3] . The atom radiates gravitons [1] , which consist of energy; however, there is no supply of energy to the atom, which is a perpetual mobile machine.

State of the art physics describe forces in very different ways; the Coulomb force is mediated by an electric field [4] [5] , the strong nuclear force is mediated by a particle (the gluon) [1] , in the Standard Model the gravitational force is mediated by a particle (the graviton) [1] and in string theory the gravitational force is mediated by one-dimensional strings [14] . In the present study these three forces are described by the same mechanism; the interaction between energy quanta. These three forces are derived from the same equation (Equation 8); however, where each operates in its own frequency domain.

In state of the art physics the photon is an elementary particle and the force needed to accelerate a body (the inertial mass) is described by a scalar field (the Higgs Field) and a scalar boson (the Higgs boson) [7] -[9] . Attempts have also been made to extend the scalar Higgs field to incorporate other areas, e.g. gravitons [10] [11] . In the present study there is no difference between the mechanism generating photons and the mechanism resulting in inertial mass; in both cases enforced energy hv is super positioned on re-emitted energy quanta and which constitutes the photon. In the present study scalar fields do not exist; the energy quantum is a vector and hence, fields are vector fields.

State of the art elementary particle physics make little distinction between energy, force and matter; all three are often described as particles. The photon, which mediates energy, is a particle [1] [6] . The gluon and the graviton, which mediate forces, are particles [1] . Matter like the electron, quarks, neutron and proton are also particles [1] . In the present study there is a clear difference between 1) energy which is mediated by energy quanta or super positioned energy hv on energy quanta, 2) forces which are created by the interaction between energy quanta which have the same frequency and 3) matter which always consist of configurations of oscillator quanta and which are amalgamated by absorbed and re-emitted energy quanta.

In the Standard Model the quark is described by its flavors: up, down, strange, charm, top, bottom, red, blue and green [1] . In the present study only three entities are introduced: energy quanta, oscillator quantum and configurations of oscillator quanta and where all three have frequency and spin. The oscillator quantum and its configurations constitute forced damped oscillators with different natural frequencies. The definitions of energy, forced damped oscillator, frequency, natural frequency and spin are well known [4] [5] .

Numerous different particles have been discovered when splitting larger particles, e.g. the neutron. It is proposed that many of these "particles" are fragments of different configurations (e.g. μq and quarks) contained in the destroyed particle. These fragments have, initially, a spin frequency; however, the spin frequency does not equal the frequency of any energy quanta. Thus, it is unable to absorb energy quanta, resulting in that it rapidly decays. It is proposed that many observed "particles", which rapidly decay, are not particles; they are fragments which do not fulfill the criteria of absorbing and re-emitting energy quanta. They should be regarded as noise. Using these criteria will probably ease the understanding of the atom.

The future will show if the atom is extremely complex, or quite simple. If it is reasonably simple, there may exist a theory which amalgamates elementary particle physics, quantum physics, general relativity, string theory, astrophysics, electromagnetic theory and the forces within chemistry into one comprehensive theory; Theory of Everything, TOE.

CITE THIS PAPER

Hans W. Giertz (2016) A Novel Model of the Atom. *Journal of Modern Physics*,**07**,413-421.

REFERENCES

1. Burgess, C. and Moore, G. (2007) The Standard Model: A Primer. Cambridge University Press, Cambridge.
2. Giertz, H.W. (2015) Journal of Modern Physics, 6, 157-167.
3. Giertz, H.W. (2015) Journal of Basic and Applied Research International, 9, 2395-3438

4. Bleaney, B.I. and Bleaney, B. (1965) Electricity and Magnetism. Oxford University Press, Amen House, London.
5. Harris, R. (2008) Modern Physics. 2nd Edition, Pearson International Edition, Addison Wesley, Boston.
6. Sipe, J.E. (1995) Physical Review A, 52, 1875-1883.
7. Higgs, P.W. (1964) Physical Review Letters, 11, 508-509.
8. Higgs, P.W. (1964) Physical Letters, 12, 132-133.
9. Englert, F. and Brout, R. (1964) Physical Review Letters, 13, 321-323.
10. Oda, I. (2010) Physics Letters B, 690, 322-327.
11. Bezrukov, F. and Shaposhnikov, M. (2008) Physics Letters B, 659, 703-706.
12. Eisberg, R. and Resnick, R. (1985) Quantum Physics of Atoms, Molecules, Solids, Nuclei, and Particles. 2nd Edition, Wiley, New York.
13. Einstein, A. (1916) Annalen der Physik, 49, 285-339.
14. Polchinski, J. (1998) String Theory. Cambridge University Press, Cambridge

CHAPTER 2

Model of the Pulsing Atom

Heinrich Ehrlich

Chemical Department, Lomonosov Moscow State University, Moscow, Russia

ABSTRACT

In this work, we reanalyzed the movement of an electron in the electrostatic field of nucleus. The trajectory of the electron's motion is an ellipse with a minor semiaxis, tending towards zero. From a mathematical point of view the movement of an electron in such an orbit will be equivalent to the oscillation of an electron. The action produced by electrons in movement between stationary points is discrete and proportional to a Planck constant. This condition sets the allowable values of the electron energy and the radius of their orbit. Electrons on the same shell perform symmetric synchronous oscillations. Their frequency is of the order of 10^{16} Hz. Most of the time the electrons are located on the periphery of the atom, periodically they simultaneously rush to the nucleus, the atom rapidly compresses and immediately decompresses, i.e. pulsates. The model gives Bohr formula for the energy of single-electron atom and suitable values of ionization potentials of the atoms of the second period of the Periodic Table.

Keywords: Atom Structure, Quantum Theory, Oscillation, Bohr-Sommerfeld Model, Ionization Potentials

1. INTRODUCTION

A little over one hundred years ago, Niels Bohr [1] introduced the planetary model of the atom—with electrons revolving in circular orbits around a nucleus. Arnold Sommerfeld extended this model to include elliptic orbits [2] . The main contradiction of these models was—that in line with Maxwell theory the electron moving with acceleration in an electrostatic field of nucleus must emit electromagnetic waves, and thus loose energy. Bohr by passed this restriction by postulating the existence of certain special orbits, along which the electron can move without electromagnetic radiation, and introduced the action quantization rule for movement along these orbits. This postulate looked (and looks) insufficiently substantiated, and it has been rejected alongside the Bohr model, with onset of the era of wave mechanics.

Yet one more option has remained unanalyzed. Within the ambit of the classical Kepler problem, reducing electron's angular momentum leads to the contraction of elliptic orbit—so at M à 0 the trajectory of the electron will aspire to a straight line.

At the point of maximum distance from the nucleus, the speed of the electron equals zero. This allows taking a new look at the radiation problem during electron's accelerating movement in the field of nucleus. Preliminary estimates show that the action, A, made by the electron during movement at atomic distances has the order of a Planck constant, h. According to Max Planck's classic quantum theory there exists a minimal value of action, h. So the action of the electron on the trajectory between two stationary points cannot not be less than h. Thus, condition A = h sets electron orbit parameters with a minimal distance from the nucleus. The electron moves in such an orbit with acceleration, nevertheless, it cannot emit radiation, because this would lead to reduction in its energy and speed, and thus reducing its action below minimum values practically possible.

This logical conclusion looks like the Bohr postulate on the existence of special orbits of an electron. Yet they differ essentially. While continuously moving in circular orbit, an electron gradually produces more and more action, and in excess of the minimum value (h) gains the ability to emit energy in the form of electromagnetic waves. The fact that this doesn't occur should certainly be postulated. In our model, movement happens between stationary points, while the lack of radiation is a direct consequence of its minimal action.

Reasoning in the same way, we reach a conclusion about the existence of a second orbit of an electron, and movement in such orbit between stationary points satisfies condition A = 2h. In this case the action A = h can transform into electromagnetic radiation, the "residual" action of an electron will be equal to the minimal value, i.e. h. Continuing with this line of reasoning, we reach a conclusion about the existence of a set of orbits of an electron in an atom meeting the general condition A = n * h. For single-electron systems (hydrogen-like atoms) the levels with n > 1 correspond to the excited states of atom. Experimental data show that the lifetime of such excited states many times exceeds the orbital period of the electron (see below for a more detailed explanation). This indicates a critical feature of the action of the Maxwell electromagnetic theory at the micro level. The imperative nature of its action peculiar to macro world, at the micro level is transformed to a probable one. An electron moving with acceleration and action exceeding the minimum value h generates a quantum of electromagnetic radiation only with a certain probability which is significantly smaller 1.

We can assume that the emission of a photon occurs at the time of passing of an electron at the minimum distance from a nucleus, the "over flight" of the nucleus, when the electron's speed and acceleration reach maximum values. And there the absorption of a photon takes place. This mechanism permits to remove one more contradiction of the Bohr model which cannot explain how the electron skips from one orbit to another at the emission or absorption of a photon.

The electron which is at the maximum distance from a nucleus cannot absorb or emit a photon. It can be removed from its orbit mechanically, for example, after impact with an external electron or other particle, including collision with other atoms.

In conclusion, let's discuss the problem of multielectron atoms, in which electrons settle into orbits which have various values of action. Though the probability of emission of a photon at the movement of an electron with $A > h$ is very small, sooner or later it will occur, and as a result all electrons have to pass into an orbit with $A = h$. But as we know from experience, electronic configurations of multielectron atoms in their basic state are steady, and in the absence of external influences it is possible to speak, in principle, about their infinite stability. This can be explained through the operation of the following mechanism. A photon is emitted when an electron moving on a "far" orbit passes near a nucleus is absorbed by the electron moving on a "near" orbit and also being at this moment near a nucleus. As a result, these electrons seemingly exchange orbits, and at the same time the electronic configuration of the atom in general remains invariable.

The time which an electron spent near a nucleus is insignificantly small in comparison with the orbital period, and during this time the vector of speed changes the direction to the opposite. So, from a mathematical point of view the movement of an electron in such an orbit will be equivalent to the oscillation of an electron along the straight line connecting it to a nucleus with a perfectly elastic reflection from the nucleus. Here after for brevity's sake we will speak about the oscillations of electrons.

2. HYDROGEN ATOM AND HYDROGEN-LIKE ATOMS

Let's now consider the motion of an electron in the nucleus field along the line connecting their centers. Since the masses of the nucleus and electron differ by three orders, it is possible to disregard the nucleus motion in the first approximation. Furthermore, we disregard the nucleus's own dimensions and the ones of the electron, and treat them as material points.

According to Newton's second law:

$$m\frac{d^2x}{dt^2} = -\frac{az}{x^2}$$

where x—distance to the nucleus, z?nuclear charge.

$$a = e^2/4\pi\varepsilon_0$$

Multiply by dx/dt and integrate:

$$\frac{m}{2}\left(\frac{dx}{dt}\right)^2 = \frac{az}{x} - E$$

where E—integration constant, which has the dimension of energy.

$$\frac{dx}{dt} = \frac{\sqrt{2}}{\sqrt{m}}\sqrt{\frac{az}{x} - E} \tag{1}$$

The action of the electron during movement from $x = R_1 = az/E$ to the nucleus ($x = 0$):

$$A = \int_{R_1}^{0} p\,dx = \int_{R_1}^{0} mv\,dx = \sqrt{2m}\int_{R_1}^{0}\sqrt{\frac{az}{x} - E}\,dx$$

$$= \sqrt{2m}\left(\sqrt{x}\sqrt{az - Ex} + \frac{az}{\sqrt{E}}\arcsin\sqrt{\frac{Ex}{az}}\right)\Bigg\{\begin{matrix}0\\R_1\end{matrix}$$

$$= \sqrt{2m}\frac{az}{\sqrt{E}}\frac{\pi}{2} = \frac{\sqrt{2m}ze^2}{8\varepsilon_o\sqrt{E}} \tag{2}$$

For the unidirectional oscillatory motion of the electron its action should be a divisible of h/2:

$$\frac{\sqrt{2mze^2}}{8\varepsilon_o\sqrt{E}} = \frac{h}{2}n$$

From this:

$$E = \frac{me^4z^2}{8\varepsilon_o^2h^2n^2} \tag{3}$$

which coincides with the Bohr equation for the energy spectrum of an electron in a hydrogen-like atom.

Maximum distance of an electron from the nucleus is given by:

$$R_n = \frac{2\varepsilon_o h^2 n^2}{\pi m e^2 z} \tag{4}$$

For the first level of the hydrogen atom R_1 = 106 pm, which is twice the Bohr radius; and the same as the diameter of a hydrogen atom according to Bohr, which is logical for this method of calculation. In our model it's meaningless to talk about the radius of the hydrogen atom, but for multielectron atoms the term acquires a clear physical meaning. Multielectron atoms possess the shape close to spherical one with the radius equal to the utmost limit of the electrons' movement.

The equation of electron motion is obtained from the differential Equation (1). Separating the variables and integrating, we obtain:

$$t = \frac{\sqrt{m}}{\sqrt{2E}}\sqrt{x}\sqrt{ax - Ex} + \frac{\sqrt{m}}{\sqrt{2}}\frac{az}{E^{3/2}}\arcsin\sqrt{1 - \frac{Ex}{az}} \tag{5}$$

The half-period of oscillation we obtain substituting in (5) extreme values $x = R_1 = az/E$ and $x = 0$:

$$\tau_1 = \frac{2\varepsilon_o^2 h^3}{e^4 z^2 m} = \frac{h}{4E}$$

For the first level of the hydrogen atom this value equals 0.76×10^{-16} s.

In general case:

$$\tau_n = \frac{2\varepsilon_o^2 h^3 n^3}{e^4 z^2 m} \tag{6}$$

This implies that most of the time an electron is located near the border of the atom: one-third of the time in the range of $(0.93 - 1)\cdot R_1$ and a half in the range of $(0.84 - 1)\cdot R_1$. Similar with the approach of quantum mechanics, we can talk about the probability of finding the electron at a certain distance from the nucleus.

On the other hand, the location of the electron can be characterized by some average distance from the nucleus:

$$x_{av} = \frac{\int_0^{\tau_1} x\,dt}{\tau_1} = 0.75 \cdot R_1 \tag{7}$$

This ratio can prove useful in calculating the characteristics of atoms and molecules because it permits approximation of some electrons in an atom or an atom in the whole by a "static" model, in which electrons are located at some fixed distances from the nucleus.

The phase of maximum distance from the nucleus is also very interesting. $R_n \sim n^2$, $\tau_n \sim n^3$. From this it follows that the electron in highest excited levels is located very far away from the nucleus and for an extremely long time, as compared to the basic level. If we monitor such an electron, we can easily take it for a free electron that is not associated with any particular atom, moving with a slow speed and exchanging energy with other "free" electrons. And suddenly this electron takes off and begins to move with ever-increasing speed to its "own" nucleus; then pushes away from it, and after some time again goes into a "free" movement state. It seems to us that this physical picture is very similar to the quantum entanglement.

3. HELIUM ATOM AND TWO-ELECTRON (2E) ATOMS

The stable oscillations of two electrons must be synchronized. There are two possible modes of oscillations: symmetric, when the electrons are moving simultaneously to/from the nucleus; and asymmetric, in which the location of one electron near the nucleus corresponds to the maximum distance from the nucleus of the second electron.

For symmetric oscillations of two electrons in 2e-atoms:

The problem is solved in explicit form:

$$m\frac{d^2x}{dt^2} = -\frac{az}{x^2} + \frac{a}{(2x)^2} \qquad (8)$$

$$\frac{m}{2}\left(\frac{dx}{dt}\right)^2 = \frac{az}{x} - \frac{a}{4x} - E = \frac{a}{x}(z - 0.25) - E$$

Hence, for helium and 2e?atoms we have: an atom radius $R_1 = R_H/(z - 0.25)$, where R_H?the radius of the hydrogen atom, z?the nuclear charge;

Energy of an electron, in eV, $E = 13.596 \cdot (z - 0.25)^2$, where 13.596 eV? energy of electron detachment from the lowest level of the hydrogen atom.

The energy of the system, consisting of two electrons (energy of 2e-atom), equals to:

$$E_2 = 2 \times 13.596 \times (z - 0.25)^2.$$

The energy required for the detachment of the first electron, or the first ionization potential, $I_1 = E_2 - E_1$, where E_1?the energy of a corresponding hydrogen-like atom, $E_1 = 13.596\ z^2$.

For helium we have $E_2 = 83.24$ eV, $I_1 = 28.83$ eV, which exceeds the experimental data by 5.4% and 17.2%, respectively. With increasing nuclear charge deviation from the literature data decreases and for neon, for example, it is +1.0% and + 2.2% respectively.

Evaluation of electrons' energy at the asymmetric oscillations gives the following values for helium: $E_2 = 90.44$ eV, $I_1 = 36.03$ eV, which is higher than the experimental data by 14.5% and 46.5%, respectively.

Thus, an option of the symmetric oscillations looks preferable. In general, the proposed model of the atom can be called pulsing. Most of the time the electrons are located on the periphery of the atom, periodically they simultaneously rush to the nucleus, the atom rapidly compresses and immediately decompresses, i.e. pulsates.

The accurate solution of the differential equation can be obtained for two-electron atoms. In principle, this can be done also for multielectron systems, but it generates some difficulties and inconveniencies. Therefore, below we have used numerical solution methods.

4. MULTIELECTRON ATOMS

Only two electron scan occupy the first energy level. Perhaps in the future it will be possible to justify this statement, yet we accept it as given, as a postulate— which follows from the totality of experimental data on the structure of atoms.

Thus, all the electrons of the atoms of the second period, with the exception of two electrons, are in the second energy level, for which A = 2·h.These electrons make synchronous symmetric oscillations, that is, have the same oscillation frequency, and approximately the same energy and the same range of oscillations (see below).In the proposed model there is no reason for the separation to s- and p-electrons, and all electrons of this energy level are equivalent.

To calculate the energy of an electron in a multielectron atom of the second period of the Periodic table we have used the equation:

$$\int_{R_2}^{0}\left(\frac{za}{x}-\frac{1}{2}E_{e-e,in}-\frac{1}{2}E_{e-e,ex}-E_j^2\right)^{1/2}dx=\frac{2h}{2\sqrt{2}\sqrt{m}} \quad (9)$$

where E_j^2 –the energy of an electron in the second level containing j electrons;

$E_{e-e,in}$ –energy of interaction of the electron with other electrons of the second level;

$E_{e-e,ex}$ –energy of interaction of the electron with the first level electrons.

The calculation algorithm is as follows.

In the first step we calculate the energy of interaction between the outer electrons, which depends on their relative position. It is convenient to characterize this interaction by parameter z', so that $1/2\cdot E_{e-e,in}=z'\cdot a/R$.

Assessment of the interaction energy with the first level electrons is carried out using a static model with an arrangement of these electrons at a distance of 0.75·R_1 from the nucleus and their approximating by a sphere with a charge of −2. Then, in accordance with the laws of electrostatics we can divide the integral (9) into two:

$$\int_{R_2}^{0.75R_1}\left(\frac{(z-1)a}{x}-\frac{z'a}{x}-E_j^2\right)^{1/2}dx+\int_{0.75R_1}^{0}\left(\frac{za}{x}-\frac{z'a}{x}-\frac{a}{0.75R_1}-E_j^2\right)^{1/2}dx=\frac{2h}{2\sqrt{2}\sqrt{m}} \tag{10}$$

where R_1—radius of the first shell. Calculations showed that the R_1 value is not dependent on the presence and the number of electrons at the second level, and coincides with the value R_1 of the corresponding helium-like (2e) atom. Substituting the values z' and R_1 into Equation (10), we find numerically the value E_j^2 and the value R_2corresponding to it.

In the second step we calculate the energy of the first level electrons, which unlike the radius depends on the number and location of the electrons at the outer level. Here again we use the static model and approximate the outer electrons by a sphere with radius 0.75·R_2 and a charge ?j. Calculations are carried out according to the equation, which is a combination of (9) and (10):

$$\int_{R_1}^{0}\left(\frac{za}{x}-\frac{z'a}{x}-\frac{\frac{j}{2}a}{0.75R_2}-E_2^1\right)^{1/2}dx=\frac{h}{2\sqrt{2}\sqrt{m}}$$

where $z' = 0.25$.

In the third step, we calculate the total energy of (je, 2e)-atom:

$$E_j = 2 * E_2^1 + j * E_j^2$$

In the fourth step, we calculate the ionization energy of (je, 2e)-atom, or (n ? j + 1)-the atom' stepwise potential containing n electrons on the filled outer level.

$$I_{n-j+1} = E_j - E_{j-1}$$

This is to say, we avoid adiabatic approximation and assume that the energy state of the inner level electrons also changes in the ionization process. The ionization energy is calculated as the energy difference between the "equilibrium" states of the atom.

The main uncertainty in this model arises from the choice of mutual spatial arrangement of electron orbits—an electron configuration. A configuration with the maximum possible distance between the electrons looks natural, since it provides a minimum energy of electron repulsion and thus a minimum of the potential energy of the atom. E.g., for 4, 6 and 8 electrons such optimal configurations are the tetrahedron, the octahedron and the cube, respectively. However, calculation of ionization energy potential of various atoms for such configurations shows substantial differences from the published (experimental and theoretical) data.

We have considered various options for the location of the electrons in space. The best results were obtained with a cubic system: eight cells are located in the corners of a cube and electrons consistently fill them with the formation of the following configurations.

This approach may look rather artificial: but, firstly, it is very similar to the ideas of Gilbert Lewis [3] and Irving Langmuir [4] about the structure of electron shells. Secondly, if we look at the configuration with five electrons, we see something like today's accepted configuration for the nitrogen atom—with two paired s-electrons and three unpaired p-electrons, and so on. Of course, this scheme is very idealized and in reality these configurations may be highly distorted (even beyond recognition), but it is useful, in particular, to consider chemical bonding.

When analyzing the configurations presented in Figure 1 we see that the electrons there are not equivalent (beside 2 and 8 electrons configurations), in particular, they have different energies of interaction with neighboring electrons. On the other hand, in the proposed model of a pulsing atom, all the electrons on the same shell should oscillate with the same frequency, i.e. should be equivalent in this sense. This can be achieved by assigning different energy to different electrons and, accordingly, a different radius of orbit. But in this case different electrons will produce different actions during one movement cycle. How will this affect the action quantization rule? We propose to use here the principle of summed action: in a system of interacting particles the summed action of particles is quantized. This is in tune with Planck's idea of quantization in the system of interacting resonators [5] . As for calculations, this principle permits to use average energy of electron repulsion and characterize the electrons on the same shell by the average energy values and the radius of their orbit - which greatly simplifies the calculations. It should be noted that differences in the characteristics of different electrons in the atom are not as great as they might seem at a first glance at the electron configuration. For example, for nitrogen atoms, the half-period of oscillations of outer shell electrons equals 25.52×10^{-18} s, average energy-78.31 eV, radius of the orbit-85.5 pm. Each electron has the energy of 76.08, 78.34, 78.34, 78.34, and 80.40 eV with the orbit radius 84.3, 85.5, 85.5, 85.5, and 86.6 pm, correspondingly, that is, the deviation from the average values does not exceed 3% for energy and 1.5% for radius.

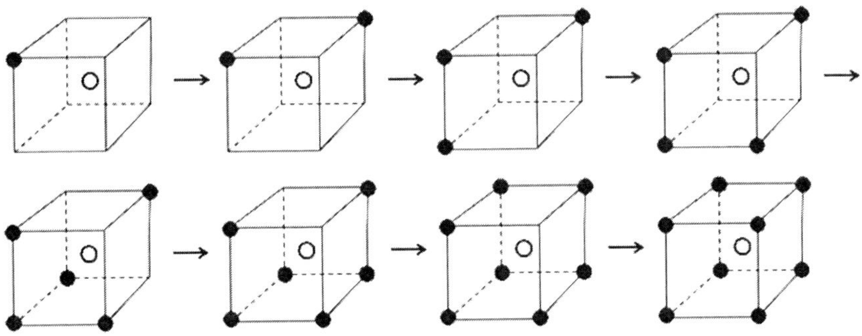

Figure 1. The order of filling of the second electron shell of an atom.

Table 1. The calculated ionization potentials (in eV) of the atoms of the second period of the Periodic Table and their deviation from the literature [6] values (in%).

Element	I_1	I_2	I_3	I_4	I_5	I_6	I_7	I_8
Li	5.27 −2.2							
Be	9.00 −3.4	18.60 +2.1						
B	10.26 +23.6	25.28 +0.5	39.46 +4.0					
C	11.45 +1.7	28.23 +15.8	48.35 +1.0	66.77 +3.5				
N	15.77 +8.5	31.33 +5.8	52.97 +11.6	78.30 +1.1	100.96 +3.1			
O	8.17 −40.0	37.39 +6.5	57.95 +5.5	84.39 +9.0	115.09 +1.0	141.77 +2.6		
F	11.35 −34.8	29.53 −15.6	67.55 +7.8	91.12 +5.6	123.06 +7.8	158.61 +0.9	189.48 +2.3	
Ne	16.28 −24.5	36.28 −11.4	58.93 −7.1	103.59 +6.7	131.52 +4.2	168.04 +6.4	208.99 +0.8	244.07 +2.1

We have calculated the total energy of the electron shells of all the elements of the second period of the Periodic table for different configurations of shells. Their deviation from published values does not exceed 1%. For ionization potentials (Table 1) the discrepancy from the published data is much higher, but overall is quite acceptable. We note also that the ionization potential is calculated as the difference between two large numbers, e.g., the first ionization potential of neon is only 0.5% of the total energy of the electron shell, so the error in its calculation must be inevitably large.

5. CONCLUSIONS

Thus, the proposed model of a pulsing atom can be summarized in the following theses.

1) An electron in an atom is considered as a material object moving along a particular trajectory.

2) The trajectory of the electron's motion is an ellipse with a minor semiaxis, tending towards zero. From a mathematical point of view the movement through such an orbit is equivalent to oscillation with a perfectly elastic repulsion from the nucleus.

3) At the point of maximum distance from the nucleus the speed of the electron drops to zero (stationary point). This distance corresponds to the major semiaxis of the elliptic trajectory. We call this the radius of the orbit. For outer shell electrons this is equivalent to the radius of the atom.

4) The action produced by electrons in movement between stationary points is discrete and proportional to a Planck constant, A = n·h, where $n = 1, 2, 3, \cdots$. This condition sets the allowable values of the electron energy and the radius of their orbit.

5) An electron moving in an orbit with A = h, does not emit electromagnetic waves. When moving in orbits with A > h, the emission of a photon is possible, and it occurs at the maximum speed of an electron's motion, near the nucleus. Emission is probabilistic in nature. A photon emitted by an electron can be absorbed by another electron which is also moving close to the nucleus.

6) Most of the time the electron is located on the periphery of the orbit (the atom), and periodically rushes to the nucleus and rebounds back. The average distance of the electron from the nucleus is 0.75·R, where R—radius of the orbit. The frequency of oscillation of the electron is of the order of 10^{16} Hz.

7) Electrons in multielectron atoms are located on the electron shells corresponding to A = n·h, where $n = 1, 2, 3, \cdots$. It is postulated that the first shell can be occupied by 2 electrons, the second—by 8 electrons, etc.

8) Electrons on the same shell perform symmetric synchronous oscillations.

9) The frequency of their oscillation is the same, and in this sense all the electrons on one shell are equivalent. Separation into s-, p-, etc., electrons is excluded.

10) To calculate the energy of multielectron atoms, the principle of summed action is used: in a system of interacting particles the summed action of particles is quantized.

CITE THIS PAPER

Ehrlich, H. (2017) Model of the Pulsing Atom. Advances in Materials Physics and Chemistry, 7, 188- 197.

REFERENCES

1. Bohr, N. (1913) On the Constitution of Atoms and Molecules. Philosophical Magazine, 26, 1-24.
2. Sommerfeld, A. (1916) Zur Quantentheorie der Spektrallinien. Annalen der Physik, 51, 1-94.
3. Lewis, G.N. (1916) The Atom and the Molecule. Journal of the American Chemical Society, 38, 762-785.
4. Langmuir, I. (1919) The Arrangement of Electrons in Atoms and Molecules. Journal of the American Chemical Society, 41, 868-934.
5. Planck, M. (1901) Uber das Gesetz der Energieverteilungim Normalspektrum. Annalen der Physik, 4, 553-563.
6. Sansonetti, J.E., Martin, W.C. and Yong, S.L. (2004) Handbook of Basic Atomic Spectroscopic Data. National Institute of Standards and Technology, Gaithersburg, USA.

CHAPTER 3

Nucleon in Nuclear Matter

Aparajita Bhattacharya*, A. Sagari, B. Chakrabarti, A. Chandra

Department of Physics, Jadavpur University, Calcutta, India

ABSTRACT

The modification of the properties of nucleon in nuclear medium has been investigated in the context of flux tube model. A nucleon has been described as diquark-quark system connected by flux tube and quasi particle diquark model has been used to describe the diquak constituting the nucleon. The modification of incompressiblity, the Roper resonance etc in the nuclear medium have been investigated. The results are compared with recent experimental and theoretical predictions. Some interesting observations are made.

Keywords: Flux Tube, Diquark, Nuclear Matter

1. INTRODUCTION

EMC effect [1] hints at the modification of the properties of the nucleon in the nuclear environment particularly swelling of nucleon nucleus. The experiment shows that the effect is more as the nuclei become larger and denser. The swelling has largely been interpreted as the effect of attraction of the neighbouring quarks to the quarks constituting the nucleon. This in turn predicts that there are fewer high momentum quarks in the nucleon when it is in nuclear matter as swelling means larger wavelength and small momentum. Recently Seely et al [2] have reported that the modification of the particle properties do not depend on the mass or density of the nucleus as a whole but only on the immediate neighbourhood within nucleus i.e. the EMC effect depends on the local environment. A number of works have been done in the context of different of models with varied dynamical theory to explain the EMC effect as the swelling of nucleon in the nuclear medium [3]. Oka et al [4] have shown from the general ground that a composite quantum system will swell (shrink) in an external attractive (repulsive) potential and have concluded that the nucleon in a nucleus swells due to the attraction of the nuclear force. Dukelsky et al [5] have investigated the EMC effect incorporating mean field acting on the valence quark of nucleon embedded in nuclear medium. They have obtained a swelling of 2 percent which is much lower than the other estimates [4,6]. Kim et al [7] have investigated the modification of nucleon form factor in nucleus. They have estimated anomalous magnetic moment from the modified form factor and found that it contributes more to the transverse structure function than the longitudinal one. Wu et al [8] have studied the modification of the nucleon properties in the Quark Meson Coupling mode (QMC) [9] which includes the effect of the motion of the quark inside the nucleus. They have studied the effective mass, radius, correlated charge density distribution of a nucleon in nuclear matter and have observed that the quark structure plays a crucial role in the description of the nuclear matter and finite nuclei. Mathieu et al [10] have studied the modification of hadron properties in the context of the flux tube model. They have suggested that nuclear binding and EMC effect are the consequences of the partial deconfinement. In the present work we have investigated the modification of the properties of a nucleon in nuclear medium in the context of the flux tube model. A diquark-quark configuration connected by a flux tube has been considered for a nucleon. The quasi-particle diquark model of diquark suggested by us has been used to estimate the diquark mass. The compressibility, Roper resonance, critical density, swelling of the nucleon in the nuclear medium has been investigated. Some interesting observations are made. In the flux tube model a nucleon is described as a quark and a diquark system connected by a flux tube [10]. The Hamiltonian for an isolated nucleon can be written as:

$$H = \frac{p^2}{2m} + \sigma r - \frac{4\alpha_s}{3} \tag{1}$$

where 'm' is the reduced mass of the system, p is the relative momentum, ‘r’ is the distance between quark and diquark, σ and α_s are the string tension and the strong coupling constant respectively. When the nucleon is in nuclear matter the Hamiltonian changes due to the effect of the neighbouring nucleons. In the presence of other nucleon the flux tubes are topologically rearranged such a way that the linear potential of the system is minimum. Considering the effect of one perturbing nucleon, the linear potential of the nucleon can be modified to [10]:

$$V_{Leff}(r) = \sigma r - \frac{\sigma\rho}{6}\frac{\pi}{24}\int I\psi(r)I^2$$

$$\left(t^2 - 4u^2\right)^2 \theta\left(t - 2u\right) d^3 r' \tag{2}$$

where r and r' are the variables that represent the length of the flux tube for the two-nucleon clusterings, t and u are functions of r and r′ as in Mathieu et al [10]. $\psi(r)$ is typical nucleon wave function, ρ is the nuclear matter density, σ is string constant. The one gluon exchange term also would be modified partially due to the presence of the other nucleon and may be represented as in [10]:

$$V_{Ceff}(r) = -\frac{4\alpha_s}{3} - \frac{4\alpha_s\rho}{18}$$

$$\int I\psi(r)I^2\left(\frac{2\pi}{3}\right)(1-z)\left(r^2 - r'^2 - 4rr'\right)d^3\, r' \tag{3}$$

where $z = \cos(\boldsymbol{R}, u)$, here R is the distance between two nucleons [10]. The Hamiltonian can thus be expressed as:

$$H = \frac{p^2}{2m} + V_{Leff}(r) + V_{Ceff}(r) \tag{4}$$

In general the contribution to the potential runs as:

$$V = \int V_{eff}(r)\, I\psi(r')\, I^2 d^3\, r' \tag{5}$$

In computing the Hamiltonian we have used the quark-diquark configuration for a nucleon assuming diquark behaving as a quasi particle suggested by us to estimate the diquark mass. With diquark within, baryon can be described as two-body (involving q-qq) system. In the present study we have considered the proton as a $[ud]_0$ (scalar) diquark and a u quark system and investigated the modification of the properties of the proton in the nuclear environment.

2. THE QUASI PARTICLE MODEL OF DIQUARK

Recently we have proposed a quasi-particle picture of the diquark [11] in analogy to the quasi-particle in usual condensed matter physics. It is well known that the quasi-particles are particle like entities arising in some system of interacting particles. They are low-lying excited states possessing the energy very close to the ground state and the properties of the system and the prpperties of the system can be obtained by investigating the properties of the quasi-particles to a considerable extent. Such states are often observed in the usual condensed matter physics. A crystal electron (with mass m) is subjected to two types of forces namely the effect of crystal field grad V and an external force F which accelerates the electron. Under the influence of the two forces, an electron in a crystal behaves like a quasi-particle whose effective mass m^* reflects the inertia of electrons which are already in a crystal field [12].

$$\frac{m}{m^*} = 1 - \frac{1}{F}\frac{\partial V}{\partial r} \tag{6}$$

We assume similar type of picture for the diquarks. Diquarks are supposed to be fundamental entities and we have considered that their masses get modified due to the two types of forces inside the hadron. One is due to the presence of a background quark field inside the hadron resembling one gluon exchange type interaction and for external force harmonic oscillator type of confinement interaction has been considered. In the context of the potential model potential has been represented as:

$$V_{ij} = -\frac{\alpha}{r} + \left(F_i, F_j\right)\left(-\frac{1}{2}Kr^2\right) \tag{7}$$

where α is the coupling constant, $F_i \cdot F_j = -\frac{2}{3}$ for qq interaction [13] and K is the strength parameter, so that,

$$V_{ij} = -\frac{\alpha}{r} + ar^2 \tag{8}$$

where $a = K/3$ The mass of the diquark in the background of the quark field gets modified in the quasiparticle approximation as it happens in the crystal lattice [12].

While retrieving the $\frac{1}{F}\frac{\partial V}{\partial r}$ part we have considered the one-gluon exchange potential for background field and for external force F we have used the confinement force [11]. Thus we come across:

$$\frac{\left(m_{qi}+m_{qj}\right)}{m_D}=1+\frac{\alpha}{2ar_D^3} \tag{9}$$

where m_D, r_D are diquark mass and radius respectively, $m_u = m_d = 360\ MeV$, $\alpha = \frac{2}{3}, \alpha_s = 0.393$ (with $\alpha_s = 0.58$), $K = 241.5\ MeV\,fm^{-2}$ [13] the diquarks masses have been estimated as $m_{ud} = 506\ MeV$. To estimate the energy we need the wave function for the nucleon. We have used the wave function for the baryon as suggested in the context of the Statistical Model suggested by us [14]. In this model, a hadron is assumed to be consists of a virtual qq(bar) in addition to the valence partners which determines the quantum number of the colourless hadron. The quarks, real and virtual, are assumed to be of same colour and flavor so that they may be regarded as identical and indistinguishable and are treated on the same footing. The indistinguishability of the valence quark with the virtual partner calls for the existence of quantum mechanical uncertainty in its available phase space. The valence quarks are assumed to be non-interacting with each other and considered to be moving almost independently in conformity with the experimental finding of asymptotic freedom. However it is considered that the valence quarks move in an average smooth background potential due to their interaction with the virtual sea. With the above considerations an expression for the probability density of the baryon in the ground state has been derived as [14],

$$\left|\varphi(r)\right|^2 = \frac{315}{64\pi r_0^{\frac{9}{2}}}\left(r_0 - r\right)^{\frac{3}{2}}\theta\left(r_0 - r\right) \tag{10}$$

where r_0 is the radius parameter of the corresponding hadron and $\theta\left(r_0 - r\right)$ represents the usual step function.

In the modified version of the statistical model, the exchange effect which arises from the tendency of quarks of like spin to keep them apart as suggested by Dirac [15] has been incorporated. The wave function is thus obtained as [16]:

$$|\varphi(r)|^2 = A(r_0 - r)^{\frac{3\mu L}{2}} \theta(r_0 - r) \tag{11}$$

The exchange effect has been incorporated through a term $\mu_L = (\alpha_{ex}/a)\frac{dp}{dr}$. The normalization constant is obtained as $A = \left[4\pi r_0^{3\mu L/2+3} B(3, 3\mu_L/2+1)\right]^{-1}$ where B is the beta function. Using these wave functions as input in the expression (2) and (3), and taking $\mu_L = 3.66$ [17] the Hamiltonians are obtained as:

$$\langle H \rangle = \frac{28.079}{r_0^2} + 0.545\sigma r_0 - 0.055\sigma r_0^4 - \frac{3\alpha_s}{r_0} + 0.312\alpha_s \rho r_0^2 \tag{12}$$

And

$$\langle H \rangle = \frac{46.205}{r_0^2} + 0.312\sigma r_0 - 0.0088\sigma r_0^4 - \frac{5.66\alpha_s}{r_0} + 0.112\alpha_s \rho r_0^2 \tag{13}$$

Neglecting the Coulomb term and minimizing the Hamiltonian with respect to radius parameter we come across the following expressions corresponding to equations (12) and (13) respectively as:

$$r_0 = \left(\frac{\sigma}{206.048}\right)^{\frac{1}{3}} \left[1 \pm (1 - 106.51\rho)^{\frac{1}{2}}\right]^{-\frac{1}{3}} \tag{14}$$

And

$$r_0 = \left(\frac{\sigma}{584.88}\right)^{\frac{1}{3}} \left[1 \pm (1 - 130.19\rho)^{\frac{1}{2}}\right]^{-\frac{1}{3}} \tag{15}$$

The compressibility of a nucleon can be expressed as:

$$K = \frac{1}{3} r_0^2 \left(\frac{d^2 E}{dr^2} \right) \tag{16}$$

The expression for Roper excitation energy runs as:

$$\Delta E = \left(\frac{K}{m_q \quad r_0^2} \right)^{\frac{1}{2}} \tag{17}$$

From the expressions (14) and (15) it has been found that ρ reaches a maximum at $\rho = \frac{\sigma}{106.516} GeV^3$ and $\rho = \frac{\sigma}{130.19} GeV^3$ respectively which corresponds to the real root of the radius parameters and can be defined as the de-confining density or the critical density ρ_c of the medium at which the nucleon ceases to exist. We have estimated the compressibility, Roper Resonance with the input of the radii has been estimated with string constant $\sigma = 0.219\ GeV^2$ [18] from Equations (14) and (15) corresponding to $\rho = 0$. The results are displayed in **Table 1** and **Table 2** corresponding to the energy expressions from Equations (12) and (13) respectively.

Table 1. Compressibility, Roper Resonance Binding Energy, Critical density with $|\varphi(r)|^2$ from (10).

$\sigma(GeV^2)$	Compressibility K_f	K_b	Roper Resonance B.E.(GeV)	P_c
0.219	0.787	0.526	0.029	$1.51\rho_0$

Table 2. Compressibility, Roper Resonance Binding Energy, Critical density with $|\varphi(r)|^2$ from (11).

$\sigma(GeV^2)$	Compressibility K_f	K_b	Roper Resonance B.E.(GeV)	P_c
0.219	0.635	0.517	0.012	$1.24\rho_0$

3. DISCUSSIONS

In the present work we have investigated the properties of the nucleon in the nuclear matter in the context of the flux tube model considering a nucleon consist of a quark and a diquark connected by a flux tube. Quasi-particle diquark model [14] is used for describing the diquark. The compressibility, Roper excitation energy have been estimated. The results obtained are found to be in the reasonable agreement with the existing predictions. The nucleon shows a swelling in the nuclear medium and the percentage increase in radius at the critical density have been estimated to be of 26 percent. In the context of the non-topological soliton model Wen et al [19] have estimated the increase in nucleon radius by 10 to 16 percent at normal nuclear matter density whereas Noble [20] has observed the increase in charge radius of nucleon to be 30 percent. Mathieu et al [10] have observed 25 percent increase in radius at deconfinement density in the context of the flux tube model. Dukelsky et al [5] have obtained the swelling to be of 2 percent much lower than the existing predictions mentioning the limitation of the model they have used. Chang et al [21] have investigated the modification of nucleon properties in the context of the Global Colour Symmetry Model and have observed that the nuclear density is 8 times the normal nuclear density at which the radius becomes infinite representing a phase transition. Achtzehnter et al [22] obtained the critical density to be of the order of normal nuclear density in the context of soliton bag model and argued that the model cannot adequately incorporate the repulsive part of the nucleon-nucleon interaction at small relative distance. The critical density obtained in the present work seems to be lower than the other existing estimates. The difference in compressibility (ΔK) between the free nucleon and nucleon in the medium at normal nuclear density ($= 0.17\, fm^{-3}\ \rho_0$) has been found to be 0.225 GeV and 0.118 GeV for two version of the Statistical Model used. The reduction of compressibility in medium is found to be substantial in the present investigation. The decrease in compressibility of the nucleon the medium shows that the nucleons in nuclear medium are more diffused than free nucleons. The Roper resonance excitation energy is more in free nucleon than when it is in nuclear medium. The difference has been obtained as 0.029 GeV and 0.012 GeV in the present work. Mathieu et al [10] obtained the value as 0.021 GeV whereas the most accepted value is of the order of 0.040 GeV. In the current investigation we have neglected the contribution from Coulomb term. It may be mentioned that the most uncertainty comes from the radius parameter of the nucleon which has a wide range of values in the literature. The string constant σ also have important role in determining the in medium properties and precise knowledge of its value is yet to be determined. It has a range of values [18]. It is interesting to observe that in the current work we have come across an expression where the radius parameter is related to the string constant. If we put radius of the proton to be 0.865 fm which is the charge radius of the proton and suggested to be the equal to the radius parameter defining it as a meso object [23], the string constant would have the value 1.26 GeV^2 which is quite high compared to the common accepted values. Compressibility also has wide range of values and is under discussion in the current literature. It has been suggested that the information on the compressibility of a system can be obtained from the dynamical properties of the size degree of freedom [24].

4. CONCLUSIONS

Aparajita Bhattacharya, A. Sagari, B. Chakrabarti, A. Chandra (quasiparticle diquark model) of the nucleon has been used to study the in medium modification of the nucleon properties. Diquarks are now being supposed to be the building block of nucleons particularly after the discovery of the exotics particles. A number of models have been suggested for diquarks [25]. We have also suggested Quasi Particle model and investigated the properties of proton in the framework of the model. Results are found to be interesting. The exchange effect of the quark spin [15] seems to reduce values of all the properties yielding comparatively larger value of the radius parameter. In the present investigation we have dealt with the modification of the nucleon properties due to medium effect through the density of the nuclear matter. However it may be pointed out that recently it has been suggested by Seely et al [2] that nucleon dependence of quark distribution of nucleon depends on the local environment not on A or the density. More experimental and theoretical efforts are needed to reveal the in medium effect on the nucleon which in turn would help us to know exact nature of the underlying interaction.

REFERENCES

1. J. J. Aubert et al, Physics Letters, Vol. 123 B, 1983.
2. J. Seely, Physical Review Letters, Vol. 103, 2009.
3. S. Sarangi and M. K. Banerjee, Physical Review C, Vol. 45, 1992.
4. M. Oka and R. D. Amado, Physical Review C, Vol. 35, 1987.
5. D. Dukelsky, Journal of Physics G, Vol. 35, 1995.
6. K.Braucr et al, Nuclear Physics A, Vol. 347, 198.
7. K. S. Kim et al, The European Physical Journal A, Vol. 8, 2000.
8. W. Wu and S. Hong, The Journal of Physical Chemistry C, Vol. 32, Suppl. II , 2008.
9. H. Muller and B. K. Jennings, Nuclear Physics A, Vol. 626, 1997.

10. P. Mathieu and P. J. S. Watson, Canadian Journal of Physics, Vol. 64, 1986. [Citation Time(s):8]
11. A. Bhattacharya et al, Physical Review C, Vol. 81, 2010.
12. A. Haug, Theoretical Solid State Physics, Vol. 1, Pergamon Press, Oxford, 1975, P. 100.
13. R. K. Bhaduri et al, Physical Review Letter, Vol. 44, 1980.
14. A. Bhattacharya et al, Nuclear Physics B, Vol. 142, 2005.
15. P. A. M. Dirac, Proceedings of the Cambridge Philosophical Society, Vol. 2b, 1930.
16. S. N. Banerjee et al, International Journal of Modern Physics, Vol. A16, 2001.
17. S. N. Banerjee et al., Had. J. Vol. 11, 1988.
18. G. L. Strobel, International Journal of Theoretical Physics, Vol. 39, 2000.
19. W. Wen and H. Shen, Physical Review C, Vol. 77, 2008.
20. J. V. Noble, Physical Review Letter, Vol. 46, 1981.
21. L. Chang et al., Nuclear Physics A, Vol. 750, 2005.
22. J. Achtzehnter and W. Scheid, Physical Review D, Vol. 32, 1985.
23. E. F. Hefter, Physical Review A. Vol. 32, 1985, p. 1201.
24. T. Meissner, Physical Review D, Vol. 39, 1989.
25. R. L. Jaffe; Physical Review D, Vol. 71, 2006.

CHAPTER 4

How Do the Nucleons Pack in an Atomic Nucleus?

Xiaodong Li*, Qijun Liu, Gongyi Li, Yihe Li, Zengyong Chu

Department of Chemistry and Biology, College of Science, National University of Defense Technology, Changsha, China

ABSTRACT

A nuclear structure model of "ring plus extra nucleon" is proposed. For nuclei larger than ^4He, protons (P) and neutrons (N) are basically bound alternatively to form a ZP + ZN ring. The ring folds with a "bond angle" of 90° for every 3 continuous nucleons to make the nucleons packed densely. Extra N('s) can bind to ring-P with the same "bond angle" and "bond distance". When 2 or more P's are geometrically available, the extra N tends to be stable. Extra P can bind with ring N in a similar way when the ratio of N/P < 1 although the binding is weaker than that of extra N. Even-Z rings, as well as normal even-even nuclei, always have superimposed gravity centers of P and N; while for odd-Z rings, as well as all odd-A (A: number of nucleon) nuclei, the centers of P and N must be eccentric. The eccentricity results in a depression of binding energy (E_B) and therefore odd and even Z dependent zigzag features of E_B/A. This can be well explained by the shift of eccentricity by extra nucleons. Symmetrical center may present in even-Z rings and normal even-even nuclei. While for odd-Z ring, only antisymmetric center (every P can find an N through the center and vice versa) is possible. Based on this model, a pair of mirror nuclei, $P_{X+n}N_X$ and P_XN_{X+n}, should be equivalent in packing structure just like black-white photo and the negative film. Therefore, an identical spin and parity was confirmed for any pair. In addition, the E_B/A difference of mirror nuclei pair is nearly a constant of 0.184n MeV. Many other facts can also be easily understood from this model, such as the neutron halo, the unusual stability sequence of ^{9}Be, ^{7}Be and ^{8}Be and so on.

Keywords: Nuclear Structure, Ring plus Extra Nucleon Model, Nuclear Symmetricity

1. INTRODUCTION

The packing model of the nucleons in an atomic nucleus always arouses curiosity of many scientists and amateurs. The model determines the shape of a nucleus as well as its various properties, such as the stability, the magnetic moment, the spin, the binding energy and radioactivity. Sophisticated theories of nuclear model were established and many facts were well explained [1] [2] . However, there is obviously still a mysterious veil on some characteristics of nuclei to be revealed, especially the packing structure of nucleons inside a nucleus. Thanks to X-ray diffraction, chemists know plenty of details about how atoms pack to build a molecule. Molecular structure establishes the basis of modern chemistry. However, since nuclei are too small and are usually embedded in electron "cloud", it is very hard to "see" them.

In nature, the equilibrium shape of a centered direction-independent object is a sphere. These objects vary vastly in composition and size, including planets, liquid droplet, bubble, most nucleoli in cell, crystalline spherulite, latex micelles, isolated atoms (at least those which can be packed in cubic lattices) and possibly some elementary particles. A plant leaf is far from a sphere because the petiole to the stem and the surface toward the sun are very direction-related. The non-sphere shape of a molecule comes from the direction-related chemical bonds. Many atomic nuclei are generally not sphere and may possess remarkable non-zero quadrupole moment [2] . That means tha the nucleons, protons (P) and neutrons (N) are bound in specific direction and are packed in a particular structure. In this paper, the authors try to propose a model that depicts the packing of nucleons in a nucleus based on the conception of covalent molecule configuration.

2. THE BINDING OF PROTON AND NEUTRON IN A NUCLEUS

From the reported data [3] -[5] , there are 7 isotopes of hydrogen, or single-P nuclides: ^{1}H, ^{2}H, ^{3}H, ^{4}H, ^{5}H, ^{6}H and ^{7}H. In addition, reported single-N nuclei include ^{2}H, ^{3}He, ^{4}Li, ^{5}Be and ^{6}B. This gives a clue that N, as well as P, can be a 6-coordinate center bound with 6 "ligands" at maximum as shown in Figure 1(a), where the proper octahedral structure seems to be a right way to ensure the ligands to be indistinguishable. Although these structures need to be further verified, especially the 90° triangular ^{3}H and ^{3}He, it is obvious that, in the case of single center, when the "coordination number" is higher than 2, the isotope will be unstable and tends to kick out the excess P or N.

Based on these structures, we assume that P can only bind with N and vice versa (considering deuterium, PN, is stable, but there is no conclusive evidence about the existence of NN and PP). The "bond distance" between bound P and N, D_{NP}, is in the range of 1 to 2 fm, and the "bond angle" for 3 continuous nucleons is 90° or 180°.

The single-center nuclei are characterized by extremely low binding energy E_B (and therefore a low E_B/A), even for the stable nuclei as ^{2}H and ^{3}He, noted in Figure 1(a) comparing with other nuclei. This is probably because of the existence of suspending open P or open N. Therefore, we assume that, for a nucleus with ratio of N:P = 1, open nucleon should be avoided and all P, as well as all N, should be equivalent. Thus, P and N should be bound alternatively to form a ring. It is obviously that the ring is not shaped like a pearl necklace on neck; instead, the ring must fold somehow, like a necklace in casket. The most possible way of folding a ring is taking a "bond angle" of 90° for every 3 continuous nucleons, to ensure the nuclide to adopt the densest packing and the most nearly sphere symmetrical shape. However, for most nuclei, the ratio of N/P is more or less than unit. Among hundreds of stable nuclei, only 12 are composed purely of a ring: ^{4}He, ^{6}Li, ^{10}B, ^{12}C , ^{14}N, ^{16}O, ^{20}Ne, ^{24}Mg, ^{28}Si, ^{32}S, ^{36}Ar and ^{40}Ca. Thus, extra N('s) or extra P('s) will be necessary to connect or insert the ring somehow in most nuclei.

3. RING STRUCTURE

Based on the "ring plus extra nucleon" hypothesis, the shape of ^{4}He should be a square, which is the smallest ring. Figure 1(b) demonstrates the shapes and E_B of nuclei of the P_2N_2 family, where the extra open N generally contributes very little to E_B, while extra open P generally gives minus contribution. Thus, in the P_2N_2 family, all the nuclei which possess open nucleon are unstable. In addition, the E_B and life time of P_2N_{2+X} are always in larger number than $P_{2+X}N_2$. This corresponds to the fact that stable nuclei present a ratio $N/P \geq 1$(except ^{3}He).

In Figure 1(b), some nuclei may have 2 or more isomers, for example, ^{5}He may have 2 isomers, ^{6}He 3, ^{7}He 6 and so on. It is reasonable to assume that favored structure should be the highest symmetry and the most compact. It should be noted that, for saturating the 6 coordinates, ^{12}He will be the possibly heaviest isotope of helium (^{10}He is the maximum found). For the same reason, for elements of $Z = 7$ to 10, ^{9}N, ^{10}O, ^{11}F and ^{12}Ne will be the possibly lightest isotopes respectively, which have not been found yet.

Figure 2 gives the possible shapes of some light rings and their stable isotopes. For odd-Z rings, although there are a few possibilities and more when Z becomes large, a common feature is that the gravity centers of P and N are separated and are eccentric to the gravity center of nuclide. This can be easily proved mathematically. In addition, one can always find one or more arrangement with antisymmetric center, through which each P can find an N by space reflection and vice versa. We assume that the most possible odd-Z ring structure is antisymmetric and the most densely packed. For ^{7}Li and other nuclei, the extra N may not likely bind to a ring P as an open N like unstable nuclides of P_2N_2 family. Instead, when 2 or more ring P's are geometrically available to share an extra N with 90 "bond angles" and same D_{PN}, the extra N tends to be stable. It seems this is the most likely way the extra N combines with the ring considering the geometric arrangement and indistinguishableness of N's. Actually, among all stable nuclei, the ratio of N/P is in the range of 1.00 to 1.54 (^{208}Pb), that means ring P is able to stably bind a little more than 0.5 extra N (or 2 ring P's share one extra N) at maximum, while no extra P can stably bind to ring N. This is the reason that, in the P_2N_2 family, all the nuclides with extra P or extra N are unstable because of the valence limitation of the square shaped ring. From the shape of the ring structure, one can predict the possibly heaviest and the lightest isotopes of the related elements. As an example, based on P_3N_3 ring, the possibly heaviest isotope of lithium is ^{16}Li and the possibly lightest isotope of aluminium is ^{16}Al. For other elements, the possibly heaviest and lightest isotopes can also be predicted if the structure of related ring is known.

The possible packing structures of some light even-Z rings and their stable isotopes are also shown in Figure 2. By contrast to odd-Z rings, all the even-Z rings possess a superimposed gravity center of P and N, which was tested by hundreds of examples without any exception and is destined to be proved mathematically. In addition, some of the structures are center-

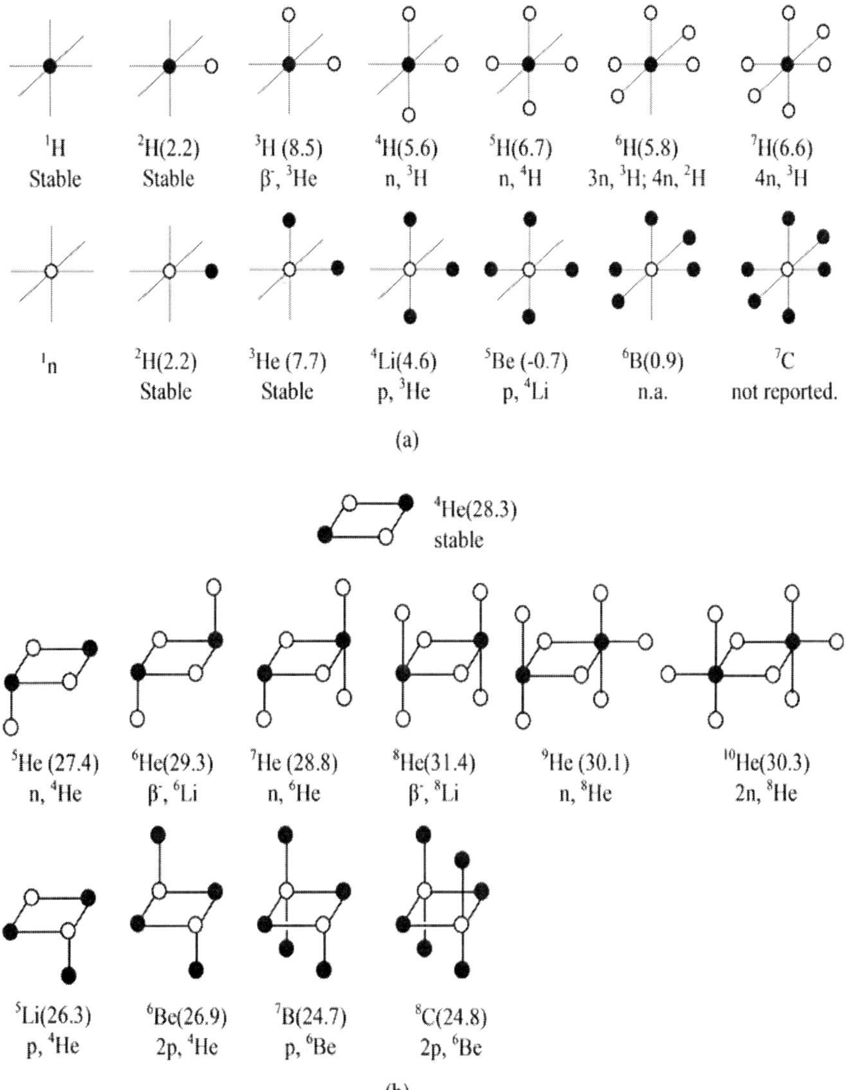

Figure 1. The possible shapes of (a) single-P nuclides (isotopes of H) and single-N nuclides (from ^{1}n to ^{7}C), (b) nuclides based on P_2N_2 ring. Black ball stands for P and white one for N (same in following figures). E_B (in MeV, in parenthesis), main decay mode and the daughter isotope are marked beside each nuclide.

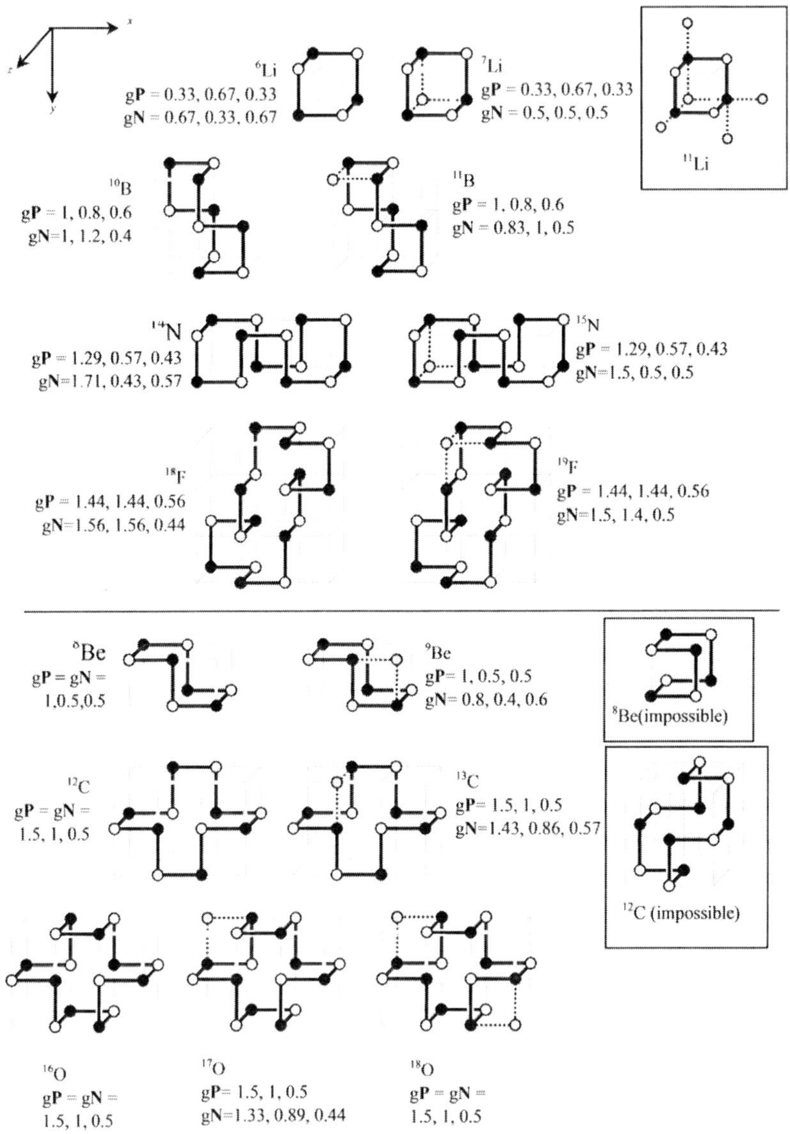

Figure 2. The possible structures of some light nuclides based on "ring plus extra nucleon" model. For odd-Z nuclides (above), the gravity centers of P and N (gP and gN) are eccentric and the rings are antisymmetrical. For even-Z nuclides (below), gP and gN are superimposed, while the rings are center symmetrical and are indistinguishable when P and N interchange. The extra N is highlighted in gray and bond in dot-line. ^{8}Be in frame is impossible because it is non center symmetrical and geometrically unavailable for any stable extra N.

symmetrical. We assume that the most possible ring structure of even-Z nuclei is the most densely packed one with center-symmetry and is indistinguishable when all P and N interchange. In Figure 2, the framed ^{12}C is marked "impossible" because it mutates to another structure when P and N interchange. It is interesting to find that ^{8}Be is composed of two P_2N_2 squares and is ready to yield into two α-particles. Only ^{9}Be, with an extra N that bridges two squares by the manner mentioned above, is stable. Therefore, as another restriction, the most possible ring structure should avoid the arrangement that can split into two normal rings.

Because of the symmetrical and geometrical restrictions of ring structure, empty open space in a large nucleus ($Z > 20$) is unavoidable. An important function of extra N is to fill the open space, making the nuclide packed more densely and establishing adequate non-bond interaction. Figure 3shows the nuclide size, characterized by radius of gyration, $[R^2]^{0.5}$ (as defined in the following equation) versus the nucleon number A. The points of stable nuclides locate around the line of slope = 1/3, or the nuclide volume is roughly proportional to A. In addition, one can find a general trend that, for rings, the radius expands related to the line with Z increases due to the open space (^{16}O is special possibly because of double magic numbers, which will be revealed in future).

From the possible structures shown in Figure 2, one can figure out that if nuclei possess too many N's to be bind properly by 2 or more ring P's, some open N will be unavoidable as a neutron halo and the nucleus will be unusually large, such as ^{11}Li in Figure 2. This is qualitatively accordant with the scattering cross-section observations for extremely N-rich nuclides [6] .

4. BINDING ENERGY PER NUCLEON

The E_B/A for all the P_xN_x rings, in Figure 4, display a zigzag feature, where even-Z rings are at the vertex and odd-Z rings at the nadir. As mentioned earlier, the fundamental difference for even-Z and odd-Z ring is that the former has superimposed gravity centers of P and N, while the later is eccentric. It is obvious that eccentricity results in lower E_B, which is possibly because the eccentric nucleus inside an atom requires more energy to spin faster to avoid the surrounding electrons to "feel" uneven.

It is well known that E_B/A of the isotopes of an element displays a specific zigzag feature, where the vertex always relates odd A for odd-Z elements, and relates even A for even-Z elements, as shown in the examples ($_{26}Fe$ and $_{29}Cu$) in Figure 4. Weizasäcker's semi-empirical mass formula treated this feature by a selective function, f_5 in an ad hoc way [2] .

In this work, the degree of eccentricity, S_g, is defined as the separation of the gravity centers of P and N, $D_{g(PN)}$, over 2 times the radius of gyration, $[R^2]^{0.5}$,

$$S_g = \frac{D_{g(PN)}}{2[R^2]^{0.5}} = \frac{D_{g(PN)}}{2\sqrt{\frac{1}{A}\sum_{i=1}^{A} D_i^2}}$$

where D_i is the length from i'th nucleon to the gravity center of the nucleus. For any even-Z ring, S_g is always zero because $D_{g(PN)} = 0$; while for odd-Z rings, S_g is higher than 0. From the definition, deuterium, ^2H, has the highest S_g of 1. For other light odd-Z rings, such as ^6Li and ^{10}B, S_g is also quite high, and therefore leads up to a very low E_B. The depression of E_B/A by eccentricity, ΔE_{ec}, is defined in Figure 4 for some simple odd-Z rings and even-odd nuclides. The relationship between S_g and ΔE_{ec} is demonstrated in Figure 5. It should be noted that this relationship is not very quantitative because ΔE_{ec} is also affected by non-bond interactions of a particular nuclide as will be mentioned later.

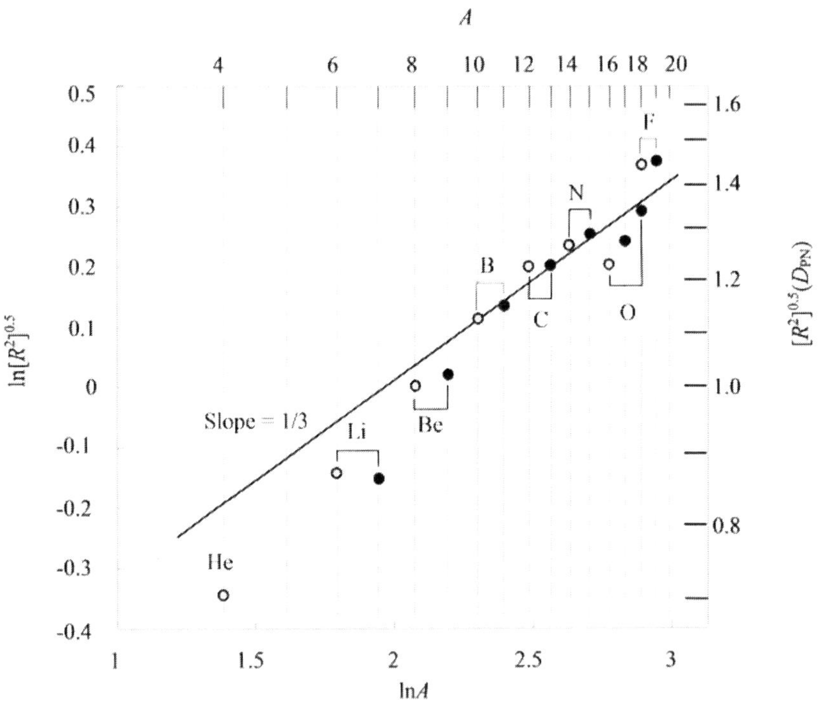

Figure 3. Radius of gyration vs A for rings (in light gray) and their stable isotopes (in black) of some light nuclei (Z = 2 to 9).

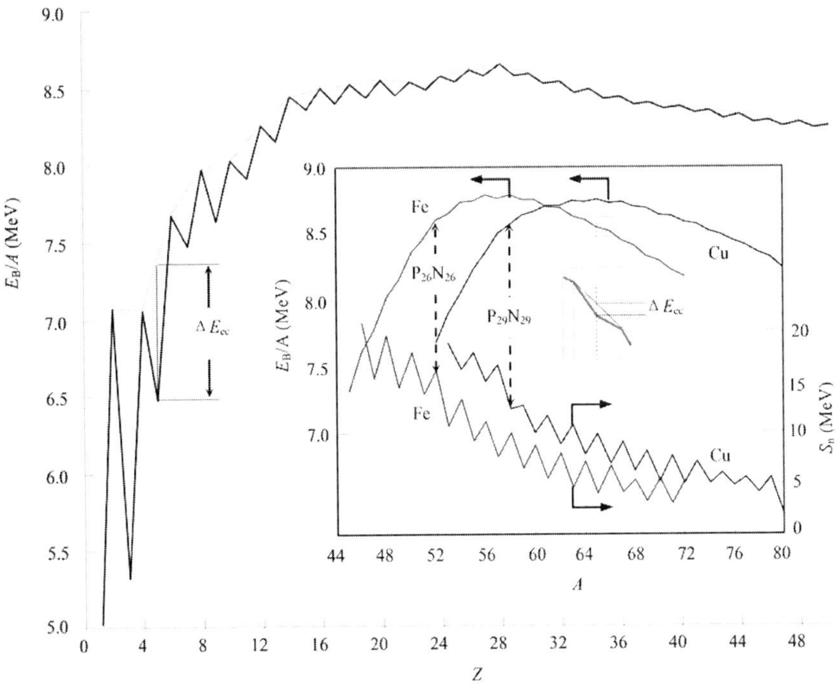

Figure 4. E_B/A of ring nuclides P_ZN_Z. The zigzag feature implies a fundamental difference between odd- and even-Z rings. The envelope line for even-Z ring is shown, with which the depression energy by eccentricity, ΔE_{ec}, for odd-Z ring is defined. Embededis zigzag feature of E_B/A (above) and S_n (below) for even-Z elements ($_{26}$Fe as an example) and odd-Z elements ($_{29}$Cu as an example). ΔE_{ec} for P_ZN_{Z+1}(even-Z) is defined.

Cohesive interactions between metallic atoms are mostly governed by their electron configuration. Since the high S_g means the high nuclear "polarity", which provides an extra interaction between isolated atoms, like polar molecules have higher intermolecular interactions than nonpolar ones. That can explain Li, Be and B respectively demonstrate the unusually highest melting point and boiling point in their group (similar electron configuration) of periodic table.

To understand the zigzag feature, Figure 6 depicts the $D_{g(PN)}$ and eccentricity S_g shift caused by extra nucleon in different cases, that results in fluctuation of E_B/A exactly as the way as the rings or isotopes of any element do (vertex for even Z at even A while odd Z at odd A).

The envelope line of E_B/A for even-Z rings in Figure 4 demonstrates that, without the energy depression by eccentricity, E_B/A increases from very low for light Z ring to maximum at $P_{28}N_{28}$, and then decreases slowly. This should be due to the variation of non-bond interaction.

For most elements, E_B/A of a ring, as basic structure, is not the highest among the isotopes, as the examples shown in Figure 4. It is because extra N, which fills the open spaces in large ring and builds adequate non-bond interaction, may lead to higher E_B/A and stability. Nevertheless, special stability of ring structure can be detected in many facts. Weizasäcker formula includes a symmetry term f_4 emphasizing P_ZN_Z is more favored in energy [2] . This can also be shown in Figure 4 that, except P-rich nuclides, P_ZN_Z always possesses the highest one-N separation energy, S_n (for most odd-Z nuclei the highest is P_ZN_{Z+1}because of lower eccentricity). That is to say, N in a ring structure is generally more stable than extra N. Since extra P is much more unstable, the normal extra nuclcons can only be N or P, excluding N and P.

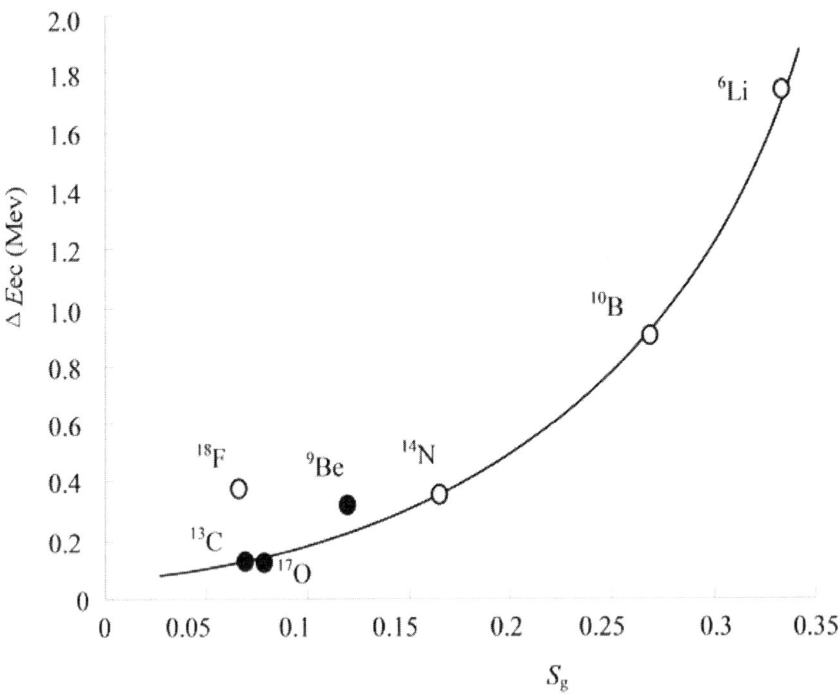

Figure 5. Relationship between degree of eccentricity, S_g, and depression energy by eccentricity, ΔE_{ec}, for eccentric nuclides. The ΔE_{ec} of odd-Z ring (in light gray) and even-Z nuclide P_ZN_{Z+1} (in black) are defined in Figure 4.

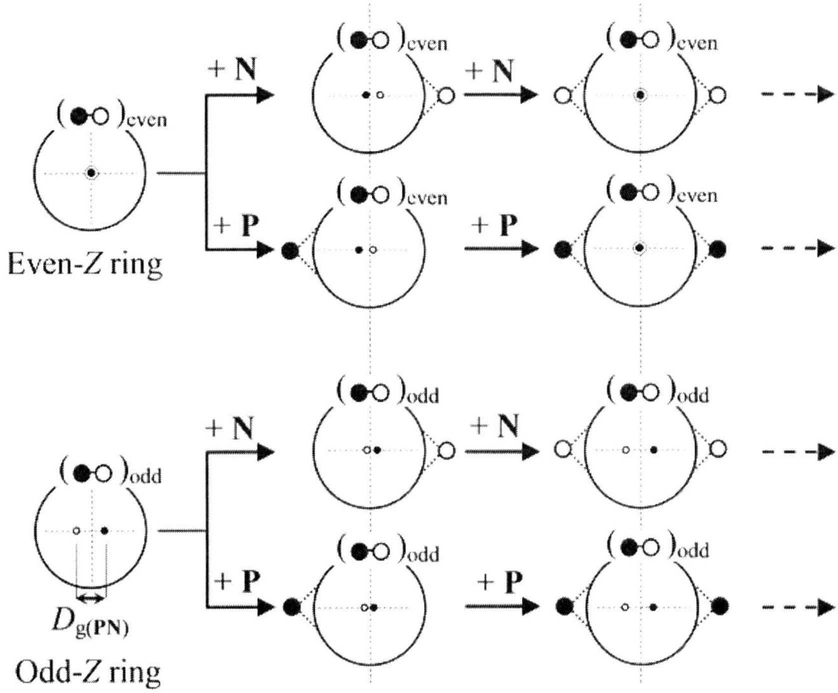

Figure 6. The explanation of E_B/A zigzag feature of nuclei. Even-Z ring has superimposed gravity centers of P (tiny black dot) and N (tiny white dot); while odd-Z ring has eccentric gravity centers. Extra N or P shifts the gravity centers and therefore $D_{g(PN)}$ as well as ΔE_{ec} fluctuates.

Basically, E_B/A of a nucleus is governed by many factors. Strong force between binding P and N in a ring plays the most important role, up to about 7 MeV (as in ^{4}He), which should be equal for all rings. The weak force acts between non-bond nucleons (P..N, N..N and P..P) in a ring that provides higher E_B/A up to 8.5 MeV (as in ^{40}Ca). The open spaces hinders non-bond interaction, and makes nuclide very unstable even E_B/A is not so low. The extra N shared by 2 or more ring P's, which contributes similar strong force, provides more weak force and, therefore, can increase E_B/A up to 8.8 MeV (as in ^{62}Ni). When ring is too large, the open spaces may not be completely filled by extra N, and E_B/A lowers gradually. Eccentricity is an important factor that decreases E_B especially for light nuclei as depicted in Figure 5. In addition, excess extra N or P can only connect to the ring as open nucleons, which have very little or minus contribution to strong force and thus result in many very unstable nuclei with very low E_B/A. Because of the complicated effects, the nuclide shapes shown in Figure 2 are only the most possible arrangements and need to be confirmed or revised by new theory considering more details of nuclear forces and new experimental observations.

5. MIRROR NUCLEI

Mirror nuclei are the pair of nuclides $P_{X+n}N_X$ and P_XN_{X+n}. Some correlations of the pair have been reported since long ago [7] -[10] . The most remarkable fact is the identical spin and parity, J^π, of mirror nuclei in ground state as summarized in Table 1, with very few exceptions or uncertainties. Since the ring is indistinguishable when P and N interchange, any nuclide with n extra N(¢s) will have a counterpart with n extra P(¢s) in similar way of binding to the ring. Thus, the shapes of the pair nuclides are similar but P and N interchange, just like a black-white photo and the negative film. This should be the basic reason for the following equation.

$$J^\pi\left(\mathbf{P}_{X\pm n}\mathbf{N}_X\right) = J^\pi\left(\mathbf{P}_X\mathbf{N}_{X\pm n}\right)$$

It is also reasonable to predict the identity of degree of eccentricity for the pair.

$$S_g\left(\mathbf{P}_{X\pm n}\mathbf{N}_X\right) = S_g\left(\mathbf{P}_X\mathbf{N}_{X\pm n}\right)$$

^{7}Be is present in the atmosphere in trace amount [11] as the only natural occurring radioisotope with the ratio of $N/P < 1$. As the mirror nuclide of ^{7}Li, ^{7}Be has an extra P and can also obtain a large energy bonus from the highest eccentricity in ^{6}Li although, as expectation, ^{7}Be is still not stable enough as ^{7}Li.

Figure 7 demonstrates the E_B/A of mirror nuclei. It is clear that the E_B/A of P_XN_{X+n} is always larger than that of $P_{X+n}N_X$. Compared with the zigzag feature of ring P_XN_X mentioned earlier, when n is odd, the curves are smooth because odd extra P or N diminishes the depression energy by eccentricity for odd-X nuclides. Of course, when n is even, the curves zigzag again. In addition, it is notable that the difference of E_B/A for any pair is nearly a constant: averagely 0.179, 0.368, 0.559, 0.763, 0.926 and 1.084 MeV for n = 1, 2, 3, 4, 5 and 6 respectively (the curves for n = 5 and 6 are omitted in Figure 7). This can be described with following equation.

$$E_B/A\left(\mathbf{P}_X\mathbf{N}_{X+n}\right) - E_B/A\left(\mathbf{P}_{X+n}\mathbf{N}_X\right) \approx 0.184n\,(\mathrm{MeV})$$

In summary, the ring plus extra nucleon model seems properly meet the structure requirement to the correlations of mirror nuclei.

Table 1. The identical spin and parity for mirror nuclei: $J^{\pi}(P_{X+n}N_X) = J^{\pi}(P_XN_{X+n})^*$.

X	$P_{X+5}N_X$	$P_{X+4}N_X$	$P_{X+3}N_X$	$P_{X+2}N_X$	$P_{X+1}N_X$	P_XN_X	P_XN_{X+1}	P_XN_{X+2}	P_XN_{X+3}	P_XN_{X+4}	P_XN_{X+5}
1		^{6}B	^{5}Be $1/2^{+}$	^{4}Li 2^{-}	^{3}He $1/2^{+}$	^{2}H 1^{+}	^{3}H $1/2^{+}$	^{4}H 2^{-}	^{5}H $1/2^{+}$	^{6}H 2^{-}	
2		^{8}C 0^{+}	^{7}B $3/2^{-}$	^{6}Be 0^{+}	^{5}Li $3/2^{-}$	^{4}He 0^{+}	^{5}He $3/2^{-}$	^{6}He 0^{+}	^{7}He $3/2^{-}$	^{8}He 0^{+}	
3		^{10}N 2^{-}	^{9}C $3/2^{-}$	^{8}B 2^{+}	^{7}Be $3/2^{-}$	^{6}Li 1^{+}	^{7}Li $3/2^{-}$	^{8}Li 2^{+}	^{9}Li $3/2^{-}$	^{10}Li $1^{-},2^{-}$	
4		^{12}O 0^{+}	^{11}N $1/2^{+}$	^{10}C 0^{+}	^{9}B $3/2^{-}$	^{8}Be 0^{+}	^{9}Be $3/2^{-}$	^{10}Be 0^{+}	^{11}Be $1/2^{+}$	^{12}Be 0^{+}	
5		^{14}F 2^{-}	^{13}O $3/2^{-}$	^{12}N 1^{+}	^{11}C $3/2^{-}$	^{10}B 3^{+}	^{11}B $3/2^{-}$	^{12}B 1^{+}	^{13}B $3/2^{-}$	^{14}B 2^{-}	
6		^{16}Ne 0^{+}	^{15}F $1/2^{+}$	^{14}O 0^{+}	^{13}N $1/2^{-}$	^{12}C 0^{+}	^{13}C $1/2^{-}$	^{14}C 0^{+}	^{15}C $1/2^{+}$	^{16}C 0^{+}	
7	^{19}Mg $1/2^{-}$	^{18}Na	^{17}Ne $1/2^{-}$	^{16}F 0^{-}?	^{15}O $1/2^{-}$	^{14}N 1^{+}	^{15}N $1/2^{-}$	^{16}N 2^{-}?	^{17}N $1/2^{-}$	^{18}N $1-$	^{19}N $1/2^{-}$
8	^{21}Al $1/2^{+}$	^{20}Mg 0^{+}	^{19}Na $5/2^{+}$	^{18}Ne 0^{+}	^{17}F $5/2^{+}$	^{16}O 0^{+}	^{17}O $5/2^{+}$	^{18}O 0^{+}	^{19}O $5/2^{+}$	^{20}O 0^{+}	^{21}O $(1/2,3/2,5/2)^{+}$
9	^{23}Si $3/2^{+}$	^{22}Al 3^{+}	^{21}Mg $5/2,3/2^{+}$	^{20}Na 2^{+}	^{19}Ne $1/2^{+}$	^{18}F 1^{+}	^{19}F $1/2^{+}$	^{20}F 2^{+}	^{21}F $5/2^{+}$	^{22}F $4^{+},3^{+}$	^{23}F $(3/2,5/2)^{+}$
10	^{25}P $1/2^{+}$?	^{24}Si 0^{+}	^{23}Al $5/2^{+}$	^{22}Mg 0^{+}	^{21}Na $3/2^{+}$	^{20}Ne 0^{+}	^{21}Ne $3/2^{+}$	^{22}Ne 0^{+}	^{23}Ne $5/2^{+}$	^{24}Ne 0^{+}	^{25}Ne $3/2^{+}$?
11	^{27}S $5/2^{+}$	^{26}P 3^{+}	^{25}Si $5/2^{+}$	^{24}Al 4^{+}	^{23}Mg $3/2^{+}$	^{22}Na 3^{+}	^{23}Na $3/2^{+}$	^{24}Na 4^{+}	^{25}Na $5/2^{+}$	^{26}Na 3^{+}	^{27}Na $5/2^{+}$
12	^{29}Cl $3/2^{+}$	^{28}S 0^{+}	^{27}P $1/2^{+}$	^{26}Si 0^{+}	^{25}Al $5/2^{+}$	^{24}Mg 0^{+}	^{25}Mg $5/2^{+}$	^{26}Mg 0^{+}	^{27}Mg $1/2^{+}$	^{28}Mg 0^{+}	^{29}Mg $3/2^{+}$
13	^{31}Ar $5/2^{+}$	^{30}Cl 3^{+}	^{29}S $5/2^{+}$	^{28}P 3^{+}	^{27}Si $5/2^{+}$	^{26}Al 5^{+}	^{27}Al $5/2^{+}$	^{28}Al 3^{+}	^{29}Al $5/2^{+}$	^{30}Al 3^{+}	^{31}Al $(3/2,5/2)^{+}$
14	^{33}K $3/2^{+}$	^{32}Ar 0^{+}	^{31}Cl $3/2^{+}$	^{30}S 0^{+}	^{29}P $1/2^{+}$	^{28}Si 0^{+}	^{29}Si $1/2^{+}$	^{30}Si 0^{+}	^{31}Si $3/2^{+}$	^{32}Si 0^{+}	^{33}Si $3/2^{+}$
15	^{35}Ca $1/2^{+}$	^{34}K 1^{+}	^{33}Ar $1/2^{+}$	^{32}Cl 1^{+}	^{31}S $1/2^{+}$	^{30}P 1^{+}	^{31}P $1/2^{+}$	^{32}P 1^{+}	^{33}P $1/2^{+}$	^{34}P 1^{+}	^{35}P $1/2^{+}$
16	^{37}Sc $7/2^{-}$	^{36}Ca 0^{+}	^{35}K $3/2^{+}$	^{34}Ar 0^{+}	^{33}Cl $3/2^{+}$	^{32}S 0^{+}	^{33}S $3/2^{+}$	^{34}S 0^{+}	^{35}S $3/2^{+}$	^{36}S 0^{+}	^{37}S $7/2^{-}$
17	^{39}Ti $3/2^{+}$	^{38}Sc 2^{-}	^{37}Ca $3/2^{+}$	^{36}K 2^{+}	^{35}Ar $3/2^{+}$	^{34}Cl 0^{+}	^{35}Cl $3/2^{+}$	^{36}Cl 2^{+}	^{37}Cl $3/2^{+}$	^{38}Cl 2^{-}	^{39}Cl $3/2^{+}$
18	^{41}V $7/2^{-}$	^{40}Ti 0^{+}	^{39}Sc $7/2^{-}$	^{38}Ca 0^{+}	^{37}K $3/2^{+}$	^{36}Ar 0^{+}	^{37}Ar $3/2^{+}$	^{38}Ar 0^{+}	^{39}Ar $7/2^{-}$	^{40}Ar 0^{+}	^{41}Ar $7/2^{-}$
19	^{43}Cr $3/2^{+}$	^{42}V 2^{-}	^{41}Ti $3/2^{+}$	^{40}Sc 4^{-}	^{39}Ca $3/2^{+}$	^{38}K 3^{+}	^{39}K $3/2^{+}$	^{40}K 4^{-}	^{41}K $3/2^{+}$	^{42}K 2^{-}	^{43}K $3/2^{+}$
20	^{45}Mn $7/2^{-}$	^{44}Cr 0^{+}	^{43}V $7/2^{-}$	^{42}Ti 0^{+}	^{41}Sc $7/2^{-}$	^{40}Ca 0^{+}	^{41}Ca $7/2^{-}$	^{42}Ca 0^{+}	^{43}Ca $7/2^{-}$	^{44}Ca 0^{+}	^{45}Ca $7/2^{-}$
21	^{47}Fe $7/2^{-}$	^{46}Mn 4^{+}	^{45}Cr $7/2^{-}$	^{44}V 2^{+}	^{43}Ti $7/2^{-}$	^{42}Sc 0^{+}	^{43}Sc $7/2^{-}$	^{44}Sc 2^{+}	^{45}Sc $7/2^{-}$	^{46}Sc 4^{+}	^{47}Sc $7/2^{-}$
22	^{49}Co $7/2^{-}$	^{48}Fe 0^{+}	^{47}Mn $5/2^{-}$	^{46}Cr 0^{+}	^{45}V $7/2^{-}$	^{44}Ti 0^{+}	^{45}Ti $7/2^{-}$	^{46}Ti 0^{+}	^{47}Ti $5/2^{-}$	^{48}Ti 0^{+}	^{49}Ti $7/2^{-}$
23	^{51}Ni $7/2^{-}$	^{50}Co 6^{+}	^{49}Fe $7/2^{-}$	^{48}Mn 4^{+}	^{47}Cr $3/2^{-}$	^{46}V 0^{+}	^{47}V $3/2^{-}$	^{48}V 4^{+}	^{49}V $7/2^{-}$	^{50}V 6^{+}	^{51}V $7/2^{-}$
24	^{53}Cu $3/2^{-}$	^{52}Ni 0^{+}	^{51}Co $7/2^{-}$	^{50}Fe 0^{+}	^{49}Mn $5/2^{-}$	^{48}Cr 0^{+}	^{49}Cr $5/2^{-}$	^{50}Cr 0^{+}	^{51}Cr $7/2^{-}$	^{52}Cr 0^{+}	^{53}Cr $3/2^{-}$
25	^{55}Zn $5/2^{-}$	^{54}Cu 3^{+}	^{53}Ni $7/2^{-}$	^{52}Co 6^{+}	^{51}Fe $5/2^{-}$	^{50}Mn 0^{+}	^{51}Mn $5/2^{-}$	^{52}Mn 6^{+}	^{53}Mn $7/2^{-}$	^{54}Mn 3^{+}	^{55}Mn $5/2^{-}$
26	^{57}Ga $1/2^{-}$	^{56}Zn 0^{+}	^{55}Cu $3/2^{-}$	^{54}Ni 0^{+}	^{53}Co $7/2^{-}$	^{52}Fe 0^{+}	^{53}Fe $7/2^{-}$	^{54}Fe 0^{+}	^{55}Fe $3/2^{-}$	^{56}Fe 0^{+}	^{57}Fe $1/2^{-}$
27	^{59}Ce $7/2^{-}$	^{58}Ga 2^{+}	^{57}Zn $7/2^{-}$	^{56}Cu 4^{+}	^{55}Ni $7/2^{-}$	^{54}Co 0^{+}	^{55}Co $7/2^{-}$	^{56}Co 4^{+}	^{57}Co $7/2^{-}$	^{58}Co 2^{+}	^{59}Co $7/2^{-}$
28	^{61}As $3/2^{-}$	^{60}Ge 0^{+}	^{59}Ga $3/2^{-}$	^{58}Zn 0^{+}	^{57}Cu $3/2^{-}$	^{56}Ni 0^{+}	^{57}Ni $3/2^{-}$	^{58}Ni 0^{+}	^{59}Ni $3/2^{-}$	^{60}Ni 0^{+}	^{61}Ni $3/2^{-}$
29		^{62}As 1^{+}	^{61}Ge $3/2^{-}$	^{60}Ga 2^{+}	^{59}Zn $3/2^{-}$	^{58}Cu 1^{+}	^{59}Cu $3/2^{-}$	^{60}Cu 2^{+}	^{61}Cu $3/2^{-}$	^{62}Cu 1^{+}	
30		^{64}Se 0^{+}	^{63}As $3/2^{-}$	^{62}Ge 0^{+}	^{61}Ga $3/2^{-}$	^{60}Zn 0^{+}	^{61}Zn $3/2^{-}$	^{62}Zn 0^{+}	^{63}Zn $3/2^{-}$	^{64}Zn 0^{+}	
31			^{65}Se $3/2^{-}$	^{64}As 0^{+}	^{63}Ge $3/2^{-}$	^{62}Ga 0^{+}	^{63}Ga $3/2^{-}$	^{64}Ga 0^{+}	^{65}Ga $3/2^{-}$		
32			^{67}Br $1/2^{-}$	^{66}Se 0^{+}	^{65}As $3/2^{-}$	^{64}Ge 0^{+}	^{65}Ge $3/2^{-}$	^{66}Ge 0^{+}	^{67}Ge $1/2^{-}$		
33			^{69}Kr $5/2^{-}$	^{68}Br 3^{+}	^{67}Se $5/2^{-}$	^{66}As 0^{+}	^{67}As $5/2^{-}$	^{68}As 3^{+}	^{69}As $5/2^{-}$		
34			^{71}Rb $5/2^{-}$	^{70}Kr 0^{+}	^{69}Br $1/2^{-}$	^{68}Se 0^{+}	^{69}Se $1/2^{-}$	^{70}Se 0^{+}	^{71}Se $5/2^{-}$		
35			^{73}Sr $1/2^{-}$	^{72}Rb 3^{+}?	^{71}Kr $5/2^{-}$	^{70}Br 0^{+}	^{71}Br $5/2^{-}$	^{72}Br 1^{+}?	^{73}Br $1/2^{-}$		
36				^{74}Sr 0^{+}	^{73}Rb $3/2^{-}$	^{72}Kr 0^{+}	^{73}Kr $3/2^{-}$	^{74}Kr 0^{+}			
37				^{76}Y	^{75}Sr $3/2^{-}$	^{74}Rb 0^{+}	^{75}Rb $3/2^{-}$	^{76}Rb 1^{-}			
38				^{78}Zr 0^{+}	^{77}Y $5/2^{+}$	^{76}Sr 0^{+}	^{77}Sr $5/2^{+}$	^{78}Sr 0^{+}			
39					^{79}Zr $5/2^{+}$	^{78}Y 0^{+}	^{79}Y $5/2^{+}$				
40					^{81}Nb $3/2^{-}$	^{80}Zr 0^{+}	^{81}Zr $3/2^{-}$				
41					^{83}Mo $3/2^{-}$?	^{82}Nb 0^{+}	^{83}Nb $5/2^{+}$?				
42					^{85}Tc $1/2^{-}$	^{84}Mo 0^{+}	^{85}Mo $1/2^{-}$				
43					^{87}Ru $1/2^{-}$	^{86}Tc 0^{+}	^{87}Tc $1/2^{-}$				
44					^{89}Rh $7/2^{+}$	^{88}Ru 0^{+}	^{89}Ru $7/2^{+}$				
45					^{91}Pd $7/2^{+}$	^{90}Rh 0^{+}	^{91}Rh $7/2^{+}$				
46					^{93}Ag $9/2^{+}$	^{92}Pd 0^{+}	^{93}Pd $9/2^{+}$				
47					^{95}Cd $9/2^{+}$	^{94}Ag 0^{+}	^{95}Ag $9/2^{+}$				
48					^{97}In $9/2^{+}$	^{96}Cd 0^{+}	^{97}Cd $9/2^{+}$				
49					^{99}Sn $9/2^{+}$	^{98}In	^{99}In $9/2^{+}$				

*Not include the mirror nuclei $n > 5$, which satisfy the identity as the cases in the table.

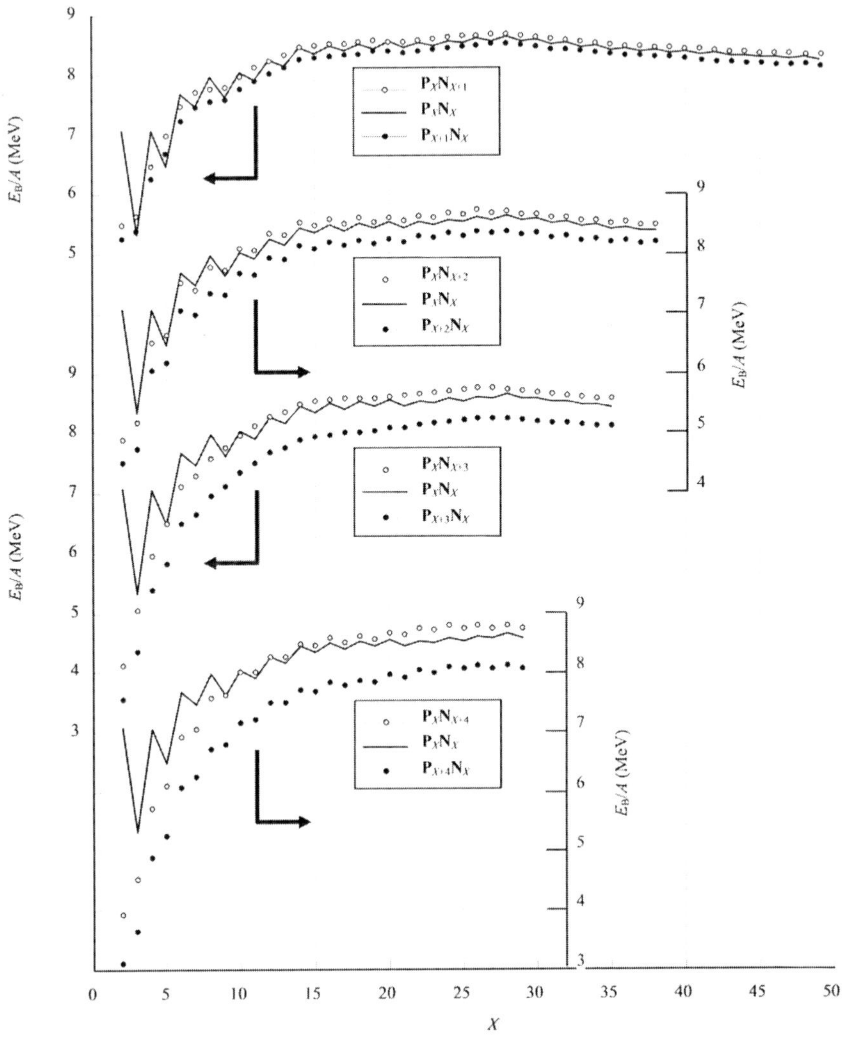

Figure 7. The comparison of E_B/A for mirror nuclei with the number of extra nucleon n = 1, 2, 3 and 4 (from top to bottom). For each case, the E_B/A difference between mirror nuclei is nearly a constant. When n = 1, 3, the curves are smooth; when n = 2, 4, they become zigzag.

6. CONCLUSION

A nucleus P_XN_Y (X, Y ≥ 2) is composed of a ring P_XN_X with P and N alternative binding plus Y − X extra N(¢s) if Y > X, or a ring P_YN_Y plus X − Y extra P(¢s) if X > Y. The ring folds with a "bond angle" of 90° for every 3 continuous nucleons. Extra N can bound to ring-P with the same "bond angle" and "bond distance". Extra P can bound to ring N in a similar way but the binding is weaker than that of extra P. When 2 or more ring P's are geometrically available, the extra N tends to be stable. Excess extra N results in open N with low E_B/A and neutron halo with large radius. Even-Z rings, as well as normal even-even nuclei, always have superimposed gravity centers of P and N; while for odd-Z rings, as well as odd-A nuclides, both centers of P and N must be eccentric. The eccentricity results in a depression of E_B. Therefore, the zigzag features of E_B/A of an element differing for odd and even Z can be simply explained by the shift of eccentricity by extra nucleons. Symmetry center may present in any even-even nuclei. While for odd-Z rings, only antisymmetry center is possible. In both cases of even and odd-Z, the rings are indistinguishable when P and N interchange. Based on the ring plus extra nucleon model, the mirror nuclei are equivalent in structure and demonstrate identical spin and parity. As the eccentricity shift, the E_B/A curves of mirror nuclei display smooth when the number of extra nucleon n is odd, and zigzag when n is even. The E_B/A difference between the pair is nearly a constant of 0.184n MeV.

ACKNOWLEDGEMENTS

Xiaodong Li, Qijun Liu, Gongyi Li, Yihe Li, Zengyong Chu The authors thank Prof. Jubo Zhu and Prof. Jianshu Luo for their effort in mathematical proof as well as many valuable discussions with Qinghua Wang, Tianjiao Hu and Zhenghua Jiang.

REFERENCES

1. Rowe, D.J. and Wood, J.L. (2010) Fundamentals of Nuclear Models, Foundational Models. World Scientific, Hackensack.
2. Martin, B.R. (2009) Nuclear and Particle Physics, an Introduction. 2nd Edition, Wiley, Hoboken.
3. Audi, G., Wapstra, A.H. and Thibault, C. (2003) The AME2003 Atomic Mass Evaluation (II). Tables, Graphs and References. Nuclear Physics A, 729, 337-676.
4. Audi, G., Bersillon, O., Blachot, J. and Wapstra, A.H. (2003) The NUBASE Evaluation of Nuclear and Decay Properties. Nuclear Physics A, 729, 3-128.
5. Holden, N.E. (2004) Table of the Isotopes. In: Lide, D.R., Ed., CRC Handbook of Chemistry and Physics, 85th Edition, CRC Press, Boca Raton.
6. Hansen, P.G. and Jonson, B. (1987) The Neutron Halo of Extremely Neutron-Rich Nuclei. Europhysics Letters, 4, 409- 414.
7. Wilson R.R. (1952) Radii of Mirror Nuclei. Physical Review, 88, 350-351.
8. Ivanter I.G. and Troitski M.A. (1964) Mass Difference of Mirror Nuclei. Journal of Experimental and Theoretical Physics, 47, 1773-1776.

9. Ajzenberg-Selove, F. (1996) Energy Levels of Light Nuclei A=13-15. Nuclear Physics A, 523, 1-196.
10. Bentley, M.A., et al. (2006) High-Spin Spectroscopy of Natural and Unnatural Parity States in the Mirror-Pair $^{45}V/^{45}Ti$. Physical Review C, 73, 1-10.
11. Usoskin, L.G. and Kovalsov, G.A. (2008) Production of cosmogenic ^{7}Be isotope in the atmosphere: Full 3-D modeling. Journal of Geophysical Research, 113, 1-12.

CHAPTER 5

Structure Mechanism of Ordinary Matter Mass Formation

Sergey N. Golubev

Laboratory of Quantum Standards, All-Russian Research Institute of Metrological Service, Moscow, Russia

ABSTRACT

The famous paradoxes of quantum mechanics are created by the fact that elementary particles exist as the alternation between two structural states with different properties. This leads to probabilistic behavior, uncertainty principle, quantum tunneling, etc. The alternation of states plays the role of the frequency generator or clock. But for the objective character of quantum interactions the length standard also should exist in nature. Such analog of the rule must be physically real and in direct sense have to participate in the of particles interactions. Just this is the main role of physical vacuum. For such role vacuum should have quasi-crystalline geometry structure. Its symmetry requires in standard views only one fundamental change. In the quasicrystalline structure of the vacuum, the virtual shells of the real particles and atomic nuclei are not diffuse "clouds", as is assumed today. Virtual environments are clearly structured and rigidly quantised shells with the geometric structure similar to fullerenes. Such shells are forming for greater than 99% of the known substance mass. Virtual particles forming such shells belong to the group of bosons and probably are just Higgs bosons existing in the ordinary matter. Chemical fullerenes form a series of discrete geometric structures. In a similar manner virtual analogues of fullerenes form a series of discrete masses, which really exist in the nature as a set of elementary particles and atomic nuclei masses.

Keywords: Quantum Mechanics, Physical Vacuum, Quasi-Crystalline Structure, Mass Formation, Dark Matter, Higgs Bosons, Gravitation

1. INTRODUCTION

One of the central unsolved problems of all physics is the existence of a discrete set of masses of elementary particles and atomic nuclei, which somehow prove to be stable. In this article, we shall find the answer and even will be able to explain rations between the masses which really exist in nature. Why protons mass is 1836 more than electrons mass? Why there are no stable particles with intermediate values of the mass? About these issues thousands of researchers thought for decades. Below we shall calculate the ratio 1:1836 just in the mind, without any calculating devices.

We will try to understand the main principle of the structural organization of quantum particles in ordinary three-dimensional geometrical space and at a normal time, but not in different variants of abstract space which are used for traditional formal descriptions of quantum processes. The article offers new realistic interpretation of quantum mechanics and realistic explanation for mass formation of ordinary matter, actually the mass of atomic nuclei and elementary particles.

2. SUBJECT OF INVESTIGATION

According to standard theory around any atomic nuclei and any elementary particle exists something like the diffuse shell or the "cloud" of virtual particles. The main subject of this article is the true structure of these vacuum shells. The analysis shows that the shells have discrete and strictly quantum geometric structure which determines important properties of the particles. Such vacuum shells are absolutely necessary and very important structure components of atomic nuclei and elementary particles.

According to the standard theory the true value of the electron charge is not observable, and its tabular value is the result of attenuation due to the partial shielding by the vacuum. The consistency of the electron charge indirectly implies that vacuum shell has some stable structure. Below it is shown that the numerical value of the famous constant 137.03605... is created by vacuum shell with geometric structure similar to fullerene C_{60}. A similar situation is also in the field of gravitation. Vacuum shells, composed of bosons and geometrically similar to the fullerenes, are creating more than 99% of the mass of atomic nuclei and elementary particles, i.e. the mass of ordinary matter. Structure and the nature role of these vacuum shells are the main subjects of this article.

3. METHODOLOGY OF THE NEW APPROACH TO STRUCTURE OF MICROCOSM

Particles with different masses have different structures. At the same time points do not have any internal structure by definition. Structural mechanism of mass creating does need to reject the view of particles as points. Today, such possibility has already appeared, but in the absence of methods for direct observations this requires a new approach to the problem.

The theory of relativity assumes that in each point of space there is a clock and the three-dimensional system of rules. In describing the macrocosm is quite enough to have the basic possibility to put a clock and a ruler to any desired point. In the formal description of the microcosm, this requirement is simply saved. In result we receive the look of a passive observer. For a participant in the events quantum measurement and interaction of particles can have an objective character only if in nature exists some physically real, not drawn on paper, analogue of clocks and rulers. Their presence seems to be impossible, but only at first glance. The operating principle of such clocks and rulers became clear after 40 years of electron microscopy research of the quantitative geometry of biological structures [1] - [8] .

In result of such approach we see the world of elementary particles already not as a collection of paradoxes, but as a rational and elegant nature device. Nature occurs to be an excellent engineer, not a fan of mathematical puzzles. We change our view not only on elementary particles, but also on ourselves, primarily on the microscopic structure of our own body.

4. OPERATION PRINCIPLE OF NATURAL FREQUENCY STANDARD AND REALISTIC INTERPRETATION OF QUANTUM MECHANICS

The natural frequency generator operates due to the fact that particles exist in the form of alternating two states with different properties. Just this alternation distinguishes elementary particles from small solids, as well as from atoms and molecules. This alternation of states creates the most famous paradoxes of quantum mechanics:

1) Probabilistic behavior. Quantum particles are involved in the physical interaction in both states but in completely different manners. Usually neighboring particles have random phase relation in alternation of states.

2) The uncertainty principle. The full range of particles properties is the sum of the properties of its two alternating forms. Because each particle moves the sum of properties can be determined only at such an interval of the particle movement when the both forms are present. This is the mathematical equivalent of the uncertainty principle standard definition.

3) The lack of classical trajectory. The line along which the particle moves in the form of spontaneous alternation of two states with different properties can be called the trajectory only conditionally. Such a line is not a classical trajectory. For this reason, the concept of classical trajectory does not apply to quantum particles.

4) The proper time of the particle. The period of oscillations with the alternation of states in terms of the Theory of Relativity corresponds to the concept of the proper time of the particle. At relativistic velocities, that quantum interconversion frequency decreases. At the speed of light, the alteration of states of photons and neutrinos disappears. The relativity theory calls this event the proper time stop of the particle.

5) Energy approach. Due to the alternation of states with different properties the force interactions between the particles attain a paradoxical specificity. That is why the energy approach is used for quantum particles instead force approach.

6) Geometry form and dimensions. The spontaneous attempts to represent the average view of two alternating states has led to the concept of particles without shape and size. Actually both alternating states of elementary particles have the size and geometric shape. At the first approximation, they are interconnected by a spatial inversion.

Four types of fundamental interactions for microcosm time scales do not appear simultaneously; instead, they belong to two tandems. Each tandem is peculiar to one of the two alternating states of a quantum particle, hence providing the main sense of the idea of the Great combination of physical interactions.

At one stage, the particle is involved in two types of interaction with a formally unlimited radius of action: electromagnetic and gravitational. At the transition to another state, the particle transforms into the proper diffraction picture. With this inversion of the geometric structure of the particle, its interactions with its theoretically unlimited radius of action transform into the interactions with a very small radius of action. Electromagnetic interaction proceeds to the weak nuclear interaction corresponding to the analytic theory of single electroweak interaction. Gravitational interactions with inversion of the structure of particles are converted into strong nuclear interactions.

According to standard theory, the electrostatic interaction between the charges is provided by the exchange of virtual photons, which are the quanta of alternating electromagnetic field. This aspect of the theory is fully consistent with our assertion of the existence of quantum particles in the form of the alternation of states with different properties. Specifically, at the scale of the microcosm, the electric charges and masses of the electron and proton are pulsating. At the stage of absence of electric charge the particles acquire the natural ability to tunnel through the electromagnetic barriers

5. OPERATION PRINCIPLE OF NATURAL LENGTH STANDARD, PROTON-ELECTRON MASS RATIO

The alternation of states of quantum particles explains the paradox of their behavior. Peculiarities of their geometric structure and mass formation are explained by the operation principle of natural length standard. The role of such three-dimensional system of length standards or rulers plays the physical vacuum with the quasi-crys- talline structure.

Ruler as the standard length requires at least two points. Therefore, a ruler cannot be inside each point of the microcosm. However, it is sufficient if each point forms part of the three-dimensional structure of natural rulers. It is impossible to obtain such rulers by returning to the classical hypothesis crystalline ether because all points of the classical ideal crystal cannot, in principle, be distinguished from each other. Therefore, the "division" of such rulers cannot, in principle, be digitised in any manner, either. The periodicity of the structure and the principal indistinguishability of its component points are, by definition, identical.

The necessary non-periodic structure was discovered by the famous British mathematician Roger Penrose approximately 50 years ago [9] . Nevertheless, for decades, three-dimensional Penrose patterns or pentagonal quasicrystals were perceived merely as entertaining mathematical puzzles. In 1984, quasicrystals were obtained in special metal alloys [10] . The Nobel Prize in Chemistry for obtaining metallic quasicrystals was awarded in 2011, approximately 30 years after the metal samples themselves and their diffraction patterns were obtained. Immediately following the discovery, the metal quasicrystals were considered "pseudoscience". Convincing evidence of the quasicrystalline of such ephemeral substance as the physical vacuum and such complex objects as living organisms require significant amounts of time to gather. The initial ideas of the quasicrystalline of the vacuum and biostructures were published in 1996 [3] [4] .

The non-periodic filling of the volume, i.e., for quasicrystal construction, requires only two varieties of elementary cells, namely, two types of rhombohedrons of a special form for which the ratio of volumetric diagonals equals the famous proportions of the golden section. This requirement is neither mystical nor accidental but is the result of the appearance of quintic symmetry axes. The presence of the axes is incompatible with the periodicity of the structure and is prohibited for classical crystals. For aperiodic quasicrystal structures, however, such symmetry is inevitable. Quasicrystals do not violate the known theorem on the prohibition of crystallographic axes of symmetry of the fifth order, but go beyond its axioms. The proof of the theorem does not assume that co-existence of two types of elementary cells is possible.

A physical vacuum with an aperiodic quasicrystal structure can be compared with the gas of crystal dislocations and considered a type of synthesis of two classical hypotheses of crystalline and gaseous ether. Such a vacuum, however, does not contradict the Theory of Relativity; moreover it is required for its applicability to the microcosm, thereby ensuring the availability of an analogue line at each point. Features of the quasicrystal symmetry that do not allow using such a vacuum as an absolute reference system are discussed in detail in books [5] [6] . For half a century, the understanding of quasicrystals has transitioned from being a mathematical puzzle to corresponding to key structure of our world.

The transition from abstract geometry to the structure of the vacuum is associated with the peculiarities of symmetry. To implement the axis of symmetry of the fifth order it is necessary to have 5, 10, 15, ... points. To implement the plane of symmetry it is necessary to have 2, 4, 6, 8, ... points. To implement the more complex combination consisting from the axis of symmetry of the fifth order and the plane of symmetry perpendicular to the axis it is necessary to have 10, 20, 30, ... points. To implement the entire set of symmetry elements of the quasi-crystall it is necessary to have 60, 120, 180, ... points. In other words, in the world with quasi-crystalline vacuum structure there are symmetry limitations for possible structures of virtual shells. Actually virtual shells of atomic nuclei and elementary particles are similar to fullerenes, of course only by the geometric structure. It must be clear that we are not talking about a hopeless attempt to extend the laws of chemistry on the atomic nucleus and the elementary particles, the similarity of the geometric structures is the result of the universality of the symmetry laws.

So, the condition for the existence of a natural standard of length is the quasicrystalline structure of the vacuum with infinitely large size. In the world based on such vacuum virtual shells of real particles and atomic nuclei acquire the structure of fullerenes. These structures have a certain finite size.

It is necessary to repeat and emphasize that for elementary particles we are talking about only one of the two alternating states. Geometric structure of another state of the same particle remains a subject for future research.

Today the foundations of the structure of fullerenes became widely known, so simple illustrations that I have presented in previous works, became redundant. But we will emphasize a number of features of these amazing chemical compounds of carbon that are important when comparing with the vacuum.

Today tools for industrial processing of diamonds are made from fullerenes. The high hardness of diamond was traditionally explained by the compactness of the atomic structure. The material from empty fullerene molecules occurs to be about 15% firmer. Such hardness and other unique properties of fullerenes are caused by the geometrical similarity of their molecules and the fundamental structures of the vacuum. In the unconscious form using of fullerenes in nanotechnology already started engineering using of physical vacuum effects.

Chemical fullerenes are formed only by carbon. It is the single element that creates chemical models of the vacuum shells, fractal increased in the size of course. The reason is associated with the numerical value of its mass and is explained in comments 2.2 to Table 2. The ability of living organisms to create analog models of vacuum structures goes much further. Despite the tremendous variety of organic substances, their carbon frame is displayed with high precision by a certain set of fragments, namely, only two crystal lattices of pure carbon: diamond and graphite. In any biological structure, such fragments of diamond and graphite lattices replace two types of rhombohedrons of a canonical quasicrystal, i.e., as a more complex quasicrystalline structure.

Direct geometrical similarity of the structures transforms biological cells in the analog models of physical vacuum world. Due to this biological cells live according to the laws of the quantum world without solving any equations.

Proton and electron mass ratio can be calculated simply in mind, without using any calculation equipment. Among all fullerenes the molecule C_{60} is distinctive. Elementary particles with a shell geometrically similar to C_{60} are the proton and the neutron. The algorithm for calculating the mass of these particles is more straightforward than expected. Real electric charge surrounded by the shell of 60 virtual particles becomes a complex of 61 particles.

Virtual particles do not have a proper weight, but the very existence of an ordered structure is possible only when specific interactions with each other and with the central core samples (i.e., with a real particle) occurs. In the system of 61 particles, only 1830 pairwise interactions are possible in a combinatorial sense. The contribution of each of them to the total mass of the proton equals the electron mass m_e that is a natural standard unit of mass. The total contribution of the shell is 1830 m_e.

The structure of the protons electric charge, the same is the structure of protons core, is not a subject of any chemical analogy but the subject of the quark theory. There are three quarks and three gluons which provide all possible combinations of interactions between quarks. Each quark mass and the dynamic equivalent of each gluon mass equal the electron mass m_e. Gluons have no resting mass; their mass is a relativistic mass equivalent. The overall proton core mass is 6 m_e and forms as the sum of the masses of three quarks with a mass m_e of each and that of the three gluons, each weighing m_e as well. The increase in the total mass of the proton to 1836 m_e is provided by a discrete virtual shell of 60 virtual particles.

The full estimated mass of a proton is 1836 m_e, with the actual weight being 1836.15 ... (m_e), with 0.008% of the calculating error. The virtual shell of the proton and neutron are identical, but for dynamic reasons, the mass of the neutron core is 8.7 m_e.

Protons shell consists of 60 virtual particles which are principally indistinguishable from each other. They belong to the group of bosons and perhaps can be identified with Higgs bosons. For the details of the comparison see below.

6. ATOMIC NUCLEI MASS FORMATION

Empirically found algorithm for calculating the proton mass (in units of m_e) in general is universal. But for composite atomic nuclei it is necessary to take into account some features. Alpha particle mass and the mass of all other atomic nuclei are formed in the process, which gives a realistic view of the nature of the strong nuclear interactions. At the confluence of the protons and neutrons in the single atomic nucleus their individual shells of 60 virtual particles are lost. Instead them arises a single common shell for the entire atomic nucleus as a single entity. According to the standard theory of strong nuclear interactions are non-Central in nature. Just such forces should result from the formation of the common shell around atomic nucleus.

The shell of an alpha particle (helium nucleus) is similar to fullerene C_{120}. Together with the core samples, they represent a set of 121 particles in which 7260 pairwise interactions are combinatorially possible. According to the empirically found algorithm they form a mass of 7260 m_e. The core mass of the alpha particle is the sum of the masses of two protons cores and two neutron cores:

$$(6+6+8.7+8.7)m_e = 29.4m_e. \tag{1}$$

The full weight of the alpha particle is the sum of the core mass and the mass generating by the shell:

$$(7260+29.4)m_e = 7289.4m_e = 4.00 \text{ amu.} \tag{2}$$

A similar mass calculation of chemical elements and their isotopes within the entire Periodic Table reveals the undeniable correlations between the structure of the atomic nuclei of the virtual shells and the real properties of the atoms themselves. The natural mechanism of mass formation is universal for all atoms of the Periodic Table of elements, as well as for the various elementary particles.

Algorithm used during protons and alpha particles mass calculating permits logical generalization. The result is an empirical equation for calculation atomic mass of all chemical elements. But the equation has important peculiarities. According to sense, the numbers of protons and neutrons, the numbers of virtual bosons in the shell are always only integer numbers. Integer equations are called Diophantus. But the name does not reflect the specifics of the equations, it reflects that physical meaning have only integer solutions of the equations. Such equations have two important features:

1) Such equations can be solved only by exhaustive search. Excluding very particular cases there are no general methods of solving.

2) Diophantine equations do not comply with the usual rule that one equation with two or three unknowns cannot be solved uniquely. Sometimes Diophantine equation with two or three unknowns has no solutions, and sometimes an integer solution is unique. Often there are several solutions, but one of them has obvious advantages in terms of symmetry or other parameters.

The calculation of the mass of the atomic nucleus using empirical Diophantine equations always includes modeling of geometric structures whose existence is demonstrated by all the independent data set. But for our purpose this is required. We are interesting not in the value of the atomic mass itself. We are interesting in the natural structure which forms just this mass value.

Algorithm of atomic mass calculation is reflected by empiric equation:

$$A = \frac{m_e}{1822.8873}\left[\frac{n(n+1)}{2} + 6Z + 8.7(A-Z) + Z\right] \tag{3}$$

A—atomic mass in atomic mass units (amu);

m_e—electron mass;

$1822.8873m_e$ = 1 amu,—standard table coefficient, ratio between different mass measurement units;

Z—number of protons in nucleus and the number of electrons in atom;

(A – Z)—number of neutrons in nucleus;

(n)—number of virtual particles in nucleus virtual covers;

$\frac{n(n+1)}{2}$ —number of all possible combinations of pairwise interactions in the system from (n + 1) particles, it means from one real particle and "n" virtual particles;

$6Zm_e$—total mass (in m_e units) of proton cores in nucleus, it means total mass of proton part in nucleus core;

$8.7(A - Z)m_e$—total mass (in m_e units) of neutron cores in nucleus, it means total mass of neutron part in nucleus core;

Zm_e—total mass of electrons in atom, including of this member transform the calculation of nucleus mass to calculation of atomic mass—compare with usual tables of atomic and isotope masses became more useful. The influence on atomic mass of interactions between nucleus and atomic electrons is not counted.

Calculation results are given in Table 1 and Table 2. We emphasize again that the reality of bosonic shells confirms not by opportunity to present atomic weight as mathematical sum in Equation (3). Real existence of nucleus virtual shells is confirmed by correlations between shell structure and real properties of chemical elements, main correlations are mentioned in comments to the tables. With that said, we should keep in mind that the most interesting calculating parameter is not the value of atomic mass, but the number of bosons in the atomic nucleus shells, that is, the parameter (n) in Equation (3) and at the same time the leftmost column in the Table 1 and Table 2.

Comments to Table 1.

Table 1. Masses of isotopes with nucleus shells with multiple to 60 numbers of virtual particles.

Number of virtual particles in atomic nuclear shell	Element	Isotope	Calculated mass	Actual mass	Calculated error, %	Occurrence of isotope, %	No. of comment
60	Proton		1836 m_e	1836.15m_e	0.08		
120	Alpha-particle		4.00	4.003	0.06		
180	Beryllium	^{9}Be	8.98	9.01	0.3	100	
240	Oxygen	^{16}O	15.93	15.99	0.4	dominate	
294	Magnium	^{24}Mg	23.99	23.985	0.01	79	1.1
300		^{25}Mg	24.88	24.986	0.4	10	1.2
306		^{26}Mg	25.97	25.983	0.03	11	
360	Chlorine	^{35}Cl	35.80	34.969	2.4	76	1.3
360		^{37}Cl	35.81	36.966	3.2	24	
420	Titanium	^{48}Ti	48.70	47.95	1.6	74	1.4
480	Cooper	^{63}Cu	63.60	62.940	1.0	69	1.5
480		^{65}Cu	63.61	64.928	2.0	31	
480	Zinc	^{64}Zn	63.61	63.929	0.5	49	1.6
540	Selenium	^{80}Se	80.48	79.917	0.7	50	
540	Bromine	^{79}Br	80.48	78.918	2.0	51	
540		^{81}Br	80.48	80.916	0.5	49	
600			Neutral vector boson				1.7
660	Tin	^{120}Sn	120.19	119.902	0.24	33	1.8
720	Neodymium	^{143}Nd	143.01	142.91	0.07	12	1.9
780	Erbium	^{168}Er	167.83	167.932	0.06	27	1.10
840			Platinum and Iridium				1.11
900			Francium and Radon				1.12
960			Einsteinium				1.13
1020	Element yet without approved name predicated by me in 1996						1.14
1080	Element which in case it will be synthesized should have mass 321.3						

1) Structure of virtual shells is based on pentagonal symmetry. Shell of 300 particles finishes the stage of fractal structure formation—appears 5 complexes with 60 particles in each. In the table for the first time appears the element with considerable weight part of not main isotopes.

2) Adding to the nucleus single neutron without any changing in shell structure increases calculated atomic mass only on neutron core mass which is 8.7m_e = 0.0048 aum. Actual differences between isotope masses can be explained only by small changes in number of particles in virtual nucleus shell. Effectiveness of such mass corrections changes with atomic mass increasing along Periodic Table. Atomic mass increasing caused by addind a single virtual particle to nucleus shell is proportional to already existing number of particles. For berillium and silenium these numbers differences in three times—180 and 540 virtual particles correspondently. But the total nucleus mass is approximately proportional to the square of virtual particles number. That is why from berillium up to silenium atomic mass increases in 9 times. And in percent form the result of single particle adding to the nucleus virtual shell is approximately inverse proportional to

the number of already existing particles. In the result 0.4% error of berillium atomic mass calculation is impossible to decrease by any correction of proposed number of particles in the shell—adding of a single particle at once changes calculated mass for 1.1%. For silenium adding of a single particle changes calculated mass only for 0.38%, so decrising calculation error on 0.4% occurs to be trivial correction. Along Periodic Tables such corrections for proposed number of virtul particles in nucleus shells became efficient just at the same level at which in nature appears elements with considerable weight part of not main isotopes (about 10%).

3) For minimization calculation error up to 0.1% you want to assume that isotope ^{37}Cl have the nucleus shell of 366 particles, isotope ^{35}Cl—of 356 particles.

4) For minimization calculation error up to 0.1% you want to assume that isotope ^{48}Ti have the nucleus shell of 417 particles, calculated mass 48.01.

5) For cooper isotopes middle mass is 63.546. Calculation error relativly this value is only 0.085%. Recall that middle mass of isotopes is always calculated with taking into account the relative occurance of isotopes. For minimization isotope masses calculation error you want to assume that isotope ^{63}Cu have the nucleus shell of 478 particles, isotope ^{65}Cu—of 485 particles. A similar reduction of error by fitting the estimated number of particles in the virtual shells further along the table, we will not comment.

6) Isotope ^{64}Zn with occurrence 49% is the main isotope of zink. Middle mass of zink isotops is 65.38 and deffers from mass of the main isotope more than one. Such great difference between middle mass and the main isotope mass occurs in the Periodic Table only few times. But Table 1 includes three such elements: zinc, selenium and tin.

7) Nuclei with shells of 600 virtual particles finish second stade of structure formation on the base of pentagonal symmetry. For nuclei with virtual shells of 300 particles calculated weight with the highest accuracy was in keeping with the magnesium isotope, relative frequency of occurrence of which is minimal. The trend continues on the nuclei with shells of 900 particles, such shells have not stable nucleuses of radon and francium. But just on the shells of 600 particles similar feature appears with maximum clarity.

Shell of 600 particles has an isotope of molybdenum ^{99}Mo, actual mass 98.908 and calculated mass 99.34, calculation error 0.4%. Half-life period is about 66 hours, this isotope is missing in nature. The same calculation mass and almost the same actual mass (98.906) has technetium isotope ^{99}Tc. The half-life period is about 200,000 years. With minimal removal of mass from the "fatal" values of half-lives of technetium isotopes ^{97}Tc and ^{98}Tc increase to approximately two and four million years, respectively. None of technetium isotopes almost never occurs in nature, although there are many isotopes in nature whose half-lives are considerably less than technetium. The lack of technetium is due to the fact that during natural nuclear reactions it is practically not formed and its natural resources are not renewed.

At first glance, the situation looks strange, because for reasons of symmetry in quasi-vacuum it is natural to expect that objects with virtual shells of 600 particles should be widespreaded. The main feature of these shells is that the main form of their existence in nature are not atomic nuclei but quite other particles—neutral vector bosons. Their mass is usually expressed in energy units; in fact it is almost identical with an atomic mass of technetium. Calculations of the mass of elementary particles, including neutral and charged vector bosons, see below. Vector bosons are the carriers of one of the four fundamental types of physical interaction—force weak nuclear interaction. In nature these are important and widespread particles, but because of the huge mass and short lifetime registration became technically possible relatively recently.

8) Isotope ^{120}Sn with occurrence 33% is the main tin isotope, see comment 1.6.

9) The most abundant neodymium isotope ^{142}Nd has occurrence 27%. On reality neodymium do not have main isotope and there is no correlation between occurrences of isotopes and exactness of coincidence calculated and actual masses.

10) The most abundant erbium isotope ^{166}Er (34%) needs for correction of virtual particles number.

11) Platinum and iridium have accurate correlation between occurrences of isotopes and exactness of coincidence calculated and actual masses. For isotope ^{195}Pt (34%) calculation error is 0.2%, for isotope ^{194}Pt (33%) calculation error is 0.3%, for isotope ^{196}Pt (25%) calculation error is 0.7%, for isotope ^{198}Pt (7%) calculation error is 1.7%.

For iridium isotope ^{193}Ir (63%) calculation error is 0.9%, for isotope ^{191}Ir (37%) calculation error is 1.9%.

12) Nuclei of francium and radon have virtual shells of 900 particles, calculated mass 223.40 in both cases. Francium have the single isotope ^{223}Fr with actual mass 223.02, calculation error 0.17%. For radon the most exactness of coincidence between calculated and actual masses have the most long-living isotope ^{222}Rn, actual mass 222.02, calculation error 0.62%.

Nuclei of francium and radon have virtual shells of 900 particles and they finish the third stage (300, 600, 900) of fracture structure formation based on pentagonal symmetry. In addition to instability there are another two pecualiarities. Firstly after francium and radon in Table 1 there are only artifcial elements. And the second peculiarity is connected with values of atomic radius—see below.

13) Virtual shell of 960 particles has einsteinium nucleus with calculated mass 254.16. In entire group of artificial transuranium elements exactly einsteinium stands out by the presence of isotopes with a relatively high half-life—about a year. For isotop ^{254}Es with a half-life of 275 days the error of mass calculation is only 0.03%. The most long-lived isotope is ^{252}Es with a half-life of 1.3 years. But the accuracy of the calculation of its mass is relatively small—error is 0.8%. Reduction of error up to 0.01% requires reducing the expected number of particles up to 956.

14) In the first publication of developing representations [4] I predicted the possibility of artificial synthesis of atoms with the shell of 1020 virtual particles and calculated mass 286.59. Then, in the calculation was neglected the difference between protons and neutrons masses, and also not included in the calculation the electrons mass. Consideration of these factors increases the calculated mass to 286.9. Predicted element has been synthesized in Dubna [11] . Its actual mass occurs to be about 288.

In the Periodic table the atomic radius varies with a distinct periodicity, which is complicated by a large number of smaller local minima and maxima. Almost all elements from the Table 1 are either directly at such minima points or tend to them. Directly in local minima are oxygen, chlorine, bromine, iridium, in the middle of two very shallow local minima are neodymium and erbium. For tin, the data are controversial, but poorly pronounced minimum, apparently, exists. On the edge of local minima are technetium, platinum, titanium, copper and zinc. For selenium the data are controversial and any interpretation is possible—selenium is the absolute champion of the Periodic table in the differences of atomic radius in measurements by different researchers.

The data on the atomic radius of radon is missing. Interpolation shows that the radius should be slightly larger than the xenon, and about the same as potassium. This severely disrupts the overall trend and is consistent with radon instability, the half-life of about 90 hours. Frances completely violates the General rule: he has the biggest radius of all the elements of the Periodic table. But since the half-life France is 22 minutes, it is the exception that just proves the rule.

From observed regularities, there is another exception of return character. There is a clear local minimum of the atomic radius, attributable to the iodine, which is not in the Table 1. But iodine also violates the General principle of the periodic table—it is lighter than the previous element, i.e., the atom falls out of the rule because it contains a lower number of neutrons.

In addition to diffraction measurements of atomic radii, there is a fundamentally different approach to estimate atomic sizes. The volume per one atom is calculated on the basis of the macroscopic density of the condensed phase and mass of one atom, expressed in macroscopic units. In the detailed reference there are such tables or graphics for at least part of the periodic table, but for independent calculation it is enough to have usual tables of density and atomic mass.

In this definition retains the overall relationship of the elements from Table 1 with local minima of the atomic radii, although some details of the picture change. In the evaluation of the atomic size on the basis of the macroscopic density there are no local minima per the halides—chlorine, bromine and iodine. But there is an additional minimum, which is magnesium. And beryllium together with boron and carbon, very little differing from each other, fall into the area of absolute minimum. In this definition of atomic volume they are the smallest atoms in the Periodic table. Not accidentally, the use of beryllium targets in many accelerator experiments. In this definition of atomic volume the hydrogen atom is much larger than atoms of beryllium or carbon. In my model this paradox has quite natural explanation: proton as the atomic nucleus of hydrogen preserves the alternation of two states with different properties and different sizes.

Fullerenes with multiple to 60 numbers of carbon atoms constitute only that part of fullerenes, which possess the highest symmetry. And chemical elements with multiple to 60 the numbers of particles in the virtual shells constitute only that part of the Periodic Table, which correspond to the minimum value of the atomic radii. Correlation between symmetry of the shell and geometric size is quite natural. Unexpected fact is the correlation with the size of the whole atom and not with the size of atomic nucleus. We must assume that the size of the shells close to the size of the entire atom and not to the size of its nucleus.

Experimental measurements of the atomic nucleus size refer to the core size. In experiments on the scattering even the existence of the virtual shells is not detected, therefore the measured sizes are not the sizes of the shell. The apparent contradiction of standard theory is poor compatibility with the official international table of atomic radii. For the most studied elements accuracy radii values is 0.01%. It seems to be obvious that such accuracy is impossible for the radius of the diffuse blurred boundaries.

In the proposed model, the tabular values of atomic radii are the radii of discrete virtual shells, whose existence in the standard theory is not known. In other words, virtual shells remain unobserved for all available devices, but just the sizes of these shells actually are summarized in the existing today official international tables of atomic radii. The appearance of this paradoxical situation is explained in the book [5] separately for different types of chemical bonds (ionic, covalent and metallic).

Physical properties of the space inside the shell and outside are different from each other. The shell creates the main part of atomic mass of the atom. In terms of the theory of relativity the space inside the shell can be considered as curved. In terms of electromagnetism, one can assume that inside the shell and outside of it the dielectric constant of vacuum is different. These changes affect the nature of propagation of electromagnetic waves inside the shells. Firstly this applies to virtual photons, carrying out interaction between the nucleus and orbital electrons.

The specific laws of chemistry are the result of differences in the properties of the valence and inner electrons. In the proposed model this fundamental basis of all of chemistry is explained by the fact that the valence electrons are located outside the atomic virtual shell and the inner electrons are inside that shell.

The general picture is the following:

1) The size of the virtual shell is comparable to the size of the whole atom. More exactly, the shell separates the zone of completely filled electron orbits from the zone of valence electrons. The size of the virtual shell with high accuracy coincides with the size of the external completely filled electron orbit.

2) The highest sublevels, d and f orbits, are always inside a virtual shell. That's why their electrons in chemical processes do not behave as valence electrons even in cases of partial filling of these orbits.

Founded algorithm atomic mass calculation applies in cases where the number of particles in the virtual shells is not multiple to 60 (Table 2).

Comments to Table 2.

1) There exist so called endometallofulleren Me C_{85}. Each molecule is the complex of 85 carbon atoms in the shell and a single metal atom in the centre. Deuterium nucleus is the structure analogue of such molecule.

2) Nucleuses with proton or neutron numbers 2, 8, 20, 28, 50, 126, 208 have improved stability and are called "magic". Special status of these numbers or discrete structures also exist on the level of elementary particles shells. But among atomic nucleuses only carbon isotope ^{12}C has the shell of "magic" number of virtual particles—208 particles. Such shell has extremely interesting peculiarity. Isotope mass ^{12}C with accuracy of approximately 2% coincides with the mass difference between two particles carriers of one of the four types of fundamental interactions: weak

nuclear forces. Such particles, as is widely agreed, are charged and neutral vector bosons.

Table 2. Masses of isotopes with nucleus shells with not multiple to 60 numbers of virtual particles.

Number of virtual particles in atomic nuclear shell	Element	Isotope	Calculated mass	Actual mass	Calculated error, %	Occurrence of isotope, %	No. of comment
85	Deuterium		$3670.7m_e$	$3671.3m_e$	0.016	Small	2.1
104	Tritium		3.01	3.016	0.25	0	
104	Helium	^{3}He	3.01	3.016	0.28	Small	
147	Lithium	^{6}Li	5.99	6.015	0.4%	7.5	
159		^{7}Li	7.01	7.016	0.1	92.5	
190	Boron	^{10}B	10.00	10.013	0.2	20	
199		^{11}B	10.96	11.009	0.4	80	
208	Carbon	^{12}C	11.98	12.000	0.20	Dominate	2.2; 2.3; 2.4
225	Nitrogen	^{14}N	14.01	14.003	0.04	Dominate	
262	Fluorine	-	19.04	18.998	0.24	100	
269	Neon	^{20}Ne	20.01	19.992	0.1	90.5	
282		^{22}Ne	21.99	21.991	0.03	9.2	
288	Natrium	-	22.93	22.990	0.26	100	
318	Aluminium	-	27.94	27.982	0.14	100	
318	Silicon	^{28}Si	27.95	27.977	0.1	92	2.5
335	Phosphorus	-	31.01	30.974	0.1	100	
340	Sulphur	^{32}S	31.94	31.972	0.1	95	
376	Kalium	^{39}K	39.04	38.964	0.2	93	
380	Argon	^{40}Ar	39.88	39.962	0.2	Dominate	
380	Calcium	^{40}Ca	39.87	39.963	0.2	Dominate	
404	Scandium	-	45.07	44.956	0.26	100	2.6
…	…	…	…	…	…	…	2.7
450	Iron	^{56}Fe	55.91	55.935	0.044	91	2.8
…	…	…	…	…	…	…	
923	Uranium	^{235}U	234.91	235.044	0.05	0.7	
929		^{238}U	237.98	238.051	0.03	Dominate	2.9

Electromagnetic resonances arise when the photon energy matches the energy difference between two levels of an electron and sometimes other particles. Resonances in the physical vacuum arise at the interaction with mass (energy) equal to the mass difference between charged and neutral vector bosons. Greater than 99% of all known substances create the virtual shell of nuclei, whose geometric structure is similar to that of fullerenes. Carbon, forming the chemical fullerenes, is the only element that creates chemical analogues of the most fundamental structures of the physical vacuum. Of course, the molecular analogues are created with a fractal increase in the absolute size. This similarity of structures is the real cause of the extraordinary properties of chemical fullerenes.

These direct geometric similarities of the structure of matter and the structure of the physical vacuum are keys to many biological problems. The possibilities of living organisms in terms of the creation of such chemical imitations or analogue models of physical vacuum shows greater potential. The uniqueness of carbon is directly related to the existence of life. For a proper understanding, it should be added that the occasionally re-emerging hypothesis concerning the possibility of life based on silicon is a misunderstanding, regardless of recent results. Silicon does not form compounds with aromatic type of connections, and without the latter, it is impossible to obtain an analogue of biochemistry.

3) Atomic nucleus of the main carbon isotope C^{12} has a shell of 208 virtual particles. A complex of 208 protons and neutrons constitutes the nucleus of lead Pb^{208}. There is also a specific fractal, the existence of which cannot be displayed as part of the conceptual apparatus of the standard theory. Among the many toxic heavy metals, lead in this regard does not stand apart. However, this element has a unique ability to penetrate into the body through intact skin. Fractality by definition signifies the presence of a geometrical similarity amongst structures, in this case, subatomic structures. A general rule, i.e., "similar dissolves similar", has been known to the field of chemistry since the times of alchemists.

4) In tandem carbon—silver surprising effect can be seen. Silver has two isotopes Ag^{107} and $Ag^{109}$with practically equal occurrences. For the formation of the conditionally averaged atomic mass of 108 units would be required shell of 624 virtual particles that is exactly 3 times more than 208 particles in the carbon shell. The atomic masses 108 and 12 units differ from each other exactly 9 times. Atomic volumes of silver and carbon are respectively 10.27 and 3.42 (diamond), i.e. differ from each other practically exactly 3 times.

In this case virtual particles in the shells of silver and carbon are identical by their characteristics. In all other cases, changes in the structure of nuclear shells lead to significant changes in the atomic mass, which corresponds to nuclear reactions and higher energies. And only between the silver and the carbon isoenergetical exchange of virtual particles in the composition of the shells became possible.

In addition to the specifics of the bactericidal properties of silver recall classic photo. Attempts to get a photo without silver failed. And attempts to substitute by other materials photographic gelatin (technically processed collagen protein) also failed. The photographic process is clearly based on the unique properties of tandem silver-carbon and is more complicated than it is common taken to think. In the standard interpretation of photographic developer is considered to be simply a chemical reducing agent of silver. But in all molecules of actually used developers there are aromatic rings. For using as reducing agents, they do not hinder, but do not help. Their constant presence in the structure of molecules developers does not correspond to a simple explanation of the photographic process.

5) Among the minerals of the earth's crust the weight part of aluminosilicates is more than the part of other minerals all together. This propensity of aluminum and silicon to the formation of joint compounds indirectly related to the identical structure of the virtual shells.

6) During the Periodic Table scandium firstly starts filling side of the subgroup. There is a real correlation with the fact that the scandium crosses important structural milestone of 400 particles in a virtual shell.

7) In Table 1 and Table 2 there are included the most important isotopes of all elements in the interval from hydrogen to scandium, without any lapses. After scandium calculation of the mass of all elements and isotopes becomes trivial fitting task (see note 1.3). In this part of periodic table I shell review only two interesting examples, the iron and uranium.

8) Iron in the periodic table corresponds to the energy optimum of nuclear forces. In this respect, it is remarkable small error in calculating the atomic mass of the main isotope of iron. The number of particles in the iron shell is exactly in the middle between the two "canonical" shells from 420 and 480 particles.

9) Transuranic elements that can be created artificially, finish periodic table on the level at which the number of virtual particles in the shell is equal to about 1000. It is just the limit for the ensemble of particles, which can be arranged according to geometric laws of clusters. This structural limitation have geometry reason and have universal application: chemical homogeneous clusters also can contain only about 1000 atoms. With further increase in the number of atoms arise notoriously heterogeneous system in the form of a mixture of microcrystals, clusters and fragments with random amorphous structure. Calculation of transuranic elements mass can be performed in the scope of virtual shells model (Table 1).

Adding to the vague end of transuranic part of the periodic table there is exactly clear undisputed border: natural part of the periodic table ends by isotope of uranium ^{238}U. A unique feature of the nucleus ^{238}U is quite remarkable value for the mass of its shell. The nucleus consists of 92 protons and 146 neutrons. Its mass is the sum of weights 92 proton cores and 146 neutron cores. In units of m_e the mass of the core ^{238}U is: $6\times 92+8.7\times 146=1822.2$. This is 99.25% of the total proton mass. Almost exact match with the full mass of the proton is not accidental. But even such accuracy estimate is understated, with respect to the discrete structure of the atomic nucleus, the accuracy is 100%. Creation of the next element means adding to ^{238}U nucleus one proton core and at least one neutron core. As a result, the boundary of the total proton mass must be crossed. But such nuclei are not formed as a result of natural processes. We will see below that the intersection of the boundary by core mass of vector bosons leads to the emergence of the phenomenon of "negative mass" (!). This structural restriction refers just the natural process of nuclei formation, but not to their existence. That is why artificial production of transuranic elements became possible.

7. MASSES OF ELEMENTARY PARTICLES

The masses of elementary particles are calculated with the same algorithm. But unlike atomic nuclei, many elementary particles have a fractal system of multiple virtual shells. The main interest in this article is not the computation of particle mass itself, but empirical attempt to find such geometric structure of bosons shells, which is forming just really existing mass values. In all cases the criterion of the correction for the geometric reconstruction is a structural alignment with other properties of the same particles.

According to the standard theory the strong interactions have such property as saturation. In my model of vacuum shells this view is clarified by empirical fact that the energy of each pairwise interaction is strictly quantized, taking the following values:

1) In the shells with the number of virtual particles 60 and over this energy is m_e. This value is by five orders of magnitude more than the characteristic energy of one chemical covalent bond. At this energy there is a saturation of bonds between virtual particles in nuclear shells. Just because of this when calculating the mass of atomic nuclei we have not met with any attenuation coefficients for the intensity of the interactions.

2) In shells of less than 60 virtual particles energy equivalent of single interaction is smaller because interactions are not saturated.

And their energy is rigidly quantized and takes the values:

- In shells of 20 - 40 particles equal to (m_e/π)
- In shells of less than 20 particles equal to $(m_e/4\pi)$.

The appearance of the number π as the coefficient of attenuation of interactions occurs also in the calculation of the electromagnetic constant 1/137 when the accuracy of calculation is 0.0001% (see below). It is no doubt that the appearance of the number π is caused by some exact structure reason. But for its understanding we need more detailed information.

7.1. Mass of Heavy Leptons

This group includes two types of particles that do not participate in strong nuclear interactions. Properties of heavy leptons are analogues of electrons or positrons, but differ from them multiple increased mass.

a) Hypothetical reconstruction gives the following structural picture. The mass of the Tau-lepton is created by the real electron mass in the role of primary generator and fractal system of three virtual shells in the role of gravitational amplification. Structure of three shells is the following:

1) The shell of 11 particles. Together with the real electric charge with mass m_e (an obvious analogue of the electron) forms a complex of 12 particles in which 66 pairwise interactions are possible combinatorial. With such a small number of particles the contribution of each interaction to the total mass the Tau lepton is not m_e, but $m_e/4\pi$. The appearance of π is the empirical fact, but we will meet it again several times. Together with the mas of the electric charge (m_e) the contribution of this complex to the total mass of the Tau lepton is:

$$m_e + (66/4\pi) m_e = 6.2521 m_e. \tag{4}$$

It is quite obvious that in ordinary space complex of 12 particles has the shape of an icosahedron with 12 vertices.

2) The shell of 20 particles, the role of core performs the first shell. In a system of 21 particles 210 pairwise interactions are possible combinatorial, contribution of each interaction is m_e/π. The contribution of this shell to the total mass of the Tau lepton is

$$(210/\pi) m_e = 66.8451 m_e. \tag{5}$$

It is quite obvious that in ordinary space the shell from 20 particles has the shape of a dodecahedron with 20 vertices.

3) The shell of 82 particles, the number 82 is one of the so-called "magic" numbers. In the role of core performs the second cover together with the first shell inside her. In the complex of 83 particles 3403 pair interactions are possible combinatorial. The contribution of each pairwise interaction in a total mass equal to m_e and coincides with the contribution of each pairwise interaction in the formation of the mass of the proton, and atomic nuclei. The whole contribution of the shell to the total mass of the Tau-lepton is 3403 m_e. Within the meaning of the algorithm this value can be only integer.

Total calculated mass of the Tau-lepton is the sum of the three masses:

$$6.2521 \cdot m_e + 66.8451 \cdot m_e + 3403 \cdot m_e = 3476.0972 \cdot m_e \quad (6)$$

Actual mass is (3477.5 ± 6)m_e. High accuracy of coincidence of the actual and estimated mass confirms the correctness of particles structural reconstruction.

Some comments about shell of 82 particles are necessary. Atomic nuclei with the number of protons or neutrons 2, 8, 20, 28, 50, 82, 128, 208 are the most stable. Actually with these numbers are connected structures of increased stability. Conventionally they are usually mentioned as "magic" numbers. It is natural to assume that similar structures exist at the level of virtual shells. Direct substitution into the corresponding Diophantine equation gives the exact coincidence of the calculated and actual mass of the Tau-lepton, which confirms the correctness of structural assumptions about the existence of a virtual shell of 82 particles.

b) The mass of the muon is known with high exactness, so its calculation will be even more convincing. Hypothetical reconstruction gives the following structural picture. Muon and Tau-lepton have only one identical structural component—it is the first shell (real electric charge and 11 virtual particles). Contribution of this component in the total mass is the same—6.2521 m_e. The third shell of 82 particles in the muon is simply absent. And the second shell is undergoing a very interesting change. The number of particles in this shell increases in such proportion that its contribution to the total mass of the muon increased by exactly 3 times, the threefold symmetry is realized not only at the level of quarks. This is achieved by increasing the number of particles from 20 particles in Tau-lepton structure up to 35 in muon structure. In a system of 21 particles (20 virtual + core) 210 pairwise interactions are possible combinatorial. But in a system of 36 particles (35 virtual + core) 630 pairwise interactions are possible combinatorial. Contribution of the second shell to the total muon mass in m_e units is: $630/\pi = 200.5352$. Total mass of the muon is the sum of these two masses:

$$6.2521 \cdot m_e + 200.5352 \cdot m_e = 206.7873 \cdot m_e \quad (7)$$

Actual mass of the muon is 206.7682m_e, the error of calculation less than 0.01%. High accuracy of coincidence of the actual and estimated mass confirms the correctness of particles structural reconstruction.

In accordance with the empirical algorithm of increasing the number of particles in the shell from 20 up to 35 increases its contribution into the formed mass exactly three times. In contrast to this, for the initial shell of 19 or 21 particles the task of increasing three times has no solution in integers. This is a typical situation with Diophantus equations and convincing illustration of the fact that the reconstruction of the particle structure is not arbitrary, as it may seem at first glance

High accuracy of coincidence between calculated and actual mass of muon once more provides the validity of virtual shells model. But in the standard model all leptons are considered to be fundamental particles, i.e., not consisting of any structural components. Calculation of heavy leptons mass proves the component structure of these particles. Another thing is that the conceptual framework of the standard model does not assume the existence of virtual shells—this is a fundamentally new type of physical objects. There is concrete correlation between leptons virtual shells structure and their non-participation in strong nuclear interactions.

Heavy leptons have the unique structural feature. As we have seen both heavy leptons form virtual shell from 11 particles. For sure in details it is not a shell around real particle with electric charge, as in all other cases. From the symmetry it is practically obvious that heavy leptons form virtual shell with direct inclusion of real particle (real electric charge), using them as the twelfth vertex of the dodecahedron.

Among the elementary particles of different types, there is a phenomenon which in current terminology is called the existence of several "generations" of particles. There are similar particles with tremendous differences in mass values. For example, electron, muon and Tau-lepton represent three generations of leptons. Standard model has no explanation for this phenomenon. Now we see that the existence of several generations of the same particles is explained by the fact that the original particle can be supplemented with different variants of virtual shells. This leads to tremendous differences in the masses and is not accompanied by changes in any other properties of the particle.

General idea of the theory of quarks ("the principle of 1/3") in a latent form is realized in the field of gravity, in the formation of muon and Tau-lepton masses, i.e. even those particles, which electric charge which is formed without participation of quarks.

In search of the general harmony of nature, the parameters of planetary orbits many times, starting with I. Kepler, were compared with the configurations of regular polyhedra, the so-called "music of the spheres". In the fractal structure of virtual shells Tau-lepton really implemented configuration in the form of an icosahedron inside the dodecahedron. But such forms exist only at the stage of electromagnetic and gravitational interactions. At the stage of existence of the strong and weak nuclear forces, the geometric shape is changing dramatically. To a first approximation the shape undergoes a geometric inversion. But the specific reconstruction of the geometric shape of the particles on the stage with existence of nuclear forces remains a subject of future investigations.

7.2. Pions Mass

a) Let us try to predict the mass of such particles, which is the simplest complex of two quarks. Its shell should consists of 40 virtual particles,—instead 60 particles in the shell of protons system with three quarks. Together with real core such particle is the complex of 41 particles with combinatorial possible 820 pairwise interactions. Their contribution to the total mass in m_e units is $820/\pi = 261.01$. Contribution to the total mass of real core is 3 m_e—two quarks and a single pairwise interaction (Figure 1(a)). Whole calculated mass is 264.01 m_e.

And now let's check the prediction of the model on the facts. Particle with calculated mass really exist. It is the neutral pion, according to standard theory it is just the simplest complex of two quarks, more exactly—quark and antiquark. Actual mass 264.15 m_e, calculation error is 0.04%.

b) The empirical reconstruction of charged pion structure gives the following picture. Charged pion in comparison with neutral pion has 3 additional quarks forming electric charge. In total core has 5 quarks and 7 pairwise interactions combinatorial possible (Figure 1(b)). Contribution to the total mass of such core is 12 m_e. Together with the shell mass 261.01 m_e whole calculated mass is 273.01 m_e. Actual mass 273.12 m_e, calculation error is 0.05%. High accuracy of coincidence of the actual and estimated mass confirms the correctness of particles structural reconstruction.

7.3. Mass of Neutral and Charged Vector Bosons, the Existence of Negative Masses

Neutral and charged vector bosons are carriers of weak nuclear forces. More precisely, according to standard theory, photon and two types of vector bosons are carriers of the unified electroweak interactions. In comparison with other particles vector bosons have giant masses.

The actual mass of neutral vector boson (Z^0 boson) is (91,187 ± 7) MeV, or (178,450 ± 15) m_e, it's about 97.9 aum. Hypothetical reconstruction gives the following structural picture. The major part of this mass is formed by the virtual shell of 600 particles. Together with the core complex from particles 601 is obtained, where 180,300 pairwise interactions are possible combinatorial. The contribution of the shell to the total weight is 180,300 m_e— a little bit more than actual mass. It may seem strange, but the central core occurs to have negative mass. Z^0 boson has the core which mass is modulo equal to the mass of the neutron:

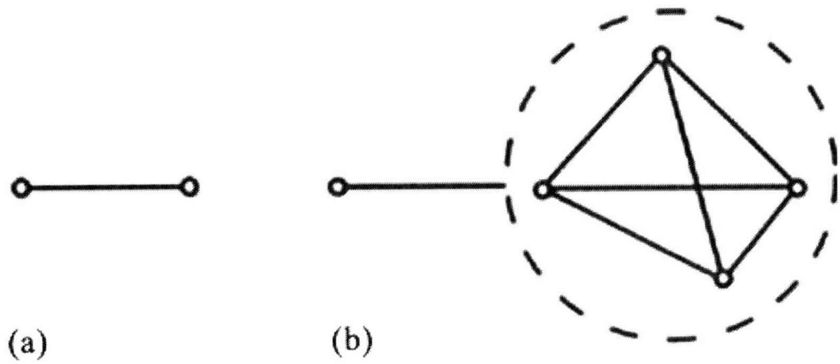

Figure 1. Structure of pion core: (a) Neutral pion, (b) Charged pion.

$$(178450 \pm 15)\cdot m_e - 180300\cdot m_e = (-1850 \pm 15)\cdot m_e \ . \quad (8)$$

This quantitative agreement cannot be accidental, and indirectly confirms the correctness of particles structural reconstruction.

The actual mass of the charged vector bosons (W^+, W^- bosons) is (80,150 ± 100) MeV, or (156,850 ± 200) m_e, it's about 86.0 aum. Hypothetical reconstruction gives the following structural picture. A virtual shell of such bosons contains 560 particles, i.e. differs from the shells of neutral Z^0 boson to double of the "magic" number (20 + 20) particles. The mass of a charged vector boson is formed as follows. In the complex of the real core and 560 virtual particles 157,080 pairwise interactions are possible combinatorial. The effect of the presence of the shell corresponds to 157,080 m_e—a little bit more than actual mass. Charged bosons also have cores with negative mass:

$$(156850 \pm 200)\cdot m_e - 157080\cdot m_e = (-230 \pm 200)\cdot m_e \ . \quad (9)$$

Probably, the core mass is modulo equal to muons mass. But for unique choice between muon and pion masses experimental measurements of the W^+, W^- bosons masses should be more exact.

Standard theory of the unified electroweak interactions (model by Salam and Weinberg) do not use the concept of negative mass, but the theory uses particles whose square seed mass are negative numbers. Accordingly such masses themselves are considered to be imaginary. Concepts of negative mass and imaginary mass have a lot in common and the differences are partly terminology. Using of negative mass concept is discussed below.

8. ELECTROMAGNETIC CONSTANT 137.03605...

It is logical to assume that the numerical value of the famous constant is also determined by quasicrystal vacuum. Hypothetical structural reconstruction involves that in addition to the gravitational shell around each elementary electric charge is formed electromagnetic shell of 60 virtual particles. In contrast with gravitation these particles are divided into two groups—"electric" and "magnetic" types. Particles are divided into groups in the same proportion as electric charge between quarks—20 and 40 particles. Real electric charge in the center is included in both groups. So there are two groups of 21 and 41 particles. The cooperative interaction between groups is the result of $21 \times 41 = 861$ pairwise interactions. Using coefficient 2π we obtain a dimensionless constant: $861/2\pi = 137{,}032406$. Because of the participation of the central particle in two groups the amendment is necessary. A more exact value of the constant is given by the empirical ratio:

$$\frac{861}{2\pi} + \frac{\pi}{861} = 137.036054774\cdots \qquad (10)$$

Exactness of the coincidence of this number with the value recommended by the "Committee on Data for Science and Technology" ("CODATA") is at the level of 0.0001% or 1 p.p.m. Actually in the experiments some metrological centers in different countries gets a value slightly more and some slightly less than the value calculated by me [12] . High accuracy of coincidence of the constants values confirms that the starting assumptions in the beginning of this section correspond to reality.

9. DISCUSSION

9.1. Dark Matter

Dark matter is not the object simply invisible from the Earth, but extremely unusually effects of gravity itself. Direct astronomical observations allow us to set the parameters of the stellar orbits. And further attempts to calculate the mass of the Central body of the galaxy lead to a paradoxical picture. It turns out that the stars on the fringes of the galaxy rotate around the Central body of suitable mass, and the stars in the central part of the same galaxy revolve around the Central body whose mass is many times smaller. Such situation is impossible with a traditional point of view and absolutely natural in the scope of virtual shells model.

If we could look at the single proton using astronomical methods, the presence of a core and of a virtual shell, we would have interpreted as existence of protons own mass 6 m_e and the concomitant existence of dark matter with a mass of 1830 m_e, i.e. 300 times greater. Dark matter is an aggregate of virtual shells of astronomical objects and with astronomical scales. Such shells really have properties of natural amplifiers of gravity, but the results can be observed only outside dimensions of the virtual shell.

Perhaps our Sun has such virtual shell. Two American space vehicles "Pioneer 10" and "Pioneer 11" at intervals of a year in the same area on the periphery of the Solar system have experienced unexplained braking. I propose that vehicles crossed the suns virtual shell and got into the zone of action of solar gravity natural amplifier.

Only astronomers can professionally and critically compare the results of observation with hypothetical explanation in the framework of the virtual shells model. Just now it is only worth to say about the possibility of unexpected approach to the largest mass in the Universe.

9.2. Higgs Bosons

All 60 virtual particles in the shell of each proton are principally indistinguishable from each other. According to generally accepted particles classification they belong to the group of bosons. The particles in the composition of virtual shells of real elementary particles and atomic nuclei actually generate more than 99% of ordinary matter mass. Namely, it is just the main property of the Higgs bosons. More detailed analysis of the properties of virtual particles from shells, as well as consideration of the complex of related issues is given in the books [5] [6] .

It is crucial that the bosons in the virtual shells do not have any own masses. As we have seen along the whole Periodic table their contribution to real mass generation depends on the number of such virtual particles in concrete shell. That is why in standard theory mass of Higgs boson itself remained unknown. At the same time in the experiment it is impossible to separate bosons own mass and its contribution to the total mass of the real particle. In such situation the experimentally measured value must radically depend on what real particles were used as a target. A unique feature of the Higgs boson is that it creates a mass of other particles, but does not have any own mass. It seems to be obvious that only particle with such combination of properties by definition may be called the Creator of mass.

Information about experimental confirmation of the Higgs bosons existence in 2012 sparked a surge of scientific interest in this particle [13] . A review of the literature on this subject is not the subject of the article. But it is important to emphasize that the proposed model of discrete bosonic shells is not a marginal alternative to the Higgs bosons. Apparently, these are two different approaches to the same problem, which future research will merge into the overall picture.

Generation of real mass is a result of interaction between virtual shells particles and specific field. In absentia it was called the Higgs field. But I am sure that fundamentally it is just alternating gravitational field. About its properties we know very small, but enough to suggest that the most intriguing modern mysteries of biology and other sciences are directly connected with alternating gravitational fields. In such fields arise very considerable strengths, evident traces of their action are obviously seen, for example, on a usual school globe [6] .

9.3. Concept of Negative Mass

Without any explanations about necessity of using the concept of negative mass long before me wrote the founder of solid state physics L. Brillouin [14] . Today it is worth to divide answers on two questions: why negative masses are necessary and why they are possible?

It is obvious that only particles with negative mass can generate attractive forces as result of exchanging interactions. The exchange of vector bosons leads to the appearance of attractive forces only because inside their structure exist components with negative mass, as we have seen above.

For elementary particles the concept of negative mass must be understood in a special sense. Existence of elementary particles not in empty space, but surrounded by physical vacuum inevitably changes the zero level of mass determining. In any real experiment it is impossible to provide selective force exerted on certain pre-selected elementary particle. Only force field in the surrounding space can be created.

The action of the gravitational field on elementary particle surrounded by vacuum conventionally can be compared to behavior in water steel balls and air bubbles. Actually air bubbles demonstrate only illusion of the negative mass, the air bubbles move against the forces of the gravitational field due to the force of Archimedes. But obvious understanding of the true situation is based on using concepts of density to water and air. To elementary particles and to physical vacuum mechanical concept of density is not applicable and adequate replacement does not exists.

In such a situation the zero level of mass determining is shifted and appears the illusion of negative mass. In vector bosons structure there are components whose interaction with the external gravitational field is weaker than that of the virtual vacuum particles. And just these structural components demonstrate the illusion of negative mass. This is a forced compromise due to lack of knowledge and imperfection of the conceptual apparatus. But compared with the representation of particles as mathematical points this seems a definite step forward.

10. CONCLUSIONS

The existence of natural standards of length and time defines the real distinguishing features of the microcosm. The standard time or frequency is created by the existence of quantum particles in the form of alternating two states with different properties, which creates the famous paradoxes—uncertainty principle, etc. Standard length is created by physical vacuum with special geometric structure similar to quasi-crystal.

In the world with such vacuum around elementary particles and atomic nuclei arise clearly structured shells, which form more than 99% of the mass of ordinary matter. Shells consist of bosons, which perhaps can be identified with the Higgs bosons.

The geometric structures of the virtual bosons shells are analogous to the fullerenes. Discrete variants of such structures create that set of masses of elementary particles and atomic nuclei that really exist in nature. In electromagnetism the virtual shell, geometrically similar to the fullerene C_{60}, determines the numerical value of the famous constant 137.03605...

CITE THIS PAPER

Sergey N. Golubev (2016) Structure Mechanism of Ordinary Matter Mass Formation. *Journal of Modern Physics*,**07**,875-891.

REFERENCES

1. Golubev, S.N. (1981) Real Crystals in Coccolithophore Skeletons. Nauka, Moscow, 164 p.
2. Golubev, S.N. (1987) Journal of General Biology, 6, 784-806.
3. Golubev, S.N. (1996) Consciousness and Physical Reality, 1, 85-92.
4. Golubev, S.N. (1996) Consciousness and Physical Reality, 1, 37-42.
5. Golubev, S.N. (2013) Quasicrystalline Structure of the Vacuum: The Key to Unlocking the Secret of Living Cells and Quantum Particles. Librokom, Moscow, 256 p.
6. Golubev, S.N. (2013) The Origin of Life and the Formation of Masses of Elementary Particles in the World of Quasicrystalline Vacuum. Folium, Moscow, 348 p.
7. Golubev, S.N. (2014) Achievements in the Life Sciences, 8, 1-9.

8. Golubev, S.N. and Golubev, S.S. (2009) Look at the Physical Position of the Microcosm from the Biologist Perspective. Dalnauka, Vladivostok, Series of the "Achievements in the Life Sciences", 245 p.
9. Penrose R. (2003) Emperor's New Mind: In Regard to Computers, Minds, and Laws of Physics. "Editorial URSS", Moscow, 384 p.
10. Schechtmann, D., Blech, S., Gratias, D. and Cahn, J.W. (1984) Physical Review Letters, 53, 1951-1953.
11. Kapitonov, I.M. (2002) Introduction to the Physics of Nuclei and Particles. "Editorial URSS", Moscow, 384 p.
12. Mohr, P.J. and Taylor, B.N. (2005) Reviews of Modern Physics, 77, 1-107.
13. Baggott, J. (2014) Higgs: The Inventation and Discovery. Centrpoligraph, Moscow, 255 p.
14. Brillouin, L. (1972) A New Look at the Theory of Relativity. Mir, Moscow, 144 p.

CHAPTER 6

Fine Structure Calculations of Atomic Data for Ar XVI

A. I. Refaie

Physics Department, Faculty of Science, Cairo University, Giza, Egypt

ABSTRACT

Fine structure energy levels, wavelengths, log gf and allowed transition probabilities (E1) have been calculated for Lithium-like Ar XVI. The optimized electrostatic parameters by a least square approach, have been used in the calculation to include the configuration interaction and relativistic effects. A total number of 69 Ar XVI levels having total angular momenta, $1/2 \leq J \leq 9/2$ of even and odd parities, orbital angular momenta $2 \leq l \leq 4$, with 546 E1 transitions for $6 \leq n \leq 10$ are considered using the relativistic effect in the Breit-Pauli method, where n is the principal quantum number. A comparison is made with the available results in literature.

Keywords: Atomic Data, Transition Probabilities, Ar XVI

1. INTRODUCTION

Identification of X-rays and EUV fine structure lines of Lithium-like Argon is needed in laboratory and in astrophysical plasma. The X-rays emitted from highly ionized atoms are proved to be important in determining the various plasma parameters [1] [2] .

Experimental data [3] - [5] have been obtained in order to be compared with the theoretical predictions. Theoretical calculations of the energy levels, radiative rates and electron impact excitation rates for transitions in Li-like ions with $12 \leq Z \leq 20$ and $2 \leq n \leq 5$ were performed [6] . Electron self-energy screening approximations have been used in multi-electron atoms [7] . Relativistic configuration-interaction calculation of the energy levels of the core-excited states of lithium-like ions was presented [8] . The energies, transition probabilities and oscillator strengths for the electric dipole transitions between the levels of the ground and the low-lying excited states have been calculated using the multi-configuration Dirac-Hartree-Fock method [9] . Theoretical calculations were performed for the K_β satellite X-rays in the active space approximation using multiconfiguration Dirac- Fock wave functions with the inclusion of the finite nuclear size effect, the Breit interaction and the quantum electrodynamic corrections [10] . The Breit-Pauli R-matrix method has been used to compute the transitions from the upper configurations (n = 4, 5) [11] . The dipole-length, dipole-velocity and dipole-acceleration absorption oscillator strengths were calculated using the energies and the multiconfiguration interaction wave functions obtained from a full core plus the correlation method, in which relativistic and mass-polarization effects on the energy, as a first-order perturbation corrections, are included [12] .

The relativistic energies, fine structures, hyperfine structures, Auger rates and widths of the core-excited states for the Li isoelectronic sequence were studied using the saddle-point variational method and the saddle-point com- plex-rotation method [13] . Electron-impact excitation for Li-like isoelectronic sequence has been obtained using the radiation and Auger damped intermediate-coupling frame transformation R-matrix approach [14] .

In the present work, fine structure energy levels, wavelengths, log gf and radiative allowed transition probabilities are presented for Ar XVI. The calculations have been performed using the Breit-Pauli (BP) approximation, for describing the relativistic interaction between electrons. That is to say, the calculations include both the relativistic Hartree-Fock (RHF) in LS coupling and relativistic correction. The interpretation of the configuration level structures are made by least-squares fit of the observed levels. The least-squares fitting procedure is used to regularize the energy levels. These adjusted fine structure energy levels are used to calculate the transition probabilities. The total number of considered Ar XVI levels is found to be 69, with 546 allowed transition rates. The configuration $1s^22s$ of Li-like ions is taken as the ground state. The energy of the other electronic configurations n_il_i is obtained by adjusting the scaling parameters E_{av} and $\zeta_i(r_i)$ and is listed in Table 1. The calculated fine structure energy levels have been tabulated and compared with the experimental and theoretical data [4] [11] . The allowed electric dipole transition probabilities have been also compared with Ref. [11] using Breit- Pauli R-matrix (BPRM) method and no corresponding experimental data have been found in the literature.

2. CALCULATION METHOD

2.1. Atomic Structure Theory

In nonrelativistic LS coupling calculations, the normal starting point is Schrödinger's equation where the Hamiltonian for a multi-electron system, in atomic units is given as [15]

$$H=\sum_{i=1}^{N}\left(-\frac{1}{2}\nabla_i^2-\frac{Z}{r_i}\right)+\sum_{i>j}^{N}\frac{1}{r_{ij}} \tag{1}$$

A stationary state of N-electrons is described by a wave function $\psi(q_1,\cdots,q_N)$, where $q_i=(r_i,s_i)$ repre- sents the space and spin coordinates of the electron i.

$$H\psi(q_1,\cdots,q_N)=E\psi(q_1,\cdots,q_N)$$

The wave function is assumed continuous with respect to the space variables and is the solution of the wave equation.

In the multiconfiguration approximation, the wave functions for a state labeled aJM$_J$ are expanded in terms of the configuration state functions

$$\psi(\alpha JM_J)=\sum_{i=1}^{M}c_i\phi(\alpha_i L_i S_i JM_j) \tag{2}$$

where a representing the configuration and also the set of quantum numbers both required to specify the state.

2.2. The Relativistic Effects

Breit-Pauli approximation, for describing the relativistic interaction between electrons in an approximate form is [16]

$$H_{BP}=H_{NR}+H_{FS}+H_{RS} \tag{3}$$

Table 1. Radial function parameters for Ar XVI ion in units of 1000 cm^{-1}.

Configuration	Parameter		Configuration	Parameter	
$1s^22s$	E_{av}	0	$1s^22p$	E_{av}	275.186
$1s^23s$	E_{av}	4181.68		ζ	17.034
$1s^24s$	E_{av}	5612.472	$1s^23p$	E_{av}	4258.086
$1s^25s$	E_{av}	6267.466		ζ	5.036
$1s^26s$	E_{av}	6620.889	$1s^24p$	E_{av}	5643.965
$1s^27s$	E_{av}	6833.028		ζ	2.12
$1s^28s$	E_{av}	6970.261	$1s^25p$	E_{av}	6283.403
$1s^29s$	E_{av}	7064.112		ζ	1.084
$1s^210s$	E_{av}	7131.113	$1s^26p$	E_{av}	6630.048
$1s^23d$	E_{av}	4288.028		ζ	0.626
	ζ	0.949	$1s^27p$	E_{av}	6838.769
$1s^24d$	E_{av}	5656.564		ζ	0.394
	ζ	0.4	$1s^28p$	E_{av}	6974.095
$1s^25d$	E_{av}	6289.829		ζ	0.264
	ζ	0.205	$1s^29p$	E_{av}	7066.799
$1s^26d$	E_{av}	6633.758		ζ	0.185
	ζ	0.118	$1s^210p$	E_{av}	7133.067
$1s^27d$	E_{av}	6841.101		ζ	0.135
	ζ	0.075	$1s^24f$	E_{av}	5658.302
$1s^28d$	E_{av}	6975.655		ζ	0.143
	ζ	0.05	$1s^25f$	E_{av}	6290.68
$1s^29d$	E_{av}	7067.894		ζ	0.073
	ζ	0.035	$1s^26f$	E_{av}	6634.243
$1s^210d$	E_{av}	7133.865		ζ	0.043
	ζ	0.025	$1s^27f$	E_{av}	6841.404
$1s^25g$	E_{av}	6291.053		ζ	0.027
	ζ	0.034	$1s^28f$	E_{av}	6975.858
$1s^26g$	E_{av}	6634.46		ζ	0.018
	ζ	0.02	$1s^29f$	E_{av}	7068.036
$1s^27g$	E_{av}	6841.541		ζ	0.013
	ζ	0.013	$1s^210f$	E_{av}	7133.969
$1s^28g$	E_{av}	6975.949		ζ	0.009
	ζ	0.008			
$1s^29g$	E_{av}	7068.1			
	ζ	0.006			
$1s^210g$	E_{av}	7134.016			
	ζ	0.004			

H_{NR} is the non-relativistic multi-electron system hamiltonian. The fine-structure operator H_{FS} is

$$H_{FS} = H_{SO} + H_{SOO} + H_{SS} \tag{4}$$

where H_{SO} is the nuclear spin-orbit term, H_{SOO} is the spin-other-orbit term and H_{SS} is the spin-spin term.

The relativistic shift operator H_{RS} commutes with L and S and can be written

$$H_{RS} = H_{MC} + H_{D_1} + H_{D_2} + H_{OO} + H_{SSC} \tag{5}$$

where H_{MC} is the mass correction term, H_{D_1} and H_{D_2} are the one and two body Darwin terms, H_{OO} is the orbit?orbit term and H_{SSC} is the spin-spin contact term.

The radial wave functions have been generated using the Relativistic Hartree-Fock (RHF) method introduced, using the computer codes (Cowan ATOMIC STRUCTURE PACKAGE) [17] . The relativistic corrections included in the differential equations are derived from the Pauli-approximation to the Dirac-Hartree?Fock equations. The mass-velocity and the Darwin operators are included in the calculations. The radial parameters have been fitted in the least square optimizing program fitting the eigen-values of the hamiltonian to the available experimental energy levels. These optimized integrals have been used to compute both the wavelengths and the transition rates.

2.3. Electric Dipole Decay Rates

The strength of a line is defined as the square of the reduced dipole matrix element [18]

$$S(\gamma J;\gamma' J') = \sum_{MM'} \left| \langle \gamma JM || \mathbf{D} || \gamma' J' M' \rangle \right|^2 \tag{6}$$

where ψ and ψ' are the wave functions composed of many basis states, the sum runs over all N electrons of the atom (or ion) and r^i is the radial position of the ith electron.

The transition probability is related to S according to

$$A_{u,l} = \frac{64\pi^4 e^2 a_o^2}{3hg_u} S(\Delta E)^3 \qquad (7)$$

where u and l represents the upper and lower levels, respectively, g_u is the statistical weight of the upper level of the transition, a_o is the Bohr radius and ΔE is the wave number of the spectral line in cm^{-1}.

The weighted oscillator strength for the transition between ψ and ψ' is defined as

$$g_u f = \frac{2}{3}(\Delta E)S \qquad (8)$$

3. RESULTS AND DISCUSSION

The diagonalization of the energy matrices with RHF values for the energy parameters leads to the theoretical predictions for the energy levels of the configurations. The computer code [17] is used to calculate energy levels, wavelengths, log gf and electric dipole transition probabilities. For even parity, the configurations $1s^2ns$, nd and ng (n = 6 - 10) have been considered while $1s^2np$ and nf (n = 6 - 10) are considered for the odd parity configurations. They correspond to total angular momenta $1/2 \leq J \leq 9/2$ of even and odd parities with n = 6 - 10 while orbital angular momenta $0 \leq L \leq 4$ and spin multiplicity $(2S + 1) = 2$.

The mentioned configurations for the experimental energy levels taken from Ref. [4] are used in the fitting procedure for Ar XVI. For $1s^2ns$ there is only one parameter (E_{av}) which is determined from the observed term $^2S_{1/2}$, for $1s^2nd$ there are two parameters (E_{av}, ζ_2) which are determined from the observed terms $^2D_{3/2}$ and $^2D_{5/2}$, for $1s^2ng$ there are two parameters (E_{av}, ζ_2) which are determined from the observed terms $^2G_{7/2}$ and $^2G_{9/2}$, for $1s^2np$ there are two parameters (E_{av}, ζ_1) which are determined from the observed terms $^2P_{1/2}$ and $^2P_{3/2}$ and for $1s^2nf$ there are two parameters (E_{av}, ζ_2) which are determined from the observed terms $^2F_{5/2}$, $^2F_{7/2}$.

The energy of the other electronic configurations are obtained by adjusting the scaling parameters E_{av} and $\zeta_i(r_i)$ are listed in Table 1.

A total of 69 fine structure energy levels for Ar XVI are listed in Table 2, as well as the values compiled by Ref. [4] and those calculated in Ref. [11] .

The calculated fine structure energy levels of the Ar XVI ion agree with the observed and calculated values within the range of 0.15% for most levels. However, ab initio calculations including relativistic effects in the Breit-

Table 2. The calculated energy levels in cm^{-1} for Ar XVI in comparison with literatures.

Configuration	E_{cal}	E [4]	E [11]	Configuration	E_{cal}	E [4]	E [11]
$1s^22s\ ^2S_{1/2}$	0	0	0	$1s^27p\ ^2P_{1/2}$	3,897,153	3,897,133	3,896,838
$1s^22p\ ^2P_{1/2}$	1,920,620	1,920,590	1,920,622	$1s^27p\ ^2P_{3/2}$	3,897,351	3,897,321	3,897,068
$1s^22p\ ^2P_{3/2}$	2,002,380	2,002,060	2,002,377	$1s^27d\ ^2D_{3/2}$	3,899,149	3,899,119	3,898,747
$1s^23s\ ^2S_{1/2}$	2,388,872	2,388,872	2,388,861	$1s^27d\ ^2D_{5/2}$	3,899,158	3,899,178	3,898,813
$1s^23p\ ^2P_{1/2}$	2,441,940	2,441,532	2,441,941	$1s^27f\ ^2F_{5/2}$	3,904,946	3,898,946	3,898,901
$1s^23p\ ^2P_{3/2}$	2,444,335	2,443,937	2,444,322	$1s^27f\ ^2F_{7/2}$	3,904,975	3,898,975	3,898,934
$1s^23d\ ^2D_{3/2}$	2,463,791	2,463,771	2,463,789	$1s^27g\ ^2G_{7/2}$	3,905,027	3,898,934	3,898,934
$1s^23d\ ^2D_{5/2}$	2,464,481	2,464,521	2,464,470	$1s^27g\ ^2G_{9/2}$	3,905,045	3,898,952	3,898,956
$1s^24s\ ^2S_{1/2}$	3,202,101	3,202,101	3,202,025	$1s^28s\ ^2S_{1/2}$	3,969,857	3,969,857	3,969,857
$1s^24p\ ^2P_{1/2}$	3,223,450	3,223,425	3,223,446	$1s^28d\ ^2D_{3/2}$	3,974,870	3,974,842	3,974,400
$1s^24p\ ^2P_{3/2}$	3,224,555	3,224,438	3,224,543	$1s^28d\ ^2D_{5/2}$	3,974,870	3,974,881	3,974,444
$1s^24d\ ^2D_{3/2}$	3,232,733	3,232,673	3,232,642	$1s^28p\ ^2P_{1/2}$	3,979,347	3,973,160	3,973,127
$1s^24d\ ^2D_{5/2}$	3,233,089	3,232,989	3,232,861	$1s^28p\ ^2P_{3/2}$	3,979,473	3,973,286	3,973,281
$1s^24f\ ^2F_{5/2}$	3,239,433	3,233,483	3,233,662	$1s^28f\ ^2F_{5/2}$	3,980,573	3,974,543	3,974,510
$1s^24f\ ^2F_{7/2}$	3,239,591	3,233,641	3,233,827	$1s^28f\ ^2F_{7/2}$	3,980,593	3,974,562	3,974,532
$1s^25s\ ^2S_{1/2}$	3,573,450	3,573,451	3,573,442	$1s^28g\ ^2G_{7/2}$	3,980,627	3,974,530	3,974,532
$1s^25p\ ^2P_{3/2}$	3,584,656	3,584,616	3,584,646	$1s^28g\ ^2G_{9/2}$	3,980,639	3,974,542	3,974,543
$1s^25d\ ^2D_{3/2}$	3,588,906	3,588,878	3,588,849	$1s^29s\ ^2S_{1/2}$	4,023,079	4,023,079	4,023,080
$1s^25d\ ^2D_{5/2}$	3,589,068	3,589,004	3,589,036	$1s^29d\ ^2D_{5/2}$	4,026,707	4,026,737	4,026,295
$1s^25p\ ^2P_{3/2}$	3,590,099	3,584,099	3,584,141	$1s^29d\ ^2D_{3/2}$	4,026,709	4,026,709	4,026,262
$1s^25f\ ^2F_{5/2}$	3,595,165	3,589,275	3,589,255	$1s^29p\ ^2P_{1/2}$	4,031,561	4,025,371	4,025,373
$1s^25f\ ^2F_{7/2}$	3,595,247	3,589,357	3,589,354	$1s^29p\ ^2P_{3/2}$	4,031,649	4,025,459	4,025,483
$1s^25g\ ^2G_{7/2}$	3,595,383	3,589,318	3,589,343	$1s^29f\ ^2F_{5/2}$	4,032,421	4,026,341	4,026,339
$1s^25g\ ^2G_{9/2}$	3,595,432	3,589,367	3,589,409	$1s^29f\ ^2F_{7/2}$	4,032,435	4,026,355	4,026,361
$1s^26s\ ^2S_{1/2}$	3,771,342	3,771,342	3,771,342	$1s^29g\ ^2G_{7/2}$	4,032,459	4,026,361	4,026,361
$1s^26p\ ^2P_{1/2}$	3,779,502	3,779,471	3,779,145	$1s^29g\ ^2G_{9/2}$	4,032,467	4,026,369	4,026,372
$1s^26p\ ^2P_{3/2}$	3,779,801	3,779,770	3,779,507	$1s^210s\ ^2S_{1/2}$	4,061,048	4,061,048	4,061,049
$1s^26d\ ^2D_{3/2}$	3,782,704	3,782,726	3,782,173	$1s^210d\ ^2D_{3/2}$	4,063,337	4,063,363	4,063,364
$1s^26d\ ^2D_{5/2}$	3,782,750	3,782,820	3,782,283	$1s^210d\ ^2D_{5/2}$	4,063,339	4,063,383	4,063,386
$1s^26f\ ^2F_{5/2}$	3,788,420	3,794,320	3,782,426	$1s^210p\ ^2P_{1/2}$	4,068,879	4,062,709	4,062,717
$1s^26f\ ^2F_{7/2}$	3,788,467	3,794,367	3,782,481	$1s^210p\ ^2P_{3/2}$	4,068,944	4,062,774	4,062,794
$1s^26g\ ^2G_{7/2}$	3,788,549	3,782,481	3,782,481	$1s^210f\ ^2F_{5/2}$	4,069,505	4,063,445	4,063,419
$1s^26g\ ^2G_{9/2}$	3,788,577	3,782,509	3,782,513	$1s^210f\ ^2F_{7/2}$	4,069,516	4,063,455	4,063,430
$1s^27s\ ^2S_{1/2}$	3,891,943	3,891,943	3,891,943	$1s^210g\ ^2G_{7/2}$	4,069,533	4,063,450	4,063,430
				$1s^210g\ ^2G_{9/2}$	4,069,539	4,063,456	4,063,441

Table 3. Wavelengths (in Å) and the transition rates (in sec^{-1}) is compared with Ref. [11] .

λ (Å)	Transition	Log gf	A_{ij} (Sec^{-1})	A_{ij} [11]
14.019	$1s^22s\ ^2S_{1/2}$ - $1s^210p\ ^2P_{1/2}$	−2.61	4.161E+10	4.000E+10
14.019	$1s^22s\ ^2S_{1/2}$ - $1s^210p\ ^2P_{3/2}$	−2.309	4.163E+10	3.972E+10
14.15	$1s^22s\ ^2S_{1/2}$ - $1s^29p\ ^2P_{3/2}$	−2.164	5.713E+10	5.442E+10
14.151	$1s^22s\ ^2S_{1/2}$ - $1s^29p\ ^2P_{1/2}$	−2.465	5.710E+10	5.483E+10
14.339	$1s^22s\ ^2S_{1/2}$ - $1s^28p\ ^2P_{1/2}$	−2.3	8.130E+10	7.795E+10
14.339	$1s^22s\ ^2S_{1/2}$ - $1s^28p\ ^2P_{3/2}$	−1.999	8.133E+10	7.733E+10
14.544	$1s^22p\ ^2P_{1/2}$ - $1s^210d\ ^2D_{3/2}$	−2.119	5.995E+10	5.942E+10
14.55	$1s^22p\ ^2P_{1/2}$ - $1s^210s\ ^2S_{1/2}$	−3.539	4.559E+09	4.468E+09
14.598	$1s^22p\ ^2P_{3/2}$ - $1s^210d\ ^2D_{3/2}$	−2.82	1.186E+10	1.167E+10
14.598	$1s^22p\ ^2P_{3/2}$ - $1s^210d\ ^2D_{5/2}$	−1.865	7.113E+10	7.037E+10
14.604	$1s^22p\ ^2P_{3/2}$ - $1s^210s\ ^2S_{1/2}$	−3.239	9.015E+09	9.165E+09
14.622	$1s^22s\ ^2S_{1/2}$ - $1s^27p\ ^2P_{3/2}$	−1.808	1.213E+11	1.150E+11
14.623	$1s^22s\ ^2S_{1/2}$ - $1s^27p\ ^2P_{1/2}$	−2.109	1.213E+11	1.160E+11
14.685	$1s^22p\ ^2P_{1/2}$ - $1s^29d\ ^2D_{3/2}$	−1.969	8.305E+10	8.230E+10
14.693	$1s^22p\ ^2P_{1/2}$ - $1s^29s\ ^2S_{1/2}$	−3.389	6.300E+09	6.171E+09
14.74	$1s^22p\ ^2P_{3/2}$ - $1s^29d\ ^2D_{3/2}$	−2.67	1.643E+10	1.616E+10
14.74	$1s^22p\ ^2P_{3/2}$ - $1s^29d\ ^2D_{5/2}$	−1.715	9.855E+10	9.748E+10
14.748	$1s^22p\ ^2P_{3/2}$ - $1s^29s\ ^2S_{1/2}$	−3.09	1.246E+10	1.266E+10
14.887	$1s^22p\ ^2P_{1/2}$ - $1s^28d\ ^2D_{3/2}$	−1.797	1.200E+11	1.188E+11
14.898	$1s^22p\ ^2P_{1/2}$ - $1s^28s\ ^2S_{1/2}$	−3.221	9.030E+09	8.879E+09
14.943	$1s^22p\ ^2P_{3/2}$ - $1s^28d\ ^2D_{3/2}$	−2.498	2.373E+10	2.334E+10
14.943	$1s^22p\ ^2P_{3/2}$ - $1s^28d\ ^2D_{5/2}$	−1.544	1.424E+11	1.407E+11
14.955	$1s^22p\ ^2P_{3/2}$ - $1s^28s\ ^2S_{1/2}$	−2.922	1.786E+10	1.822E+10
15.082	$1s^22s\ ^2S_{1/2}$ - $1s^26p\ ^2P_{3/2}$	−1.58	1.926E+11	1.815E+11
15.084	$1s^22s\ ^2S_{1/2}$ - $1s^26p\ ^2P_{1/2}$	−1.882	1.925E+11	1.832E+11
15.191	$1s^22p\ ^2P_{1/2}$ - $1s^27d\ ^2D_{3/2}$	−1.597	1.829E+11	1.809E+11
15.209	$1s^22p\ ^2P_{1/2}$ - $1s^27s\ ^2S_{1/2}$	−3.025	1.361E+10	1.347E+10
15.25	$1s^22p\ ^2P_{3/2}$ - $1s^27d\ ^2D_{3/2}$	−2.297	3.615E+10	3.555E+10
15.25	$1s^22p\ ^2P_{3/2}$ - $1s^27d\ ^2D_{5/2}$	−1.343	2.168E+11	2.143E+11
15.269	$1s^22p\ ^2P_{3/2}$ - $1s^27s\ ^2S_{1/2}$	−2.726	2.690E+10	2.760E+10
15.685	$1s^22p\ ^2P_{1/2}$ - $1s^26d\ ^2D_{3/2}$	−1.354	3.000E+11	2.965E+11
15.717	$1s^22p\ ^2P_{1/2}$ - $1s^26s\ ^2S_{1/2}$	−2.79	2.190E+10	2.188E+10
15.748	$1s^22p\ ^2P_{3/2}$ - $1s^26d\ ^2D_{3/2}$	−2.055	5.930E+10	5.830E+10
15.748	$1s^22p\ ^2P_{3/2}$ - $1s^26d\ ^2D_{5/2}$	−1.1	3.558E+11	3.512E+11
15.78	$1s^22p\ ^2P_{3/2}$ - $1s^26s\ ^2S_{1/2}$	−2.491	4.328E+10	4.483E+10

Continued

33.882	$1s^2 3s\ ^2S_{1/2} - 1s^2 10p\ ^2P_{3/2}$	−2.038	1.329E+10	1.294E+10
33.884	$1s^2 3s\ ^2S_{1/2} - 1s^2 10p\ ^2P_{1/2}$	−2.34	1.329E+10	1.309E+10
34.66	$1s^2 3s\ ^2S_{1/2} - 1s^2 9p\ ^2P_{3/2}$	−1.882	1.821E+10	1.772E+10
34.663	$1s^2 3s\ ^2S_{1/2} - 1s^2 9p\ ^2P_{1/2}$	−2.183	1.821E+10	1.793E+10
34.713	$1s^2 3p\ ^2P_{1/2} - 1s^2 10d\ ^2D_{3/2}$	−1.81	2.141E+10	2.134E+10
34.746	$1s^2 3p\ ^2P_{1/2} - 1s^2 10s\ ^2S_{1/2}$	−2.973	2.943E+09	2.720E+09
34.803	$1s^2 3p\ ^2P_{3/2} - 1s^2 10d\ ^2D_{5/2}$	−1.556	2.550E+10	2.545E+10
34.804	$1s^2 3p\ ^2P_{3/2} - 1s^2 10d\ ^2D_{3/2}$	−2.511	4.248E+09	4.235E+09
34.837	$1s^2 3p\ ^2P_{3/2} - 1s^2 10s\ ^2S_{1/2}$	−2.673	5.840E+09	5.560E+09
35.12	$1s^2 3d\ ^2D_{3/2} - 1s^2 10f\ ^2F_{5/2}$	−1.613	2.198E+10	2.196E+10
35.13	$1s^2 3d\ ^2D_{3/2} - 1s^2 10p\ ^2P_{3/2}$	−4.08	1.123E+08	1.090E+08
35.133	$1s^2 3d\ ^2D_{3/2} - 1s^2 10p\ ^2P_{1/2}$	−3.381	1.123E+09	1.137E+09
35.149	$1s^2 3d\ ^2D_{5/2} - 1s^2 10f\ ^2F_{7/2}$	−1.458	2.349E+10	2.343E+10
35.15	$1s^2 3d\ ^2D_{5/2} - 1s^2 10f\ ^2F_{5/2}$	−2.759	1.566E+09	1.556E+09
35.16	$1s^2 3d\ ^2D_{5/2} - 1s^2 10p\ ^2P_{3/2}$	−3.127	1.008E+09	9.924E+08
35.527	$1s^2 3p\ ^2P_{1/2} - 1s^2 9d\ ^2D_{3/2}$	−1.649	2.963E+10	2.951E+10
35.574	$1s^2 3p\ ^2P_{1/2} - 1s^2 9s\ ^2S_{1/2}$	−2.811	4.074E+09	3.775E+09
35.621	$1s^2 3p\ ^2P_{3/2} - 1s^2 9d\ ^2D_{5/2}$	−1.395	3.525E+10	3.520E+10
35.622	$1s^2 3p\ ^2P_{3/2} - 1s^2 9d\ ^2D_{3/2}$	−2.35	5.875E+09	5.858E+09
35.67	$1s^2 3p\ ^2P_{3/2} - 1s^2 9s\ ^2S_{1/2}$	−2.511	8.085E+09	7.719E+09
35.81	$1s^2 3s\ ^2S_{1/2} - 1s^2 8p\ ^2P_{3/2}$	−1.701	2.588E+10	2.516E+10
35.815	$1s^2 3s\ ^2S_{1/2} - 1s^2 8p\ ^2P_{1/2}$	−2.002	2.587E+10	2.546E+10
35.953	$1s^2 3d\ ^2D_{3/2} - 1s^2 9f\ ^2F_{5/2}$	−1.441	3.117E+10	3.113E+10
35.968	$1s^2 3d\ ^2D_{3/2} - 1s^2 9p\ ^2P_{3/2}$	−3.914	1.570E+08	1.526E+08
35.971	$1s^2 3d\ ^2D_{3/2} - 1s^2 9p\ ^2P_{1/2}$	−3.215	1.570E+09	1.591E+09
35.983	$1s^2 3d\ ^2D_{5/2} - 1s^2 9f\ ^2F_{7/2}$	−1.286	3.330E+10	3.322E+10
35.984	$1s^2 3d\ ^2D_{5/2} - 1s^2 9f\ ^2F_{5/2}$	−2.587	2.220E+09	2.206E+09
35.998	$1s^2 3d\ ^2D_{5/2} - 1s^2 9p\ ^2P_{3/2}$	-2.961	1.409E+09	1.389E+09
36.731	$1s^2 3p\ ^2P_{1/2} - 1s^2 8d\ ^2D_{3/2}$	−1.462	4.268E+10	4.249E+10
36.802	$1s^2 3p\ ^2P_{1/2} - 1s^2 8s\ ^2S_{1/2}$	−2.622	5.875E+09	5.466E+09
36.831	$1s^2 3p\ ^2P_{3/2} - 1s^2 8d\ ^2D_{5/2}$	−1.208	5.078E+10	5.069E+10
36.833	$1s^2 3p\ ^2P_{3/2} - 1s^2 8d\ ^2D_{3/2}$	−2.162	8.463E+09	8.438E+09
36.905	$1s^2 3p\ ^2P_{3/2} - 1s^2 8s\ ^2S_{1/2}$	−2.323	1.165E+10	1.118E+10
37.186	$1s^2 3d\ ^2D_{3/2} - 1s^2 8f\ ^2F_{5/2}$	−1.238	4.648E+10	4.644E+10
37.208	$1s^2 3d\ ^2D_{3/2} - 1s^2 8p\ ^2P_{3/2}$	−3.719	2.298E+08	2.235E+08
37.213	$1s^2 3d\ ^2D_{3/2} - 1s^2 8p\ ^2P_{1/2}$	−3.02	2.297E+09	2.330E+09
37.218	$1s^2 3d\ ^2D_{5/2} - 1s^2 8f\ ^2F_{5/2}$	−2.384	3.312E+09	3.293E+09

Continued

37.218	$1s^23d\,^2D_{5/2} - 1s^28f\,^2F_{7/2}$	−1.083	4.968E+10	4.957E+10
37.24	$1s^23d\,^2D_{5/2} - 1s^28p\,^2P_{3/2}$	−2.766	2.063E+09	2.034E+09
37.632	$1s^23s\,^2S_{1/2} - 1s^27p\,^2P_{3/2}$	−1.486	3.848E+10	3.733E+10
37.641	$1s^23s\,^2S_{1/2} - 1s^27p\,^2P_{1/2}$	−1.787	3.846E+10	3.781E+10
38.641	$1s^23p\,^2P_{1/2} - 1s^27d\,^2D_{3/2}$	−1.237	6.475E+10	6.443E+10
38.751	$1s^23p\,^2P_{3/2} - 1s^27d\,^2D_{5/2}$	−0.983	7.705E+10	7.688E+10
38.754	$1s^23p\,^2P_{3/2} - 1s^27d\,^2D_{3/2}$	−1.937	1.284E+10	1.280E+10
38.76	$1s^23p\,^2P_{1/2} - 1s^27s\,^2S_{1/2}$	−2.396	8.925E+09	8.364E+09
38.874	$1s^23p\,^2P_{3/2} - 1s^27s\,^2S_{1/2}$	−2.096	1.769E+10	1.709E+10
39.143	$1s^23d\,^2D_{3/2} - 1s^27f\,^2F_{5/2}$	−0.989	7.445E+10	7.432E+10
39.178	$1s^23d\,^2D_{5/2} - 1s^27f\,^2F_{7/2}$	−0.834	7.955E+10	7.933E+10
39.179	$1s^23d\,^2D_{3/2} - 1s^27p\,^2P_{3/2}$	−3.484	3.565E+08	3.476E+08
39.179	$1s^23d\,^2D_{5/2} - 1s^27f\,^2F_{5/2}$	−2.135	5.303E+09	5.271E+09
39.188	$1s^23d\,^2D_{3/2} - 1s^27p\,^2P_{1/2}$	−2.785	3.564E+09	3.625E+09
39.216	$1s^23d\,^2D_{5/2} - 1s^27p\,^2P_{3/2}$	−2.53	3.200E+09	3.165E+09
40.838	$1s^23s\,^2S_{1/2} - 1s^26p\,^2P_{3/2}$	−1.217	6.063E+10	5.864E+10
40.854	$1s^23s\,^2S_{1/2} - 1s^26p\,^2P_{1/2}$	−1.518	6.055E+10	5.947E+10
42.007	$1s^23p\,^2P_{1/2} - 1s^26d\,^2D_{3/2}$	−0.952	1.054E+11	1.047E+11
42.136	$1s^23p\,^2P_{3/2} - 1s^26d\,^2D_{5/2}$	−0.698	1.254E+11	1.251E+11
42.141	$1s^23p\,^2P_{3/2} - 1s^26d\,^2D_{3/2}$	−1.653	2.089E+10	2.084E+10
42.233	$1s^23p\,^2P_{1/2} - 1s^26s\,^2S_{1/2}$	−2.109	1.456E+10	1.378E+10
42.368	$1s^23p\,^2P_{3/2} - 1s^26s\,^2S_{1/2}$	−1.809	2.885E+10	2.817E+10
42.598	$1s^23d\,^2D_{3/2} - 1s^26f\,^2F_{5/2}$	−0.665	1.324E+11	1.322E+11
42.638	$1s^23d\,^2D_{5/2} - 1s^26f\,^2F_{7/2}$	−0.511	1.414E+11	1.411E+11
42.641	$1s^23d\,^2D_{5/2} - 1s^26f\,^2F_{5/2}$	−1.812	9.427E+09	9.380E+09
42.667	$1s^23d\,^2D_{3/2} - 1s^26p\,^2P_{3/2}$	3.182	6.023E+08	5.893E+08
42.684	$1s^23d\,^2D_{3/2} - 1s^26p\,^2P_{1/2}$	−2.483	6.015E+09	6.145E+09
42.71	$1s^23d\,^2D_{5/2} - 1s^26p\,^2P_{3/2}$	−2.228	5.405E+09	5.365E+09
65.761	$1s^24s\,^2S_{1/2} - 1s^210p\,^2P_{3/2}$	−1.803	6.068E+09	5.950E+09
65.77	$1s^24s\,^2S_{1/2} - 1s^210p\,^2P_{1/2}$	−2.104	6.065E+09	6.039E+09
67.025	$1s^24p\,^2P_{1/2} - 1s^210d\,^2D_{3/2}$	−1.577	9.823E+09	9.805E+09
67.147	$1s^24p\,^2P_{1/2} - 1s^210s\,^2S_{1/2}$	−2.579	1.948E+09	2.130E+09
67.165	$1s^24p\,^2P_{3/2} - 1s^210d\,^2D_{5/2}$	−1.323	1.171E+10	1.174E+10
67.168	$1s^24p\,^2P_{3/2} - 1s^210d\,^2D_{3/2}$	−2.277	1.952E+09	1.957E+09
67.291	$1s^24p\,^2P_{3/2} - 1s^210s\,^2S_{1/2}$	−2.279	3.871E+09	4.338E+09
67.66	$1s^24d\,^2D_{3/2} - 1s^210f\,^2F_{5/2}$	−1.256	1.348E+10	1.345E+10

Continued

67.697	$1s^24d\ ^2D_{5/2} - 1s^210p\ ^2P_{3/2}$	−3.475	1.219E+08	1.242E+08
67.704	$1s^24d\ ^2D_{5/2} - 1s^210f\ ^2F_{7/2}$	−1.101	1.441E+10	1.437E+10
67.705	$1s^24d\ ^2D_{5/2} - 1s^210f\ ^2F_{5/2}$	−2.402	9.610E+08	9.563E+08
67.706	$1s^24d\ ^2D_{3/2} - 1s^210p\ ^2P_{1/2}$	−2.776	1.218E+09	1.288E+09
67.743	$1s^24d\ ^2D_{3/2} - 1s^210p\ ^2P_{3/2}$	−2.521	1.095E+09	1.129E+09
67.751	$1s^24f\ ^2F_{5/2} - 1s^210g\ ^2G_{7/2}$	−1.269	9.780E+09	9.779E+09
67.756	$1s^24f\ ^2F_{5/2} - 1s^210d\ ^2D_{5/2}$	−4.398	9.677E+06	9.517E+06
67.759	$1s^24f\ ^2F_{5/2} - 1s^210d\ ^2D_{3/2}$	−3.252	2.032E+08	2.032E+08
67.773	$1s^24f\ ^2F_{7/2} - 1s^210g\ ^2G_{9/2}$	−1.156	1.014E+10	1.012E+10
67.774	$1s^24f\ ^2F_{7/2} - 1s^210d\ ^2D_{5/2}$	−2.7	4.828E+08	1.918E+08
67.774	$1s^24f\ ^2F_{7/2} - 1s^210g\ ^2G_{7/2}$	−2.7	3.621E+08	3.605E+08
68.756	$1s^24s\ ^2S_{1/2} - 1s^29p\ ^2P_{3/2}$	−1.628	8.308E+09	8.120E+09
68.769	$1s^24s\ ^2S_{1/2} - 1s^29p\ ^2P_{1/2}$	−1.929	8.300E+09	8.246E+09
70.126	$1s^24p\ ^2P_{1/2} - 1s^29d\ ^2D_{3/2}$	−1.398	1.357E+10	1.352E+10
70.279	$1s^24p\ ^2P_{3/2} - 1s^29d\ ^2D_{5/2}$	−1.144	1.617E+10	1.620E+10
70.283	$1s^24p\ ^2P_{3/2} - 1s^29d\ ^2D_{3/2}$	−2.098	2.695E+09	2.701E+09
70.31	$1s^24p\ ^2P_{1/2} - 1s^29s\ ^2S_{1/2}$	−2.396	2.714E+09	2.964E+09
70.468	$1s^24p\ ^2P_{3/2} - 1s^29s\ ^2S_{1/2}$	−2.095	5.390E+09	6.039E+09
70.819	$1s^24d\ ^2D_{3/2} - 1s^29f\ ^2F_{5/2}$	−1.065	1.907E+10	1.903E+10
70.867	$1s^24d\ ^2D_{5/2} - 1s^29f\ ^2F_{7/2}$	−0.911	2.040E+10	2.034E+10
70.869	$1s^24d\ ^2D_{5/2} - 1s^29f\ ^2F_{5/2}$	−2.212	1.360E+09	1.353E+09
70.875	$1s^24d\ ^2D_{5/2} - 1s^29p\ ^2P_{3/2}$	−3.285	1.722E+08	1.756E+08
70.889	$1s^24d\ ^2D_{3/2} - 1s^29p\ ^2P_{1/2}$	−2.586	1.721E+09	1.821E+09
70.918	$1s^24f\ ^2F_{5/2} - 1s^29g\ ^2G_{7/2}$	−1.058	1.450E+10	1.448E+10
70.926	$1s^24d\ ^2D_{3/2} - 1s^29p\ ^2P_{3/2}$	−2.331	1.547E+09	1.596E+09
70.926	$1s^24f\ ^2F_{5/2} - 1s^29d\ ^2D_{5/2}$	−4.198	1.402E+07	1.378E+07
70.931	$1s^24f\ ^2F_{5/2} - 1s^29d\ ^2D_{3/2}$	−3.052	2.943E+08	2.942E+08
70.942	$1s^24f\ ^2F_{7/2} - 1s^29g\ ^2G_{9/2}$	−0.946	1.502E+10	1.500E+10
70.944	$1s^24f\ ^2F_{7/2} - 1s^29g\ ^2G_{7/2}$	−2.49	5.364E+08	5.341E+08
70.952	$1s^24f\ ^2F_{7/2} - 1s^29d\ ^2D_{5/2}$	−2.897	2.800E+08	2.777E+08
73.435	$1s^24s\ ^2S_{1/2} - 1s^28p\ ^2P_{3/2}$	−1.419	1.178E+10	1.146E+10
73.456	$1s^24s\ ^2S_{1/2} - 1s^28p\ ^2P_{1/2}$	−1.72	1.177E+10	1.165E+10
74.977	$1s^24p\ ^2P_{1/2} - 1s^28d\ ^2D_{3/2}$	−1.183	1.948E+10	1.939E+10
75.15	$1s^24p\ ^2P_{3/2} - 1s^28d\ ^2D_{5/2}$	−0.928	2.322E+10	2.323E+10
75.157	$1s^24p\ ^2P_{3/2} - 1s^28d\ ^2D_{3/2}$	−1.883	3.870E+09	3.875E+09
75.278	$1s^24p\ ^2P_{1/2} - 1s^28s\ ^2S_{1/2}$	−2.174	3.947E+09	4.307E+09

Continued

75.458	$1s^24p\,^2P_{3/2} - 1s^28s\,^2S_{1/2}$	−1.874	7.835E+09	8.776E+09
75.766	$1s^24d\,^2D_{3/2} - 1s^28f\,^2F_{5/2}$	−0.834	2.837E+10	2.832E+10
75.82	$1s^24d\,^2D_{5/2} - 1s^28f\,^2F_{7/2}$	−0.68	3.034E+10	3.027E+10
75.823	$1s^24d\,^2D_{5/2} - 1s^28f\,^2F_{5/2}$	−1.981	2.022E+09	2.015E+09
75.857	$1s^24d\,^2D_{3/2} - 1s^28p\,^2P_{3/2}$	−3.054	2.558E+08	2.610E+08
75.878	$1s^24f\,^2F_{5/2} - 1s^28g\,^2G_{7/2}$	−0.798	2.308E+10	2.306E+10
75.88	$1s^24d\,^2D_{3/2} - 1s^28p\,^2P_{1/2}$	−2.356	2.554E+09	2.707E+09
75.89	$1s^24f\,^2F_{5/2} - 1s^28d\,^2D_{5/2}$	−3.952	2.157E+07	2.116E+07
75.898	$1s^24f\,^2F_{5/2} - 1s^28d\,^2D_{3/2}$	−2.806	4.528E+08	4.517E+08
75.904	$1s^24f\,^2F_{7/2} - 1s^28g\,^2G_{9/2}$	−0.685	2.391E+10	2.387E+10
75.906	$1s^24f\,^2F_{7/2} - 1s^28g\,^2G_{7/2}$	−2.229	8.538E+08	8.504E+08
75.915	$1s^24d\,^2D_{5/2} - 1s^28p\,^2P_{3/2}$	−2.101	2.296E+09	2.373E+09
75.919	$1s^24f\,^2F_{7/2} - 1s^28d\,^2D_{5/2}$	−2.651	4.307E+08	4.264E+08
81.533	$1s^24s\,^2S_{1/2} - 1s^27p\,^2P_{3/2}$	−1.158	1.744E+10	1.685E+10
81.573	$1s^24s\,^2S_{1/2} - 1s^27p\,^2P_{1/2}$	−1.459	1.741E+10	1.715E+10
83.393	$1s^24p\,^2P_{1/2} - 1s^27d\,^2D_{3/2}$	−0.911	2.940E+10	2.917E+10
83.601	$1s^24p\,^2P_{3/2} - 1s^27d\,^2D_{5/2}$	−0.657	3.502E+10	3.497E+10
83.615	$1s^24p\,^2P_{3/2} - 1s^27d\,^2D_{3/2}$	−1.612	5.835E+09	5.836E+09
83.95	$1s^24p\,^2P_{1/2} - 1s^27s\,^2S_{1/2}$	−1.891	6.080E+09	6.624E+09
84.175	$1s^24p\,^2P_{3/2} - 1s^27s\,^2S_{1/2}$	−1.591	1.206E+10	1.350E+10
84.361	$1s^24d\,^2D_{3/2} - 1s^27f\,^2F_{5/2}$	−0.539	4.517E+10	4.509E+10
84.425	$1s^24d\,^2D_{5/2} - 1s^27f\,^2F_{7/2}$	−0.384	4.828E+10	4.821E+10
84.432	$1s^24d\,^2D_{5/2} - 1s^27f\,^2F_{5/2}$	−1.685	3.218E+09	3.210E+09
84.496	$1s^24f\,^2F_{5/2} - 1s^27g\,^2G_{7/2}$	−0.455	4.091E+10	4.084E+10
84.519	$1s^24f\,^2F_{5/2} - 1s^27d\,^2D_{5/2}$	−3.634	3.612E+07	3.534E+07
84.527	$1s^24f\,^2F_{7/2} - 1s^27g\,^2G_{9/2}$	−0.343	4.239E+10	4.229E+10
84.531	$1s^24d\,^2D_{3/2} - 1s^27p\,^2P_{3/2}$	−2.76	4.060E+08	4.152E+08
84.531	$1s^24f\,^2F_{7/2} - 1s^27g\,^2G_{7/2}$	−1.887	1.514E+09	1.507E+09
84.533	$1s^24f\,^2F_{5/2} - 1s^27d\,^2D_{3/2}$	−2.488	7.583E+08	7.540E+08
84.555	$1s^24f\,^2F_{7/2} - 1s^27d\,^2D_{5/2}$	−2.333	7.215E+08	7.119E+08
84.573	$1s^24d\,^2D_{3/2} - 1s^27p\,^2P_{1/2}$	−2.061	4.054E+09	4.306E+09
84.602	$1s^24d\,^2D_{5/2} - 1s^27p\,^2P_{3/2}$	−1.806	3.645E+09	3.775E+09
98.243	$1s^24s\,^2S_{1/2} - 1s^26p\,^2P_{3/2}$	−0.804	2.710E+10	2.590E+10
98.333	$1s^24s\,^2S_{1/2} - 1s^26p\,^2P_{1/2}$	−1.106	2.703E+10	2.642E+10
100.833	$1s^24p\,^2P_{1/2} - 1s^26d\,^2D_{3/2}$	−0.541	4.718E+10	4.661E+10
101.127	$1s^24p\,^2P_{3/2} - 1s^26d\,^2D_{5/2}$	−0.287	5.613E+10	5.598E+10
101.158	$1s^24p\,^2P_{3/2} - 1s^26d\,^2D_{3/2}$	−1.241	9.348E+09	9.349E+09

Continued

102.14	$1s^24p\ ^2P_{1/2}$ - $1s^26s\ ^2S_{1/2}$	−1.499	1.013E+10	1.104E+10
102.229	$1s^24d\ ^2D_{3/2}$ - $1s^26f\ ^2F_{5/2}$	−0.128	7.915E+10	7.893E+10
102.318	$1s^24d\ ^2D_{5/2}$ - $1s^26f\ ^2F_{7/2}$	0.026	8.459E+10	8.442E+10
102.334	$1s^24d\ ^2D_{5/2}$ - $1s^26f\ ^2F_{5/2}$	−1.275	5.637E+09	5.624E+09
102.418	$1s^24f\ ^2F_{5/2}$ - $1s^26g\ ^2G_{7/2}$	0.039	8.699E+10	8.684E+10
102.461	$1s^24f\ ^2F_{7/2}$ - $1s^26g\ ^2G_{9/2}$	0.152	9.010E+10	8.993E+10
102.47	$1s^24f\ ^2F_{7/2}$ - $1s^26g\ ^2G_{7/2}$	−1.392	3.216E+09	3.207E+09
102.473	$1s^24p\ ^2P_{3/2}$ - $1s^26s\ ^2S_{1/2}$	−1.2	2.005E+10	2.251E+10
102.474	$1s^24f\ ^2F_{5/2}$ - $1s^26d\ ^2D_{5/2}$	−3.186	6.893E+07	6.761E+07
102.505	$1s^24f\ ^2F_{5/2}$ - $1s^26d\ ^2D_{3/2}$	−2.04	1.446E+09	1.442E+09
102.526	$1s^24f\ ^2F_{7/2}$ - $1s^26d\ ^2D_{5/2}$	−1.886	1.376E+09	1.362E+09
102.628	$1s^24d\ ^2D_{3/2}$ - $1s^26p\ ^2P_{3/2}$	−2.348	7.103E+08	7.214E+08
102.727	$1s^24d\ ^2D_{3/2}$ - $1s^26p\ ^2P_{1/2}$	−1.649	7.085E+09	7.485E+09
102.733	$1s^24d\ ^2D_{5/2}$ - $1s^26p\ ^2P_{3/2}$	−1.394	6.373E+09	6.558E+09
115.518	$1s^25s\ ^2S_{1/2}$ - $1s^210p\ ^2P_{3/2}$	−1.571	3.353E+09	3.231E+09
115.545	$1s^25s\ ^2S_{1/2}$ - $1s^210p\ ^2P_{1/2}$	−1.873	3.350E+09	3.286E+09
117.439	$1s^25p\ ^2P_{1/2}$ - $1s^210d\ ^2D_{3/2}$	−1.355	5.335E+09	5.304E+09
117.655	$1s^25p\ ^2P_{3/2}$ - $1s^210d\ ^2D_{5/2}$	−1.101	6.368E+09	6.367E+09
117.663	$1s^25p\ ^2P_{3/2}$ - $1s^210d\ ^2D_{3/2}$	−2.055	1.061E+09	1.063E+09
117.814	$1s^25p\ ^2P_{1/2}$ - $1s^210s\ ^2S_{1/2}$	−2.241	1.380E+09	1.472E+09
118.04	$1s^25p\ ^2P_{3/2}$ - $1s^210s\ ^2S_{1/2}$	−1.941	2.745E+09	2.997E+09
118.423	$1s^25d\ ^2D_{3/2}$ - $1s^210f\ ^2F_{5/2}$	−0.991	8.095E+09	8.085E+09
118.491	$1s^25d\ ^2D_{5/2}$ - $1s^210f\ ^2F_{7/2}$	−0.836	8.659E+09	8.648E+09
118.495	$1s^25d\ ^2D_{5/2}$ - $1s^210f\ ^2F_{5/2}$	−2.137	5.772E+08	5.760E+08
118.538	$1s^25d\ ^2D_{5/2}$ - $1s^210p\ ^2P_{3/2}$	−3.029	1.111E+08	1.139E+08
118.558	$1s^25f\ ^2F_{5/2}$ - $1s^210g\ ^2G_{7/2}$	−0.844	8.498E+09	8.488E+09
118.566	$1s^25d\ ^2D_{3/2}$ - $1s^210p\ ^2P_{1/2}$	−2.33	1.110E+09	1.177E+09
118.574	$1s^25f\ ^2F_{5/2}$ - $1s^210d\ ^2D_{5/2}$	−3.726	1.486E+07	1.462E+07
118.583	$1s^25f\ ^2F_{5/2}$ - $1s^210d\ ^2D_{3/2}$	−2.58	3.120E+08	3.116E+08
118.591	$1s^25f\ ^2F_{7/2}$ - $1s^210g\ ^2G_{9/2}$	−0.731	8.805E+09	8.791E+09
118.594	$1s^25f\ ^2F_{7/2}$ - $1s^210g\ ^2G_{7/2}$	−2.275	3.144E+08	3.135E+08
118.61	$1s^25d\ ^2D_{3/2}$ - $1s^210p\ ^2P_{3/2}$	−2.075	9.978E+08	1.034E+09
118.61	$1s^25f\ ^2F_{7/2}$ - $1s^210d\ ^2D_{5/2}$	−2.425	2.968E+08	2.944E+08
118.622	$1s^25g\ ^2G_{7/2}$ - $1s^210f\ ^2F_{7/2}$	−4.545	1.690E+06	1.664E+06
118.626	$1s^25g\ ^2G_{7/2}$ - $1s^210f\ ^2F_{5/2}$	−3.114	6.082E+07	6.044E+07
118.643	$1s^25g\ ^2G_{9/2}$ - $1s^210f\ ^2F_{7/2}$	−3.001	5.911E+07	5.853E+07
125.09	$1s^25s\ ^2S_{1/2}$ - $1s^29p\ ^2P_{3/2}$	−1.366	4.585E+09	4.402E+09

Continued

125.133	$1s^25s\ ^2S_{1/2} - 1s^29p\ ^2P_{3/2}$	−1.668	4.580E+09	4.481E+09
127.304	$1s^25p\ ^2P_{1/2} - 1s^29d\ ^2D_{3/2}$	−1.146	7.355E+09	7.297E+09
127.554	$1s^25p\ ^2P_{3/2} - 1s^29d\ ^2D_{5/2}$	−0.891	8.775E+09	8.766E+09
127.568	$1s^25p\ ^2P_{3/2} - 1s^29d\ ^2D_{3/2}$	−1.846	1.462E+09	1.464E+09
127.911	$1s^25p\ ^2P_{1/2} - 1s^29s\ ^2S_{1/2}$	−2.022	1.940E+09	2.069E+09
128.178	$1s^25p\ ^2P_{3/2} - 1s^29s\ ^2S_{1/2}$	−1.721	3.855E+09	4.214E+09
128.454	$1s^25d\ ^2D_{3/2} - 1s^29f\ ^2F_{5/2}$	−0.771	1.143E+10	1.141E+10
128.531	$1s^25d\ ^2D_{5/2} - 1s^29f\ ^2F_{7/2}$	−0.616	1.222E+10	1.221E+10
128.539	$1s^25d\ ^2D_{5/2} - 1s^29f\ ^2F_{5/2}$	−1.917	8.145E+08	8.132E+08
128.609	$1s^25f\ ^2F_{5/2} - 1s^29g\ ^2G_{7/2}$	−0.604	1.255E+10	1.253E+10
128.635	$1s^25f\ ^2F_{5/2} - 1s^29d\ ^2D_{5/2}$	−3.485	2.197E+07	2.162E+07
128.639	$1s^25d\ ^2D_{3/2} - 1s^29p\ ^2P_{3/2}$	−2.801	1.592E+08	1.630E+08
128.647	$1s^25f\ ^2F_{7/2} - 1s^29g\ ^2G_{9/2}$	−0.491	1.300E+10	1.298E+10
128.649	$1s^25f\ ^2F_{5/2} - 1s^29d\ ^2D_{3/2}$	−2.339	4.613E+08	4.606E+08
128.651	$1s^25f\ ^2F_{7/2} - 1s^29g\ ^2G_{7/2}$	−2.035	4.644E+08	4.630E+08
128.677	$1s^25f\ ^2F_{7/2} - 1s^29d\ ^2D_{5/2}$	−2.185	4.390E+08	4.353E+08
128.685	$1s^25d\ ^2D_{3/2} - 1s^29p\ ^2P_{1/2}$	−2.103	1.590E+09	1.685E+09
128.686	$1s^25g\ ^2G_{7/2} - 1s^29f\ ^2F_{7/2}$	−4.285	2.611E+06	2.573E+06
128.693	$1s^25g\ ^2G_{7/2} - 1s^29f\ ^2F_{5/2}$	−2.854	9.400E+07	9.345E+07
128.711	$1s^25g\ ^2G_{9/2} - 1s^29f\ ^2F_{7/2}$	−2.741	9.135E+07	9.050E+07
128.724	$1s^25d\ ^2D_{5/2} - 1s^29p\ ^2P_{3/2}$	−1.848	1.430E+09	1.481E+09
141.491	$1s^25s\ ^2S_{1/2} - 1s^28p\ ^2P_{3/2}$	−1.109	6.480E+09	6.189E+09
141.57	$1s^25s\ ^2S_{1/2} - 1s^28p\ ^2P_{1/2}$	−1.41	6.470E+09	6.307E+09
144.246	$1s^25p\ ^2P_{1/2} - 1s^28d\ ^2D_{3/2}$	−0.882	1.052E+10	1.041E+10
144.559	$1s^25p\ ^2P_{3/2} - 1s^28d\ ^2D_{5/2}$	0.628	1.254E+10	1.251E+10
144.585	$1s^25p\ ^2P_{3/2} - 1s^28d\ ^2D_{3/2}$	−1.582	2.089E+09	2.090E+09
145.361	$1s^25p\ ^2P_{1/2} - 1s^28s\ ^2S_{1/2}$	−1.742	2.859E+09	3.050E+09
145.705	$1s^25p\ ^2P_{3/2} - 1s^28s\ ^2S_{1/2}$	1.442	5.675E+09	6.216E+09
145.709	$1s^25d\ ^2D_{3/2} - 1s^28f\ ^2F_{5/2}$	0.491	1.692E+10	1.689E+10
145.804	$1s^25d\ ^2D_{5/2} - 1s^28f\ ^2F_{7/2}$	0.336	1.809E+10	1.807E+10
145.818	$1s^25d\ ^2D_{5/2} - 1s^28f\ ^2F_{5/2}$	1.637	1.206E+09	1.204E+09
145.901	$1s^25f\ ^2F_{5/2} - 1s^28g\ ^2G_{7/2}$	−0.295	1.986E+10	1.984E+10
145.948	$1s^25f\ ^2F_{7/2} - 1s^28g\ ^2G_{9/2}$	−0.182	2.058E+10	2.055E+10
145.949	$1s^25f\ ^2F_{5/2} - 1s^28d\ ^2D_{5/2}$	−3.176	3.483E+07	3.429E+07
145.956	$1s^25f\ ^2F_{7/2} - 1s^28g\ ^2G_{7/2}$	−1.726	7.348E+08	7.331E+08
145.975	$1s^25f\ ^2F_{5/2} - 1s^28d\ ^2D_{3/2}$	−2.03	7.310E+08	7.304E+08
146.003	$1s^25g\ ^2G_{7/2} - 1s^28f\ ^2F_{7/2}$	3.947	4.419E+06	4.356E+06

Continued

125.133	$1s^25s\ ^2S_{1/2} - 1s^29p\ ^2P_{3/2}$	−1.668	4.580E+09	4.481E+09
127.304	$1s^25p\ ^2P_{1/2} - 1s^29d\ ^2D_{3/2}$	−1.146	7.355E+09	7.297E+09
127.554	$1s^25p\ ^2P_{3/2} - 1s^29d\ ^2D_{5/2}$	−0.891	8.775E+09	8.766E+09
127.568	$1s^25p\ ^2P_{3/2} - 1s^29d\ ^2D_{3/2}$	−1.846	1.462E+09	1.464E+09
127.911	$1s^25p\ ^2P_{1/2} - 1s^29s\ ^2S_{1/2}$	−2.022	1.940E+09	2.069E+09
128.178	$1s^25p\ ^2P_{3/2} - 1s^29s\ ^2S_{1/2}$	−1.721	3.855E+09	4.214E+09
128.454	$1s^25d\ ^2D_{3/2} - 1s^29f\ ^2F_{5/2}$	−0.771	1.143E+10	1.141E+10
128.531	$1s^25d\ ^2D_{5/2} - 1s^29f\ ^2F_{7/2}$	−0.616	1.222E+10	1.221E+10
128.539	$1s^25d\ ^2D_{5/2} - 1s^29f\ ^2F_{5/2}$	−1.917	8.145E+08	8.132E+08
128.609	$1s^25f\ ^2F_{5/2} - 1s^29g\ ^2G_{7/2}$	−0.604	1.255E+10	1.253E+10
128.635	$1s^25f\ ^2F_{5/2} - 1s^29d\ ^2D_{5/2}$	−3.485	2.197E+07	2.162E+07
128.639	$1s^25d\ ^2D_{3/2} - 1s^29p\ ^2P_{3/2}$	−2.801	1.592E+08	1.630E+08
128.647	$1s^25f\ ^2F_{7/2} - 1s^29g\ ^2G_{9/2}$	−0.491	1.300E+10	1.298E+10
128.649	$1s^25f\ ^2F_{5/2} - 1s^29d\ ^2D_{3/2}$	−2.339	4.613E+08	4.606E+08
128.651	$1s^25f\ ^2F_{7/2} - 1s^29g\ ^2G_{7/2}$	−2.035	4.644E+08	4.630E+08
128.677	$1s^25f\ ^2F_{7/2} - 1s^29d\ ^2D_{5/2}$	−2.185	4.390E+08	4.353E+08
128.685	$1s^25d\ ^2D_{3/2} - 1s^29p\ ^2P_{1/2}$	−2.103	1.590E+09	1.685E+09
128.686	$1s^25g\ ^2G_{7/2} - 1s^29f\ ^2F_{7/2}$	−4.285	2.611E+06	2.573E+06
128.693	$1s^25g\ ^2G_{7/2} - 1s^29f\ ^2F_{5/2}$	−2.854	9.400E+07	9.345E+07
128.711	$1s^25g\ ^2G_{9/2} - 1s^29f\ ^2F_{7/2}$	−2.741	9.135E+07	9.050E+07
128.724	$1s^25d\ ^2D_{5/2} - 1s^29p\ ^2P_{3/2}$	−1.848	1.430E+09	1.481E+09
141.491	$1s^25s\ ^2S_{1/2} - 1s^28p\ ^2P_{3/2}$	−1.109	6.480E+09	6.189E+09
141.57	$1s^25s\ ^2S_{1/2} - 1s^28p\ ^2P_{1/2}$	−1.41	6.470E+09	6.307E+09
144.246	$1s^25p\ ^2P_{1/2} - 1s^28d\ ^2D_{3/2}$	−0.882	1.052E+10	1.041E+10
144.559	$1s^25p\ ^2P_{3/2} - 1s^28d\ ^2D_{5/2}$	−0.628	1.254E+10	1.251E+10
144.585	$1s^25p\ ^2P_{3/2} - 1s^28d\ ^2D_{3/2}$	−1.582	2.089E+09	2.090E+09
145.361	$1s^25p\ ^2P_{1/2} - 1s^28s\ ^2S_{1/2}$	−1.742	2.859E+09	3.050E+09
145.705	$1s^25p\ ^2P_{3/2} - 1s^28s\ ^2S_{1/2}$	−1.442	5.675E+09	6.216E+09
145.709	$1s^25d\ ^2D_{3/2} - 1s^28f\ ^2F_{5/2}$	−0.491	1.692E+10	1.689E+10
145.804	$1s^25d\ ^2D_{5/2} - 1s^28f\ ^2F_{7/2}$	−0.336	1.809E+10	1.807E+10
145.818	$1s^25d\ ^2D_{5/2} - 1s^28f\ ^2F_{5/2}$	−1.637	1.206E+09	1.204E+09
145.901	$1s^25f\ ^2F_{5/2} - 1s^28g\ ^2G_{7/2}$	−0.295	1.986E+10	1.984E+10
145.948	$1s^25f\ ^2F_{7/2} - 1s^28g\ ^2G_{9/2}$	−0.182	2.058E+10	2.055E+10
145.949	$1s^25f\ ^2F_{5/2} - 1s^28d\ ^2D_{5/2}$	−3.176	3.483E+07	3.429E+07
145.956	$1s^25f\ ^2F_{7/2} - 1s^28g\ ^2G_{7/2}$	−1.726	7.348E+08	7.331E+08
145.975	$1s^25f\ ^2F_{5/2} - 1s^28d\ ^2D_{3/2}$	−2.03	7.310E+08	7.304E+08
146.003	$1s^25g\ ^2G_{7/2} - 1s^28f\ ^2F_{7/2}$	−3.947	4.419E+06	4.356E+06

Continued

146.004	$1s^25f\ ^2F_{7/2} - 1s^28d\ ^2D_{5/2}$	−1.875	6.958E+08	6.903E+08
146.016	$1s^25g\ ^2G_{7/2} - 1s^28f\ ^2F_{5/2}$	−2.516	1.590E+08	1.582E+08
146.036	$1s^25g\ ^2G_{9/2} - 1s^28f\ ^2F_{7/2}$	2.403	1.545E+08	1.532E+08
146.048	$1s^25d\ ^2D_{3/2} - 1s^28p\ ^2P_{3/2}$	−2.511	2.412E+08	2.467E+08
146.133	$1s^25d\ ^2D_{3/2} - 1s^28p\ ^2P_{1/2}$	−1.812	2.408E+09	2.550E+09
146.158	$1s^25d\ ^2D_{5/2} - 1s^28p\ ^2P_{3/2}$	−1.557	2.166E+09	2.241E+09
174.978	$1s^25s\ ^2S_{1/2} - 1s^27p\ ^2P_{3/2}$	−0.759	9.493E+09	8.971E+09
175.159	$1s^25s\ ^2S_{1/2} - 1s^27p\ ^2P_{1/2}$	−1.06	9.465E+09	9.163E+09
178.996	$1s^25p\ ^2P_{1/2} - 1s^27d\ ^2D_{3/2}$	−0.521	1.569E+10	1.544E+10
179.459	$1s^25p\ ^2P_{3/2} - 1s^27d\ ^2D_{5/2}$	−0.267	1.868E+10	1.859E+10
179.519	$1s^25p\ ^2P_{3/2} - 1s^27d\ ^2D_{3/2}$	−1.221	3.110E+09	3.108E+09
181.216	$1s^25d\ ^2D_{3/2} - 1s^27f\ ^2F_{5/2}$	−0.105	2.658E+10	2.652E+10
181.353	$1s^25d\ ^2D_{5/2} - 1s^27f\ ^2F_{7/2}$	0.05	2.843E+10	2.839E+10
181.384	$1s^25d\ ^2D_{5/2} - 1s^27f\ ^2F_{5/2}$	−1.251	1.893E+09	1.893E+09
181.496	$1s^25f\ ^2F_{5/2} - 1s^27g\ ^2G_{7/2}$	0.137	3.473E+10	3.468E+10
181.562	$1s^25f\ ^2F_{7/2} - 1s^27g\ ^2G_{9/2}$	0.25	3.598E+10	3.593E+10
181.581	$1s^25f\ ^2F_{7/2} - 1s^27g\ ^2G_{7/2}$	−1.294	1.285E+09	1.282E+09
181.584	$1s^25p\ ^2P_{1/2} - 1s^27s\ ^2S_{1/2}$	−1.353	4.491E+09	4.796E+09
181.606	$1s^25f\ ^2F_{5/2} - 1s^27d\ ^2D_{5/2}$	−2.741	6.118E+07	6.022E+07
181.661	$1s^25g\ ^2G_{7/2} - 1s^27f\ ^2F_{7/2}$	−3.468	8.600E+06	8.471E+06
181.667	$1s^25f\ ^2F_{5/2} - 1s^27d\ ^2D_{3/2}$	−1.595	1.284E+09	1.282E+09
181.691	$1s^25f\ ^2F_{7/2} - 1s^27d\ ^2D_{5/2}$	−1.44	1.222E+09	1.212E+09
181.691	$1s^25g\ ^2G_{7/2} - 1s^27f\ ^2F_{5/2}$	−2.037	3.095E+08	3.075E+08
181.711	$1s^25g\ ^2G_{9/2} - 1s^27f\ ^2F_{7/2}$	−1.924	3.008E+08	2.979E+08
182.002	$1s^25d\ ^2D_{3/2} - 1s^27p\ ^2P_{3/2}$	2.105	3.953E+08	4.032E+08
182.122	$1s^25p\ ^2P_{3/2} - 1s^27s\ ^2S_{1/2}$	−1.053	8.900E+09	9.777E+09
182.172	$1s^25d\ ^2D_{5/2} - 1s^27p\ ^2P_{3/2}$	−1.151	3.548E+09	3.663E+09
182.198	$1s^25d\ ^2D_{3/2} - 1s^27p\ ^2P_{1/2}$	−1.407	3.939E+09	4.170E+09
195.219	$1s^26s\ ^2S_{1/2} - 1s^210p\ ^2P_{3/2}$	−1.322	2.086E+09	1.988E+09
195.296	$1s^26s\ ^2S_{1/2} - 1s^210p\ ^2P_{1/2}$	−1.623	2.084E+09	2.025E+09
198.253	$1s^26p\ ^2P_{1/2} - 1s^210d\ ^2D_{3/2}$	−1.116	3.248E+09	3.208E+09
198.598	$1s^26p\ ^2P_{3/2} - 1s^210d\ ^2D_{5/2}$	−0.862	3.877E+09	3.861E+09
198.623	$1s^26p\ ^2P_{3/2} - 1s^210d\ ^2D_{3/2}$	−1.816	6.460E+08	6.451E+08
199.326	$1s^26p\ ^2P_{1/2} - 1s^210s\ ^2S_{1/2}$	−1.907	1.039E+09	1.095E+09
199.7	$1s^26p\ ^2P_{3/2} - 1s^210s\ ^2S_{1/2}$	−1.607	2.067E+09	2.229E+09
199.852	$1s^26d\ ^2D_{3/2} - 1s^210f\ ^2F_{5/2}$	−0.734	5.137E+09	5.130E+09
199.957	$1s^26d\ ^2D_{5/2} - 1s^210f\ ^2F_{7/2}$	−0.579	7.327E+09	5.492E+09

Continued

199.97	$1s^26d\ ^2D_{5/2}$ - $1s^210f\ ^2F_{5/2}$	−1.88	3.663E+08	3.661E+08
200.061	$1s^26f\ ^2F_{5/2}$ - $1s^210g\ ^2G_{7/2}$	−0.524	6.239E+09	6.233E+09
200.107	$1s^26f\ ^2F_{5/2}$ - $1s^210d\ ^2D_{3/2}$	−3.205	1.733E+07	1.709E+07
200.113	$1s^26f\ ^2F_{7/2}$ - $1s^210g\ ^2G_{9/2}$	−0.411	6.464E+09	6.458E+09
200.121	$1s^26f\ ^2F_{7/2}$ - $1s^210g\ ^2G_{7/2}$	−1.955	2.309E+08	2.304E+08
200.132	$1s^26f\ ^2F_{5/2}$ - $1s^210d\ ^2D_{5/2}$	−2.059	3.638E+08	3.636E+08
200.166	$1s^26f\ ^2F_{7/2}$ - $1s^210d\ ^2D_{5/2}$	−1.904	3.463E+08	3.439E+08
200.171	$1s^26g\ ^2G_{7/2}$ - $1s^210f\ ^2F_{7/2}$	−3.791	3.365E+06	3.322E+06
200.178	$1s^26d\ ^2D_{3/2}$ - $1s^210p\ ^2P_{3/2}$	−2.625	9.860E+07	1.006E+08
200.184	$1s^26g\ ^2G_{7/2}$ - $1s^210f\ ^2F_{5/2}$	−2.36	1.211E+08	1.205E+08
200.207	$1s^26g\ ^2G_{9/2}$ - $1s^210f\ ^2F_{7/2}$	−2.247	1.177E+08	1.168E+08
200.259	$1s^26d\ ^2D_{3/2}$ - $1s^210p\ ^2P_{1/2}$	−1.927	9.850E+08	1.037E+09
200.297	$1s^26d\ ^2D_{5/2}$ - $1s^210p\ ^2P_{3/2}$	−1.671	8.858E+08	9.131E+08
224.214	$1s^26s\ ^2S_{1/2}$ - $1s^29p\ ^2P_{3/2}$	−1.066	2.848E+09	2.701E+09
224.354	$1s^26s\ ^2S_{1/2}$ - $1s^29p\ ^2P_{1/2}$	−1.368	2.842E+09	2.755E+09
228.092	$1s^26p\ ^2P_{1/2}$ - $1s^29d\ ^2D_{3/2}$	−0.856	4.463E+09	4.397E+09
228.536	$1s^26p\ ^2P_{3/2}$ - $1s^29d\ ^2D_{5/2}$	−0.602	5.323E+09	5.296E+09
228.582	$1s^26p\ ^2P_{3/2}$ - $1s^29d\ ^2D_{3/2}$	−1.556	8.868E+08	8.853E+08
230.048	$1s^26p\ ^2P_{1/2}$ - $1s^29s\ ^2S_{1/2}$	−1.629	1.479E+09	1.560E+09
230.186	$1s^26d\ ^2D_{3/2}$ - $1s^29f\ ^2F_{5/2}$	−0.463	7.223E+09	7.211E+09
230.32	$1s^26d\ ^2D_{5/2}$ - $1s^29f\ ^2F_{7/2}$	−0.308	1.030E+10	7.722E+09
230.343	$1s^26d\ ^2D_{5/2}$ - $1s^29f\ ^2F_{5/2}$	−1.609	5.148E+08	5.148E+08
230.453	$1s^26f\ ^2F_{5/2}$ - $1s^29g\ ^2G_{7/2}$	−0.234	9.168E+09	9.157E+09
230.518	$1s^26f\ ^2F_{7/2}$ - $1s^29g\ ^2G_{9/2}$	−0.121	9.499E+09	9.490E+09
230.532	$1s^26f\ ^2F_{7/2}$ - $1s^29g\ ^2G_{7/2}$	−1.665	3.391E+08	3.387E+08
230.536	$1s^26f\ ^2F_{5/2}$ - $1s^29d\ ^2D_{5/2}$	−2.9	2.632E+07	2.595E+07
230.547	$1s^26p\ ^2P_{3/2}$ - $1s^29s\ ^2S_{1/2}$	−1.329	2.939E+09	3.178E+09
230.583	$1s^26f\ ^2F_{5/2}$ - $1s^29d\ ^2D_{3/2}$	1.754	5.523E+08	5.521E+08
230.604	$1s^26g\ ^2G_{7/2}$ - $1s^29f\ ^2F_{7/2}$	3.463	5.394E+06	5.325E+06
230.615	$1s^26f\ ^2F_{7/2}$ - $1s^29d\ ^2D_{5/2}$	−1.599	5.258E+08	5.222E+08
230.627	$1s^26g\ ^2G_{7/2}$ - $1s^29f\ ^2F_{5/2}$	−2.032	1.942E+08	1.932E+08
230.651	$1s^26g\ ^2G_{9/2}$ - $1s^29f\ ^2F_{7/2}$	−1.919	1.886E+08	1.872E+08
230.781	$1s^26d\ ^2D_{3/2}$ - $1s^29p\ ^2P_{3/2}$	−2.337	1.440E+08	1.467E+08
230.929	$1s^26d\ ^2D_{3/2}$ - $1s^29p\ ^2P_{1/2}$	−1.639	1.437E+09	1.513E+09
230.939	$1s^26d\ ^2D_{5/2}$ - $1s^29p\ ^2P_{3/2}$	−1.384	1.293E+09	1.332E+09
275.562	$1s^25s\ ^2S_{1/2}$ - $1s^26p\ ^2P_{3/2}$	−0.196	1.398E+10	1.289E+10

Continued

276.277	$1s^25s\ ^2S_{1/2}$ - $1s^26p\ ^2P_{3/2}$	−0.498	1.387E+10	1.324E+10
283.016	$1s^26s\ ^2S_{1/2}$ - $1s^28p\ ^2P_{3/2}$	−0.717	3.990E+09	3.752E+09
283.333	$1s^26s\ ^2S_{1/2}$ - $1s^28p\ ^2P_{1/2}$	−1.019	3.977E+09	3.836E+09
284.688	$1s^25p\ ^2P_{1/2}$ - $1s^26d\ ^2D_{3/2}$	0.065	2.392E+10	2.327E+10
285.77	$1s^25p\ ^2P_{3/2}$ - $1s^26d\ ^2D_{5/2}$	0.319	2.838E+10	2.812E+10
286.012	$1s^25p\ ^2P_{3/2}$ - $1s^26d\ ^2D_{3/2}$	−0.636	4.718E+09	4.710E+09
288.885	$1s^26p\ ^2P_{1/2}$ - $1s^28d\ ^2D_{3/2}$	−0.5	6.320E+09	6.197E+09
289.566	$1s^26p\ ^2P_{3/2}$ - $1s^28d\ ^2D_{5/2}$	−0.246	7.532E+09	7.477E+09
289.671	$1s^26p\ ^2P_{3/2}$ - $1s^28d\ ^2D_{3/2}$	−1.2	1.254E+09	1.251E+09
290.162	$1s^25d\ ^2D_{3/2}$ - $1s^26f\ ^2F_{5/2}$	0.526	4.433E+10	4.418E+10
290.468	$1s^25d\ ^2D_{5/2}$ - $1s^26f\ ^2F_{7/2}$	0.68	4.735E+10	4.734E+10
290.593	$1s^25d\ ^2D_{5/2}$ - $1s^26f\ ^2F_{5/2}$	−0.621	3.152E+09	3.160E+09
290.802	$1s^25f\ ^2F_{5/2}$ - $1s^26g\ ^2G_{7/2}$	0.851	7.003E+10	6.990E+10
290.943	$1s^25f\ ^2F_{7/2}$ - $1s^26g\ ^2G_{9/2}$	0.964	7.251E+10	7.246E+10
291.019	$1s^25f\ ^2F_{7/2}$ - $1s^26g\ ^2G_{7/2}$	−0.58	2.588E+09	2.588E+09
291.254	$1s^25f\ ^2F_{5/2}$ - $1s^26d\ ^2D_{5/2}$	−2.02	1.252E+08	1.230E+08
291.257	$1s^25g\ ^2G_{7/2}$ - $1s^26f\ ^2F_{7/2}$	−2.67	2.103E+07	2.071E+07
291.384	$1s^25g\ ^2G_{7/2}$ - $1s^26f\ ^2F_{5/2}$	−1.239	7.558E+08	7.517E+08
291.388	$1s^25g\ ^2G_{9/2}$ - $1s^26f\ ^2F_{7/2}$	−1.126	7.348E+08	7.283E+08
291.472	$1s^25f\ ^2F_{7/2}$ - $1s^26d\ ^2D_{5/2}$	−0.719	2.498E+09	2.476E+09
291.505	$1s^25f\ ^2F_{5/2}$ - $1s^26d\ ^2D_{3/2}$	−0.874	2.623E+09	2.620E+09
292.191	$1s^26d\ ^2D_{3/2}$ - $1s^28f\ ^2F_{5/2}$	−0.09	1.058E+10	1.056E+10
292.391	$1s^26d\ ^2D_{5/2}$ - $1s^28f\ ^2F_{7/2}$	0.064	1.508E+10	1.131E+10
292.444	$1s^26d\ ^2D_{5/2}$ - $1s^28f\ ^2F_{5/2}$	−1.237	7.537E+08	7.544E+08
292.594	$1s^26f\ ^2F_{5/2}$ - $1s^28g\ ^2G_{7/2}$	0.168	1.433E+10	1.431E+10
292.689	$1s^26f\ ^2F_{7/2}$ - $1s^28g\ ^2G_{9/2}$	0.28	1.484E+10	1.483E+10
292.721	$1s^26f\ ^2F_{7/2}$ - $1s^28g\ ^2G_{7/2}$	−1.264	5.299E+08	5.296E+08
292.785	$1s^26f\ ^2F_{5/2}$ - $1s^28d\ ^2D_{5/2}$	−2.475	4.345E+07	4.286E+07
292.848	$1s^26g\ ^2G_{7/2}$ - $1s^28f\ ^2F_{7/2}$	−3.003	9.660E+06	9.541E+06
292.892	$1s^26f\ ^2F_{5/2}$ - $1s^28d\ ^2D_{3/2}$	−1.329	9.115E+08	9.115E+08
292.902	$1s^26g\ ^2G_{7/2}$ - $1s^28f\ ^2F_{5/2}$	−1.571	3.477E+08	3.460E+08
292.913	$1s^26f\ ^2F_{7/2}$ - $1s^28d\ ^2D_{5/2}$	−1.174	8.680E+08	8.621E+08
292.925	$1s^26g\ ^2G_{9/2}$ - $1s^28f\ ^2F_{7/2}$	−1.459	3.379E+08	3.353E+08
293.393	$1s^26p\ ^2P_{3/2}$ - $1s^28s\ ^2S_{1/2}$	−1.241	2.223E+09	2.348E+09
293.393	$1s^25d\ ^2D_{3/2}$ - $1s^26p\ ^2P_{3/2}$	−1.429	7.208E+08	7.322E+08
293.559	$1s^26d\ ^2D_{3/2}$ - $1s^28p\ ^2P_{3/2}$	−1.936	2.244E+08	2.284E+08

Continued

293.815	$1s^26d\ ^2D_{5/2}$ - $1s^28p\ ^2P_{3/2}$	−0.982	2.014E+09	2.074E+09
293.835	$1s^25d\ ^2D_{5/2}$ - $1s^26p\ ^2P_{3/2}$	−0.476	6.458E+09	6.655E+09
293.901	$1s^26d\ ^2D_{3/2}$ - $1s^28p\ ^2P_{1/2}$	−1.237	2.236E+09	2.357E+09
294.204	$1s^26p\ ^2P_{3/2}$ - $1s^28s\ ^2S_{1/2}$	−0.941	4.409E+09	4.786E+09
294.205	$1s^25d\ ^2D_{3/2}$ - $1s^26p\ ^2P_{1/2}$	−0.732	7.150E+09	7.586E+09
295.359	$1s^25p\ ^2P_{1/2}$ - $1s^26s\ ^2S_{1/2}$	−0.698	7.665E+09	8.220E+09
296.785	$1s^25p\ ^2P_{3/2}$ - $1s^26s\ ^2S_{1/2}$	−0.399	1.511E+10	1.680E+10
333.214	$1s^27s\ ^2S_{1/2}$ - $1s^210p\ ^2P_{3/2}$	−1.028	1.407E+09	1.330E+09
333.439	$1s^27s\ ^2S_{1/2}$ - $1s^210p\ ^2P_{1/2}$	−1.33	1.405E+09	1.358E+09
338.464	$1s^27p\ ^2P_{1/2}$ - $1s^210d\ ^2D_{3/2}$	−0.833	2.141E+09	2.102E+09
339.07	$1s^27p\ ^2P_{3/2}$ - $1s^210d\ ^2D_{5/2}$	−0.578	2.555E+09	2.535E+09
339.143	$1s^27p\ ^2P_{3/2}$ - $1s^210d\ ^2D_{3/2}$	−1.532	4.255E+08	4.240E+08
341.342	$1s^27d\ ^2D_{3/2}$ - $1s^210f\ ^2F_{5/2}$	−0.442	3.445E+09	3.438E+09
341.522	$1s^27d\ ^2D_{5/2}$ - $1s^210f\ ^2F_{7/2}$	−0.288	3.685E+09	3.683E+09
341.559	$1s^27d\ ^2D_{5/2}$ - $1s^210f\ ^2F_{5/2}$	−1.589	2.455E+08	2.456E+08
341.603	$1s^27p\ ^2P_{1/2}$ - $1s^210s\ ^2S_{1/2}$	−1.539	8.255E+08	8.653E+08
341.7	$1s^27f\ ^2F_{5/2}$ - $1s^210g\ ^2G_{7/2}$	−0.202	4.489E+09	4.483E+09
341.787	$1s^27f\ ^2F_{7/2}$ - $1s^210g\ ^2G_{9/2}$	−0.089	4.654E+09	4.646E+09
341.81	$1s^27f\ ^2F_{7/2}$ - $1s^210g\ ^2G_{7/2}$	−1.633	1.661E+08	1.659E+08
341.833	$1s^27f\ ^2F_{5/2}$ - $1s^210d\ ^2D_{5/2}$	−2.706	1.870E+07	1.846E+07
341.908	$1s^27f\ ^2F_{5/2}$ - $1s^210d\ ^2D_{3/2}$	−1.56	3.925E+08	3.923E+08
341.912	$1s^27g\ ^2G_{7/2}$ - $1s^210f\ ^2F_{7/2}$	−3.169	4.829E+06	4.769E+06
341.943	$1s^27f\ ^2F_{7/2}$ - $1s^210d\ ^2D_{5/2}$	−1.405	3.737E+08	3.713E+08
341.949	$1s^27g\ ^2G_{7/2}$ - $1s^210f\ ^2F_{5/2}$	−1.738	1.738E+08	1.729E+08
341.978	$1s^27g\ ^2G_{9/2}$ - $1s^210f\ ^2F_{7/2}$	−1.625	1.689E+08	1.676E+08
342.294	$1s^27p\ ^2P_{3/2}$ - $1s^210s\ ^2S_{1/2}$	−1.239	1.641E+09	1.762E+09
342.295	$1s^27d\ ^2D_{3/2}$ - $1s^210p\ ^2P_{3/2}$	−2.206	8.858E+07	9.008E+07
342.514	$1s^27d\ ^2D_{5/2}$ - $1s^210p\ ^2P_{3/2}$	−1.252	7.958E+08	8.175E+08
342.532	$1s^27d\ ^2D_{3/2}$ - $1s^210p\ ^2P_{1/2}$	−1.507	8.840E+08	9.275E+08
427.599	$1s^27s\ ^2S_{1/2}$ - $1s^29p\ ^2P_{3/2}$	−0.68	1.907E+09	1.789E+09
428.108	$1s^27s\ ^2S_{1/2}$ - $1s^29p\ ^2P_{1/2}$	−0.981	1.901E+09	1.830E+09
435.794	$1s^27p\ ^2P_{1/2}$ - $1s^29d\ ^2D_{3/2}$	−0.478	2.918E+09	2.852E+09
436.752	$1s^27p\ ^2P_{3/2}$ - $1s^29d\ ^2D_{5/2}$	−0.224	3.478E+09	3.445E+09
436.919	$1s^27p\ ^2P_{3/2}$ - $1s^29d\ ^2D_{3/2}$	−1.179	5.790E+08	5.767E+08
440.487	$1s^27d\ ^2D_{3/2}$ - $1s^29f\ ^2F_{5/2}$	−0.077	4.800E+09	4.789E+09
440.764	$1s^27d\ ^2D_{5/2}$ - $1s^29f\ ^2F_{7/2}$	0.078	5.133E+09	5.132E+09
440.849	$1s^27d\ ^2D_{5/2}$ - $1s^29f\ ^2F_{5/2}$	−1.223	3.420E+08	3.425E+08

Continued

441.044	$1s^27f\ ^2F_{5/2} - 1s^29g\ ^2G_{7/2}$	0.182	6.521E+09	6.513E+09
441.175	$1s^27f\ ^2F_{7/2} - 1s^29g\ ^2G_{9/2}$	0.295	6.757E+09	6.752E+09
441.227	$1s^27f\ ^2F_{7/2} - 1s^29g\ ^2G_{7/2}$	−1.249	2.413E+08	2.412E+08
441.349	$1s^27f\ ^2F_{5/2} - 1s^29d\ ^2D_{5/2}$	−2.287	2.943E+07	2.906E+07
441.413	$1s^27g\ ^2G_{7/2} - 1s^29f\ ^2F_{7/2}$	−2.721	8.130E+06	8.033E+06
441.499	$1s^27g\ ^2G_{7/2} - 1s^29f\ ^2F_{5/2}$	−1.29	2.925E+08	2.912E+08
441.519	$1s^27f\ ^2F_{5/2} - 1s^29d\ ^2D_{3/2}$	−1.141	6.175E+08	6.175E+08
441.524	$1s^27g\ ^2G_{9/2} - 1s^29f\ ^2F_{7/2}$	−1.177	2.844E+08	2.823E+08
441.532	$1s^27f\ ^2F_{7/2} - 1s^29d\ ^2D_{5/2}$	−0.987	5.880E+08	5.844E+08
442.669	$1s^27d\ ^2D_{3/2} - 1s^29p\ ^2P_{3/2}$	−1.807	1.327E+08	1.349E+08
442.993	$1s^27p\ ^2P_{1/2} - 1s^29s\ ^2S_{1/2}$	−1.152	1.198E+09	1.258E+09
443.035	$1s^27d\ ^2D_{5/2} - 1s^29p\ ^2P_{3/2}$	−0.853	1.191E+09	1.224E+09
443.214	$1s^27d\ ^2D_{3/2} - 1s^29p\ ^2P_{1/2}$	−1.109	1.322E+09	1.389E+09
444.156	$1s^27p\ ^2P_{3/2} - 1s^29s\ ^2S_{1/2}$	−0.852	2.377E+09	2.563E+09
458.554	$1s^26s\ ^2S_{1/2} - 1s^27p\ ^2P_{3/2}$	−0.154	5.558E+09	5.114E+09
459.8	$1s^26s\ ^2S_{1/2} - 1s^27p\ ^2P_{1/2}$	−0.457	5.510E+09	5.256E+09
472.662	$1s^26p\ ^2P_{1/2} - 1s^27d\ ^2D_{3/2}$	0.081	8.993E+09	8.711E+09
474.351	$1s^26p\ ^2P_{3/2} - 1s^27d\ ^2D_{5/2}$	0.335	1.068E+10	1.055E+10
474.771	$1s^26p\ ^2P_{3/2} - 1s^27d\ ^2D_{3/2}$	−0.62	1.775E+09	1.768E+09
481.299	$1s^26d\ ^2D_{3/2} - 1s^27f\ ^2F_{5/2}$	0.52	1.588E+10	1.582E+10
481.769	$1s^26d\ ^2D_{5/2} - 1s^27f\ ^2F_{7/2}$	0.674	2.262E+10	1.696E+10
481.987	$1s^26d\ ^2D_{5/2} - 1s^27f\ ^2F_{5/2}$	−0.627	1.129E+09	1.133E+09
482.272	$1s^26f\ ^2F_{5/2} - 1s^27g\ ^2G_{7/2}$	0.822	2.381E+10	2.378E+10
482.486	$1s^26f\ ^2F_{7/2} - 1s^27g\ ^2G_{9/2}$	0.935	2.467E+10	2.466E+10
482.618	$1s^26f\ ^2F_{7/2} - 1s^27g\ ^2G_{7/2}$	−0.609	8.803E+08	8.812E+08
483.011	$1s^26g\ ^2G_{7/2} - 1s^27f\ ^2F_{7/2}$	2.242	2.049E+07	2.022E+07
483.049	$1s^26f\ ^2F_{5/2} - 1s^27d\ ^2D_{5/2}$	−1.771	8.065E+07	7.938E+07
483.221	$1s^26g\ ^2G_{9/2} - 1s^27f\ ^2F_{7/2}$	0.698	7.160E+08	7.105E+08
483.23	$1s^26g\ ^2G_{7/2} - 1s^27f\ ^2F_{5/2}$	−0.811	7.363E+08	7.332E+08
483.397	$1s^26f\ ^2F_{7/2} - 1s^27d\ ^2D_{5/2}$	0.471	1.610E+09	1.597E+09
483.485	$1s^26f\ ^2F_{5/2} - 1s^27d\ ^2D_{3/2}$	−0.626	1.689E+09	1.689E+09
486.887	$1s^26d\ ^2D_{3/2} - 1s^27p\ ^2P_{3/2}$	1.267	3.803E+08	3.861E+08
487.591	$1s^26d\ ^2D_{5/2} - 1s^27p\ ^2P_{3/2}$	−0.313	3.408E+09	3.508E+09
488.293	$1s^26d\ ^2D_{3/2} - 1s^27p\ ^2P_{1/2}$	−0.569	3.771E+09	3.992E+09
491.144	$1s^26p\ ^2P_{1/2} - 1s^27s\ ^2S_{1/2}$	−0.588	3.572E+09	3.789E+09
493.421	$1s^26p\ ^2P_{3/2} - 1s^27s\ ^2S_{1/2}$	−0.289	7.045E+09	7.743E+09

Continued

613.972	$1s^28s\ ^2S_{1/2} - 1s^210p\ ^2P_{3/2}$	−0.645	1.003E+09	9.395E+08
614.735	$1s^28s\ ^2S_{1/2} - 1s^210p\ ^2P_{1/2}$	−0.946	9.990E+08	9.616E+08
625.015	$1s^28p\ ^2P_{3/2} - 1s^210d\ ^2D_{5/2}$	−0.457	1.490E+09	1.454E+09
626.314	$1s^28p\ ^2P_{1/2} - 1s^210d\ ^2D_{3/2}$	−0.203	1.777E+09	1.759E+09
626.565	$1s^28p\ ^2P_{3/2} - 1s^210d\ ^2D_{3/2}$	−1.157	2.958E+08	2.945E+08
631.432	$1s^28d\ ^2D_{3/2} - 1s^210f\ ^2F_{5/2}$	−0.063	2.410E+09	2.405E+09
631.802	$1s^28d\ ^2D_{5/2} - 1s^210f\ ^2F_{7/2}$	0.091	2.576E+09	2.579E+09
631.93	$1s^28d\ ^2D_{5/2} - 1s^210f\ ^2F_{5/2}$	−1.21	1.717E+08	1.721E+08
632.178	$1s^28f\ ^2F_{5/2} - 1s^210g\ ^2G_{7/2}$	0.193	3.251E+09	3.249E+09
632.351	$1s^28f\ ^2F_{7/2} - 1s^210g\ ^2G_{9/2}$	0.305	3.368E+09	3.369E+09
632.428	$1s^28f\ ^2F_{7/2} - 1s^210g\ ^2G_{7/2}$	−1.239	1.203E+08	1.204E+08
632.635	$1s^28f\ ^2F_{5/2} - 1s^210d\ ^2D_{5/2}$	−2.145	1.990E+07	1.969E+07
632.694	$1s^28g\ ^2G_{7/2} - 1s^210f\ ^2F_{7/2}$	−2.522	6.260E+06	6.201E+06
632.822	$1s^28g\ ^2G_{7/2} - 1s^210f\ ^2F_{5/2}$	−1.091	2.252E+08	2.247E+08
632.846	$1s^28g\ ^2G_{9/2} - 1s^210f\ ^2F_{7/2}$	−0.978	2.190E+08	2.178E+08
632.886	$1s^28f\ ^2F_{7/2} - 1s^210d\ ^2D_{5/2}$	−0.844	3.977E+08	3.959E+08
632.89	$1s^28f\ ^2F_{5/2} - 1s^210d\ ^2D_{3/2}$	−0.999	4.175E+08	4.181E+08
634.701	$1s^28d\ ^2D_{3/2} - 1s^210p\ ^2P_{3/2}$	−1.704	8.183E+07	8.317E+07
635.204	$1s^28d\ ^2D_{5/2} - 1s^210p\ ^2P_{3/2}$	−0.75	7.345E+08	7.546E+08
635.518	$1s^28d\ ^2D_{3/2} - 1s^210p\ ^2P_{1/2}$	−1.006	8.150E+08	8.557E+08
635.802	$1s^28p\ ^2P_{1/2} - 1s^210s\ ^2S_{1/2}$	−1.077	6.905E+08	7.230E+08
637.406	$1s^28p\ ^2P_{3/2} - 1s^210s\ ^2S_{1/2}$	−0.777	1.371E+09	1.473E+09
708.221	$1s^27s\ ^2S_{1/2} - 1s^28p\ ^2P_{3/2}$	−0.115	2.550E+09	2.345E+09
710.211	$1s^27s\ ^2S_{1/2} - 1s^28p\ ^2P_{1/2}$	−0.418	2.528E+09	2.411E+09
728.835	$1s^27p\ ^2P_{1/2} - 1s^28d\ ^2D_{3/2}$	0.1	3.955E+09	3.823E+09
731.321	$1s^27p\ ^2P_{3/2} - 1s^28d\ ^2D_{5/2}$	0.354	4.698E+09	4.636E+09
731.988	$1s^27p\ ^2P_{3/2} - 1s^28d\ ^2D_{3/2}$	−0.6	7.808E+08	7.772E+08
741.661	$1s^27d\ ^2D_{3/2} - 1s^28f\ ^2F_{5/2}$	0.522	6.730E+09	6.711E+09
742.343	$1s^27d\ ^2D_{5/2} - 1s^28f\ ^2F_{7/2}$	0.677	7.191E+09	7.198E+09
742.688	$1s^27d\ ^2D_{5/2} - 1s^28f\ ^2F_{5/2}$	−0.624	4.787E+08	4.809E+08
743.066	$1s^27f\ ^2F_{5/2} - 1s^28g\ ^2G_{7/2}$	0.808	9.715E+09	9.706E+09
743.376	$1s^27f\ ^2F_{7/2} - 1s^28g\ ^2G_{9/2}$	0.921	1.006E+10	1.007E+10
743.585	$1s^27f\ ^2F_{7/2} - 1s^28g\ ^2G_{7/2}$	−0.623	3.590E+08	3.598E+08
744.187	$1s^27g\ ^2G_{7/2} - 1s^28f\ ^2F_{7/2}$	−1.984	1.561E+07	1.543E+07
744.3	$1s^27f\ ^2F_{5/2} - 1s^28d\ ^2D_{5/2}$	−1.596	5.082E+07	5.013E+07
744.501	$1s^27g\ ^2G_{9/2} - 1s^28f\ ^2F_{7/2}$	−0.44	5.455E+08	5.421E+08
744.534	$1s^27g\ ^2G_{7/2} - 1s^28f\ ^2F_{5/2}$	−0.553	5.610E+08	5.592E+08

Continued

744.82	$1s^27f\ ^2F_{7/2}$ - $1s^28d\ ^2D_{5/2}$	-0.296	1.014E+09	1.008E+09
744.992	$1s^27f\ ^2F_{5/2}$ - $1s^28d\ ^2D_{3/2}$	-0.451	1.064E+09	1.066E+09
750.54	$1s^27d\ ^2D_{3/2}$ - $1s^28p\ ^2P_{3/2}$	-1.143	2.128E+08	2.160E+08
751.592	$1s^27d\ ^2D_{5/2}$ - $1s^28p\ ^2P_{3/2}$	-0.19	1.907E+09	1.962E+09
752.776	$1s^27d\ ^2D_{3/2}$ - $1s^28p\ ^2P_{1/2}$	-0.446	2.109E+09	2.230E+09
758.232	$1s^27p\ ^2P_{3/2}$ - $1s^28s\ ^2S_{1/2}$	-0.499	1.838E+09	1.940E+09
761.646	$1s^27p\ ^2P_{1/2}$ - $1s^28s\ ^2S_{1/2}$	-0.2	3.627E+09	3.963E+09
1034.869	$1s^28s\ ^2S_{1/2}$ - $1s^29p\ ^2P_{3/2}$	-0.079	1.298E+09	1.194E+09
1037.853	$1s^28s\ ^2S_{1/2}$ - $1s^29p\ ^2P_{1/2}$	-0.381	1.287E+09	1.228E+09
1063.712	$1s^28p\ ^2P_{1/2}$ - $1s^29d\ ^2D_{3/2}$	0.122	1.950E+09	1.881E+09
1067.211	$1s^28p\ ^2P_{3/2}$ - $1s^29d\ ^2D_{5/2}$	0.375	2.317E+09	2.283E+09
1068.208	$1s^28p\ ^2P_{3/2}$ - $1s^29d\ ^2D_{3/2}$	-0.579	3.850E+08	3.829E+08
1081.898	$1s^28d\ ^2D_{3/2}$ - $1s^29f\ ^2F_{5/2}$	0.531	3.223E+09	3.215E+09
1082.843	$1s^28d\ ^2D_{5/2}$ - $1s^29f\ ^2F_{7/2}$	0.685	3.444E+09	3.449E+09
1083.36	$1s^28d\ ^2D_{5/2}$ - $1s^29f\ ^2F_{5/2}$	-0.616	2.293E+08	2.305E+08
1083.849	$1s^28f\ ^2F_{5/2}$ - $1s^29g\ ^2G_{7/2}$	0.803	4.514E+09	4.512E+09
1084.274	$1s^28f\ ^2F_{7/2}$ - $1s^29g\ ^2G_{9/2}$	0.916	4.675E+09	4.680E+09
1084.586	$1s^28f\ ^2F_{7/2}$ - $1s^29g\ ^2G_{7/2}$	-0.628	1.669E+08	1.673E+08
1085.465	$1s^28g\ ^2G_{7/2}$ - $1s^29f\ ^2F_{7/2}$	-1.802	1.116E+07	1.105E+07
1085.693	$1s^28f\ ^2F_{5/2}$ - $1s^29d\ ^2D_{5/2}$	-1.463	3.250E+07	3.210E+07
1085.911	$1s^28g\ ^2G_{9/2}$ - $1s^29f\ ^2F_{7/2}$	-0.258	3.901E+08	3.882E+08
1085.985	$1s^28g\ ^2G_{7/2}$ - $1s^29f\ ^2F_{5/2}$	-0.371	4.012E+08	4.004E+08
1086.432	$1s^28f\ ^2F_{7/2}$ - $1s^29d\ ^2D_{5/2}$	-0.162	6.487E+08	6.455E+08
1086.725	$1s^28f\ ^2F_{5/2}$ - $1s^29d\ ^2D_{3/2}$	-0.317	6.808E+08	6.820E+08
1095.158	$1s^28d\ ^2D_{3/2}$ - $1s^29p\ ^2P_{3/2}$	-1.044	1.256E+08	1.275E+08
1096.656	$1s^28d\ ^2D_{5/2}$ - $1s^29p\ ^2P_{3/2}$	-0.09	1.126E+09	1.158E+09
1098.5	$1s^28d\ ^2D_{3/2}$ - $1s^29p\ ^2P_{1/2}$	-0.346	1.245E+09	1.315E+09
1107.647	$1s^28p\ ^2P_{1/2}$ - $1s^29s\ ^2S_{1/2}$	-0.425	1.022E+09	1.074E+09
1112.523	$1s^28p\ ^2P_{3/2}$ - $1s^29s\ ^2S_{1/2}$	-0.126	2.017E+09	2.194E+09
1448.806	$1s^29s\ ^2S_{1/2}$ - $1s^210p\ ^2P_{3/2}$	-0.045	7.163E+08	6.582E+08
1453.066	$1s^29s\ ^2S_{1/2}$ - $1s^210p\ ^2P_{1/2}$	-0.347	7.100E+08	6.775E+08
1487.797	$1s^29p\ ^2P_{1/2}$ - $1s^210d\ ^2D_{3/2}$	0.143	1.048E+09	1.009E+09
1492.551	$1s^29p\ ^2P_{3/2}$ - $1s^210d\ ^2D_{5/2}$	0.397	1.245E+09	1.226E+09
1493.972	$1s^29p\ ^2P_{3/2}$ - $1s^210d\ ^2D_{3/2}$	-0.557	2.070E+08	2.057E+08
1512.646	$1s^29d\ ^2D_{3/2}$ - $1s^210f\ ^2F_{5/2}$	0.542	1.692E+09	1.688E+09
1513.921	$1s^29d\ ^2D_{5/2}$ - $1s^210f\ ^2F_{7/2}$	0.696	1.809E+09	1.811E+09
1514.651	$1s^29d\ ^2D_{5/2}$ - $1s^210f\ ^2F_{5/2}$	-0.605	1.204E+08	1.210E+08

Continued

1515.277	$1s^2 9f\ ^2F_{5/2} - 1s^2 10g\ ^2G_{7/2}$	0.804	2.313E+09	2.312E+09
1515.845	$1s^2 9f\ ^2F_{7/2} - 1s^2 10g\ ^2G_{9/2}$	0.917	2.396E+09	2.398E+09
1516.29	$1s^2 9f\ ^2F_{7/2} - 1s^2 10g\ ^2G_{7/2}$	−0.628	8.549E+07	8.575E+07
1517.52	$1s^2 9g\ ^2G_{7/2} - 1s^2 10f\ ^2F_{7/2}$	−1.062	7.881E+06	7.807E+06
1517.904	$1s^2 9f\ ^2F_{5/2} - 1s^2 10d\ ^2D_{5/2}$	−1.355	2.130E+07	2.105E+07
1518.131	$1s^2 9g\ ^2G_{9/2} - 1s^2 10f\ ^2F_{7/2}$	−0.118	2.755E+08	2.742E+08
1518.254	$1s^2 9g\ ^2G_{7/2} - 1s^2 10f\ ^2F_{5/2}$	−0.231	2.833E+08	2.828E+08
1518.921	$1s^2 9f\ ^2F_{7/2} - 1s^2 10d\ ^2D_{5/2}$	−0.054	4.252E+08	4.232E+08
1519.375	$1s^2 9f\ ^2F_{5/2} - 1s^2 10d\ ^2D_{3/2}$	−0.209	4.460E+08	4.470E+08
1531.547	$1s^2 9d\ ^2D_{3/2} - 1s^2 10p\ ^2P_{3/2}$	−0.961	7.773E+07	7.884E+07
1533.602	$1s^2 9d\ ^2D_{5/2} - 1s^2 10p\ ^2P_{3/2}$	−0.008	6.965E+08	7.156E+08
1536.308	$1s^2 9d\ ^2D_{3/2} - 1s^2 10p\ ^2P_{3/2}$	0.264	7.700E+08	8.122E+08
1550.414	$1s^2 9p\ ^2P_{1/2} - 1s^2 10s\ ^2S_{1/2}$	−0.361	6.040E+08	6.332E+08
1557.121	$1s^2 9p\ ^2P_{3/2} - 1s^2 10s\ ^2S_{1/2}$	−0.062	1.192E+09	1.293E+09
10557.487	$1s^2 6s\ ^2S_{1/2} - 1s^2 6p\ ^2P_{3/2}$	−0.379	6.250E+06	1.713E+07
11720.307	$1s^2 6s\ ^2S_{1/2} - 1s^2 6p\ ^2P_{1/2}$	−0.725	4.569E+06	1.313E+07
16839.408	$1s^2 7s\ ^2S_{1/2} - 1s^2 7p\ ^2P_{3/2}$	−0.31	2.880E+06	7.532E+06
18701.027	$1s^2 7s\ ^2S_{1/2} - 1s^2 7p\ ^2P_{1/2}$	−0.656	2.104E+06	5.746E+06
24047.416	$1s^2 6p\ ^2P_{1/2} - 1s^2 6d\ ^2D_{3/2}$	−0.768	4.920E+05	1.143E+06
25215.587	$1s^2 8s\ ^2S_{1/2} - 1s^2 8p\ ^2P_{3/2}$	−0.25	1.473E+06	3.738E+06
28010.42	$1s^2 8s\ ^2S_{1/2} - 1s^2 8p\ ^2P_{1/2}$	−0.597	1.075E+06	2.841E+06
28449.899	$1s^2 6p\ ^2P_{3/2} - 1s^2 6d\ ^2D_{5/2}$	−0.586	3.565E+05	8.673E+05
31068.444	$1s^2 6p\ ^2P_{3/2} - 1s^2 6d\ ^2D_{3/2}$	−1.578	4.563E+04	1.156E+05
35984.154	$1s^2 9s\ ^2S_{1/2} - 1s^2 9p\ ^2P_{3/2}$	−0.198	8.155E+05	2.025E+06
38252.606	$1s^2 7p\ ^2P_{1/2} - 1s^2 7d\ ^2D_{3/2}$	−0.688	2.339E+05	5.559E+05
39980.809	$1s^2 9s\ ^2S_{1/2} - 1s^2 9p\ ^2P_{1/2}$	−0.545	5.945E+05	1.536E+06
45258.064	$1s^2 7p\ ^2P_{3/2} - 1s^2 7d\ ^2D_{5/2}$	−0.506	1.695E+05	4.232E+05
49449.846	$1s^2 10s\ ^2S_{1/2} - 1s^2 10p\ ^2P_{3/2}$	−0.152	4.805E+05	1.175E+06
54948.074	$1s^2 10s\ ^2S_{1/2} - 1s^2 10p\ ^2P_{1/2}$	−0.499	3.503E+05	8.903E+05
57160.79	$1s^2 8p\ ^2P_{1/2} - 1s^2 8d\ ^2D_{3/2}$	−0.621	1.221E+05	2.919E+05
67636.072	$1s^2 8p\ ^2P_{3/2} - 1s^2 8d\ ^2D_{5/2}$	−0.439	8.840E+04	2.220E+05
81453.058	$1s^2 9p\ ^2P_{1/2} - 1s^2 9d\ ^2D_{3/2}$	−0.565	6.850E+04	1.644E+05
96394.741	$1s^2 9p\ ^2P_{3/2} - 1s^2 9d\ ^2D_{5/2}$	−0.382	4.960E+04	1.252E+05
111775.431	$1s^2 10p\ ^2P_{1/2} - 1s^2 10d\ ^2D_{3/2}$	−0.515	4.080E+04	9.796E+04
132266.211	$1s^2 10p\ ^2P_{3/2} - 1s^2 10d\ ^2D_{5/2}$	−0.333	2.955E+04	7.471E+04

Pauli R-matrix (BPRM) method by Nahar [11] show a good agreement with the present calculations of fine structure energy levels.

The wavelengths λ (in units of Å), log gf and radiative transition probabilities for spontaneous emission (in units of 10^8 s^{-1}) for 546 electric dipole allowed transitions E1 are obtained from BP calculations. Table 3 shows the comparison of the present calculations with calculated in the literature [11] . The calculated transition probabilities in the soft X-ray region in the range 14.019 - 98.333Å is within the range 0.017% - 10% accuracy those calculated in Ref. [11] using the Breit-Pauli R-matrix method. The transition probabilities in the EUV region 100.833 - 761.646 Å show a very good agreement within 0.0065% - 10% with the available calculated [11] values. In the far UV 1034.869 - 1557.121 Å, the calculated transition probabilities is in agreement with the calculated values [11] within the range 0.22% - 8.8% accuracy. The transition probabilities in the near infra red and mid-infra red regions 10,557.487 - 132,266.211 Å show an agreement within 57% - 65% with the calculated [11] results. The calculated transition probability of the spectral line 49,449.846 Å of the $1s^2 10s\,^2S_{1/2}$ - $1s^2 10p\,^2P_{3/2}$ transition shows 59% discrepancy with the calculated results [11] . Since in the present work, the COWAN code [17] , taking into account the relativistic and configuration interaction effects has been used. There is a difference in the present results using the Breit-Pauli approximation and the Breit-Pauli R-matrix (BPRM) method in the near infra red and mid infra red regions. This discrepancy may due to the close coupling (CC) approximation using R-matrix method which may not accurate in the infra red regions. The data of transition probabilities for a great number of lines for Ar XVI is not available. Therefore, there is an urgent needs to the experimental data to be compared with the theoretical results. This experimental support will show the best methods that give accurate results.

4. CONCLUSION

Calculations have been carried out for fine structure energy levels, wavelengths and allowed transition probabilities in lithium-like Ar XVI including the relativistic effect using the Breit-Pauli approximation. Fine structure energy levels obtained from Breit-Pauli are assessed to be more accurate by 0.15% than the observed and calculated values. The transition probabilities show a good agreement in the soft X-rays, EUV and far UV regions with almost all available calculated values. A slight discrepancy appears in the near infra red and mid infra red regions in the calculated results. No experimental data for the transition probabilities exist in literatures for comparison. However, the present results show that there is an urgent need for experimental data in order to identify the spectral lines from astrophysics and the controlled thermonuclear fusion research.

CITE THIS PAPER

A. I. Refaie (2015) Fine Structure Calculations of Atomic Data for Ar XVI. *Journal of Modern Physics*,**06**,1609-1630.

REFERENCES

1. 1. Schlesser, S., Boucard, S., Covita, D.S., dos Santos, J.M.F., Fuhrmann, H., Gotta, D., Gruber, A., Hennebach, M., Hirtl, A., Indelicato, P., Le Bigot, E.-O., Simons, L.M., Stingelin, L., Trassinelli, M., Veloso, J.F.C.A., Wasser, A. and Zmeskal, J. (2013) Physical Review A, 88, Article ID: 022503.
2. Natarajan, L. (2013) Physical Review A, 88, Article ID: 052522.
3. Guerra, M., Amaro, P., Szabo, C.I., Gumberidze, A., Indelicato, P. and Santos, J.P. (2013) Journal of Physics B, 46, Article ID: 065701.
4. Saloman, E.B. (2010) Journal of Physical and Chemical Reference Data, 39, Article ID: 033101. http://physics.nist.gov/cgi-bin/ASD/energy1.pl
5. Lepson, J.K., Beiersdorfer, P., Behar, E. and Kahn, S.M. (2003) The Astrophysical Journal, 590, 604-617.
6. Aggarwal, K.M. and Keenan, F.P. (2013) Atomic Data and Nuclear Data Tables, 99, 156-248.
7. Lowe, J.A., Chantler, C.T. and Grant, I.P. (2013) Radiation Physics and Chemistry, 85, 118-123.
8. Yerokhin, V.A. and Surzhykov, A. (2012) Physical Review A, 86, Article ID: 042507.
9. Liu, S.-Z., Xie, L.-Y., Ding, X.-B. and Dong, C.-Z. (2012) Acta Physica Sinica, 61, Article ID: 093106.
10. Natarajan, L., Natarajan, A. and Kadrekar, R. (2010) Physical Review A, 82, Article ID: 062514.
11. Nahar, S.N. (2002) Astronomy & Astrophysics, 389, 716-728. http:// cds.u - strasbg.fr/cgi-bin/tipbase/tipbase
12. Hu, M.-H. and Wang, Z.-W. (2009) Chinese Physics B, 18, 2244-2249.
13. Zhu, J.J., Gou, B.C. and Wang, Y.D. (2008) Journal of Physics B, 41, Article ID: 065702.
14. Liang, G.Y. and Badnell, N.R. (2011) Astronomy & Astrophysics, 528, A69.
15. Sobel'man, I.I. (1979) Introduction to the Theory of Atomic Spectra. International Series of Monographs in National Philosophy, Pergamon Press, Oxford.
16. Fischer, C.F., Brage, T. and Jönsson, P. (2000) Computational Atomic Structure. Institute of Physics Publishing, Bristol and Philadelphia.
17. Cowan, R.D. (1981) The Theory of Atomic Structure and Spectra. University of California Press, Berkeley. http://www.tcd.ie/Physics/People /Cormac.Mc Guinness/Cowan/
18. Sobel'man, I.I. (1979) Atomic Spectra and Radiative Transitions. Springer, Berlin.

CHAPTER 7

The Atomic Regular Polyhedron Electronic Shell

Zilong Kong

Wufangfa Computer Technology Service Department, Guangzhou, China

ABSTRACT

The periodic table of elements is arranged based on a series of regular polyhedron. The stability of inert gas atoms can be explained by the distribution of electrons, as well as their motion and magnetic force structure. A magnetic force regular octahedron is proposed. It is a unique configuration that best satisfies the convergence of electrons moving in the same direction within regular polyhedra. In the case of an electrostatic force crust, the formal electron spin accounts for the crusts intrinsic magnetic moment exceeding the speed of light. If one is to consider that the electron has a magnetic outer layer and an electrostatic inner layer, then the question can be solved and abovementioned inference can provide the basis for magnetic force and momentum for the regular octahedron model. The electron periphery has twenty-petal adsorptive substances; the existence of adsorptive substance causes the magnetic force greater than the electrostatic force. Each electronic shell in the regular polyhedron is in accordance with the electron configuration of periodic table of elements; the kinetic track of each electron is a surface of regular polyhedron. The magnetic properties of iron, cobalt, and nickel can be explained by the regular dodecahedron electronic shell of an atom. The electron orbit converged from reverse direction can explain diamond. The adsorptive substances found in atomic nuclei and electrons are defined as magnetic particles called magnetons. The thermodynamic magneton theory can be better explained when it is analyzed using principles of thermodynamics, superconductivity, viscosity, and even in the creation of glass. The structure of the light is a helical line.

Keywords: Atomic Structure; Regular Polyhedron; Electronic Shell; Periodic Table of Elements; Light; Diamond

1. INTRODUCTION

The study of quantum mechanics entails quantifying atom from energy and momentum, wave properties of particle and other aspects, and various electron configurations. Based on the structure of hydrogen atom, a single electron system is initially developed. The present method is still confined to the energy level, as in a unitary approach. Solid and liquid do not have characteristic spectrum, and they lacked an item of information in energy difference. This article regards the inert gases atom as the initiation point for multiple electron system. The magnetic force structure, in which the electrons interact, is analyzed for the formation of electron configuration, using the simple geometric model, which provides the obvious details.

The octet rule [1] describes the most appearance probability of electrons of stable inert gases atoms configuration situated at the corners of the cube, and this may regard as which indicates that the electrons are close to resting. Now this article discusses the state at which electrons are moving: if one is to disregard that each electron motion must be around the atomic nucleus, changing its motion to a circle around the line that connects the nucleus with eight vertices of cube, then a configuration of eight electrons with stable motion is gained. Its symmetry is unique in the magnetic force structure.

If the electron has an electrostatic force crust, the formal electron spin accounts for the crusts intrinsic magnetic moment exceeding the speed of light, which was found by Wolfgang Pauli.

Compton wavelength λ_C of electron should be explained by a substance.

2. THE UNIQUE PROPERTIES OF THE REGULAR OCTAHEDRON CONFIGURATION

When two moving electrons converge, their electrostatic and magnetic forces interact. The electrostatic force produced between the electrons is repulsion; when the direction of motion of two electrons is the same, it brings about an attractive magnetic force. Thus if the direction of motion is the same when two electrons converge, this becomes a special condition, since the stability of the system must be maintained.

When the motion track of electrons is projected around the eight vertices on the six surfaces of cube (regular hexahedron), and the direction of motion of electron is drawn (**Figure 1**), the condition wherein electrons with the same direction of motion converge is satisfied. In the unfurled map, the same number of cube faces denotes synchronization of the track of motion of the electrons projected on these two faces. Because the electron track from an edge comes into a face and comes out from the other edge, the number of entry edges is equal to that of the exit edges, thus the total number of edges of the face must be an even number. All regular polyhedra have only five kinds, with the number of the faces taken separately as 4, 6, 8, 12, 20. This is a mathematical theorem (platonic solids, see Appendix 1). The number of edges of each face in a regular hexahedron is an even number, and other polyhedron is an odd number, thus such structure is unique.

Only the projection of electron motion is reflected above the regular hexahedron. If the circle track of electron motion is to be regarded as a face, then the distances from the nucleus to each of the faces are equal. Eight electrons can form a regular octahedron (**Figure 2**), which only reflects the track of motion of electrons. Electrons converge at the edges of the regular octahedron, satisfying the condition at which electrons converge with the same direction of motion. In the unfurled map, the edges of the same number denote synchronization of the track of motion of electrons at the four edges. Because the electron track from an edge comes into a vertex and comes out from the outer edge, the number of entry edges is equal to that of the exit edges, such that the number of edges of vertex in the regular octahedron must be an even number (see Appendix 1), and other polyhedron is an odd number. This again explains the unique property of the structure. The dual polyhedron of the regular octahedron is a regular hexahedron, which is why this result is obtained.

3. THE MAGNETIC MOMENT AND ADSORPTIVE SUBSTANCE OF ELECTRON

Usually, the magnetic force between two electrons is far less than the electrostatic force. Their difference is v_c^2/c^2 factor, where v_c is the speed of electron, and c is the speed of light. Then the magnetic force of electrons that converge on a regular octahedron is too small that it can be ignored, but the fact is not of circumstance. The electron has an intrinsic magnetic moment, if the electron is a small sphere with a crust of electrostatic force, where the charge collects on the surface of the

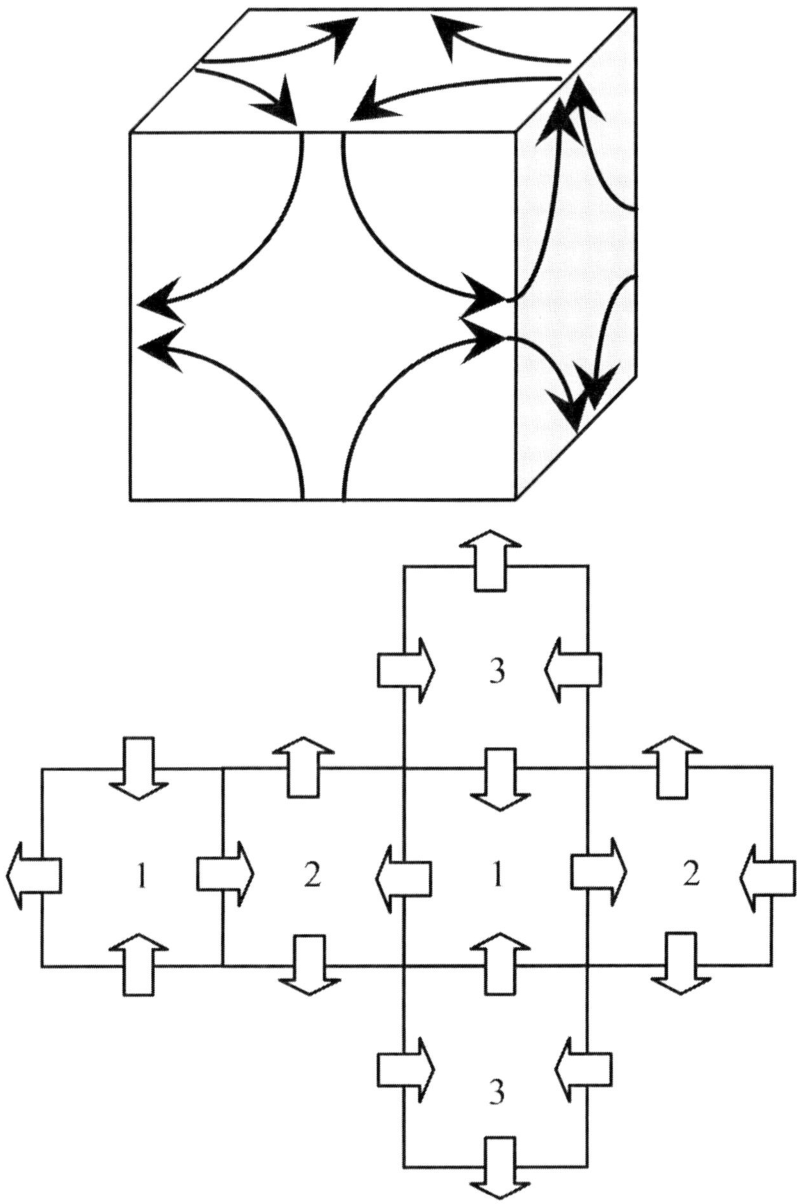

Figure 1. The projection and the unfurled map on the regular hexahedron.

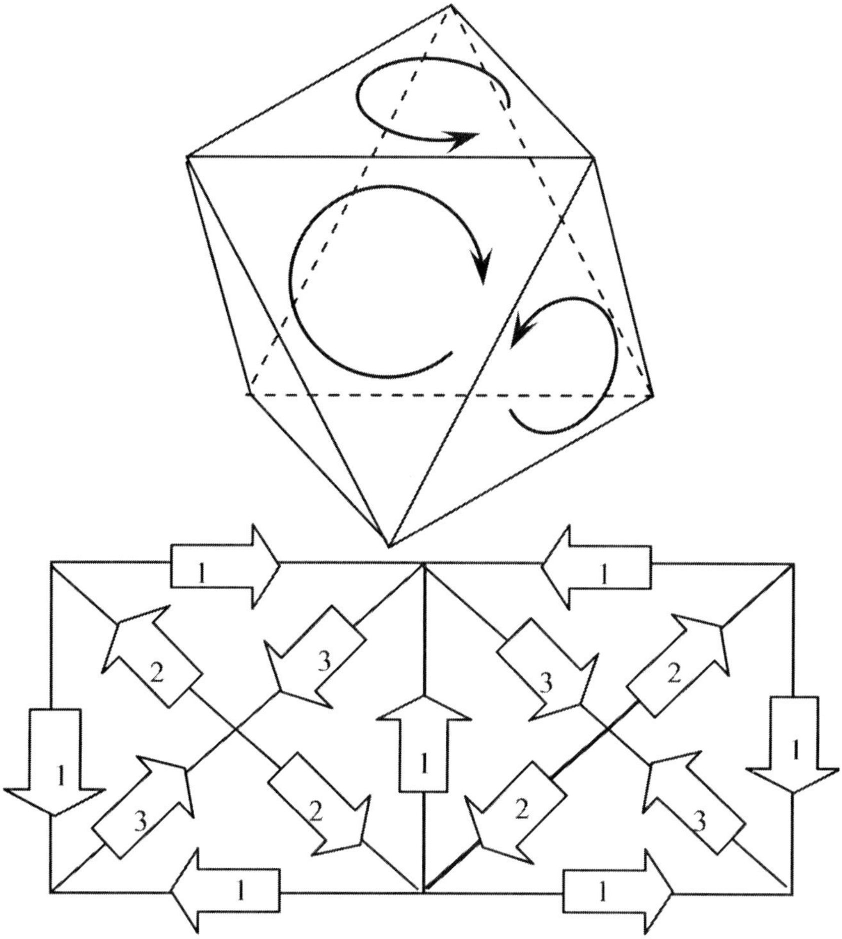

Figure 2. The track on regular octahedron and unfurled map.

sphere and the spinning produces magnetic moment; then the radius of this electrical property sphere is too small, and the intrinsic magnetic moment is too large: if the value of the intrinsic magnetic moment has to be approximated, the surface speed of sphere will have to greatly exceed the speed of light, many literature referred to this problem. Thus, in terms of distance, the magnetic force and the electrostatic force between two electrons can be compared. Otherwise, if some theory treated the electron as a point particle, this issue may be disregarded, or slide over this question.

Now and here assume that the electron has a magnetic outer layer and an electrostatic force inner layer. The electron exterior is surrounded with a layer of magnetic substance that brings about magnetic force; this layer substance is attracted by the electron that joins and rotates about the electron, producing magnetic force and moment. This is the adsorptive substance of electron: its scope radius is a lot greater than the inner layer radius of the electron. When an electron interacts with another particle outside the scope of the adsorptive substance, it produces electrostatic force, magnetic force, and magnetic moment; if another particle enters into the scope of the adsorptive substance, the adsorptive substance is destroyed or changed, the magnetic force and magnetic moment of the adsorptive substance are eliminated or weakened, and the electrostatic force of electron takes over. This assumption is in contrast with that of exceeding the speed of light with a great magnetic radius of the outer layer of the electron, the magnetic radius r_{cm} of electron being the scope radius of the electron adsorptive substance (**Figure 3**).

The electron magnetic radius r_{cm} and the magnetic force of the adsorptive substance can be calculated below. The adsorptive substance charge q moves in a circle with velocity v and radius r_i, the formative magnetic moment μ is equal to current i multiplied by the circumferential area S,

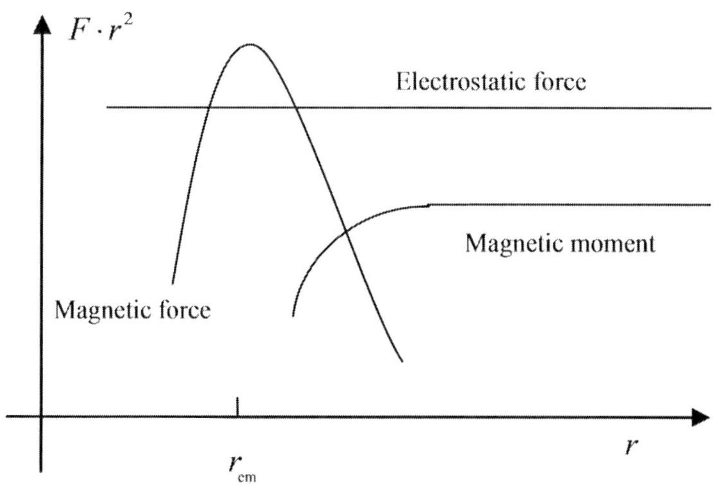

Figure 3. Comparison of electrostatic force, magnetic force and magnetic moment.

$$\mu = iS = q\frac{v}{2\pi r_i}\cdot \pi r_i^2 = \frac{1}{2}qr_i v \quad (1)$$

$$r_i = \frac{2\mu}{qv} \quad .(2)$$

The adsorptive substance and the electron have equal quantity charge. If the motion of the adsorptive substance is the same as that of electron spin, it will have an electrostatic force; because the adsorptive substance has only magnetic force but not electrostatic force. Thus the adsorptive substance and electron spin axis are in the same plane, the adsorptive substance symmetrically surrounding the electron side and moving in circle with radius r_{a}, and tangential to the inner layer of electron (**Figure 4**). When the inner layer radius of electron is ignored, $r_{\mathrm{cm}} \approx 2r_i \approx 2r_{\mathrm{a}}$. If not,

$$r_{\mathrm{cm}} = r_i + r_{\mathrm{a}} = \frac{2\mu}{qv} + r_{\mathrm{a}} \quad (3)$$

Substituted with electron magnetic moment μ_{e} and electron charge e the speed of the adsorptive substance being the speed of light c, r_{a} is Compton wavelength λ_{C} of electron, the magnetic radius of electron can be computed as follows:

$$r_{\mathrm{cm}} = \frac{2\mu_{\mathrm{e}}}{ec} + \lambda_{\mathrm{C}} = 7.72766\times 10^{-13}\,\mathrm{m} \quad (4)$$

The distance of the adsorptive substance from the electron spin axis is:

$$r_{\mathrm{ca}} = r_i - r_{\mathrm{a}} = \frac{2\mu_{\mathrm{e}}}{ec} - \lambda_{\mathrm{C}} = 4.475\times 10^{-16}\,\mathrm{m} \quad (5)$$

Using this method one can also obtain the distance of a proton adsorptive substance from a proton spin axis:

$$r_{\mathrm{pa}} = \frac{2\mu_{\mathrm{p}}}{ec} - \lambda_{\mathrm{C}}^{\mathrm{proton}} = 3.771\times 10^{-16}\,\mathrm{m} \quad (6)$$

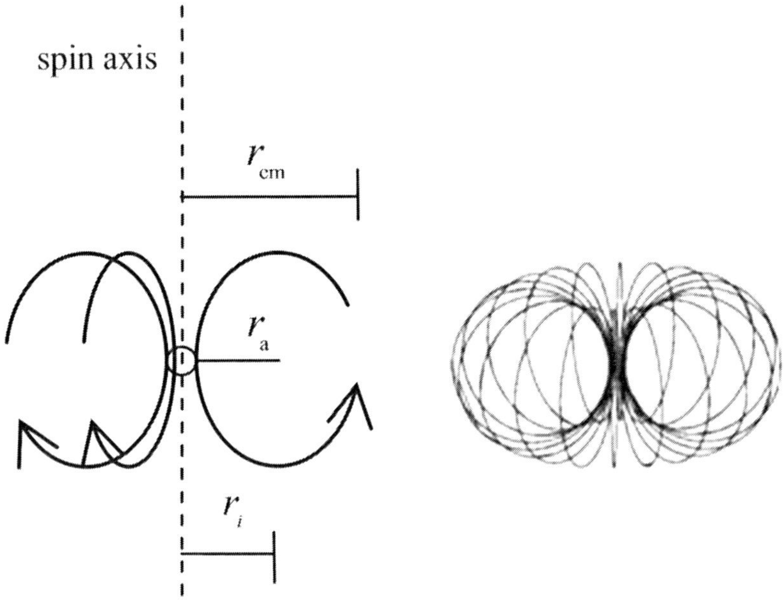

Figure 4. The motion of the electron adsorptive substance.

where r_{ca} is on the same order of magnitude as r_{pa}; from here one can also see that r_a is Compton wavelength λ_C.

Along the circle of the electron spin direction, the ring of the electron adsorptive substance is discontinuous. Suppose p is a positive integer, the adsorptive substance is composed of p circle and the number of petals of electron is p.

The electron magnetic moment reveals the quantity of electron adsorptive substance. The atomic nucleus also has an adsorptive substance that can be released, and the electron outside the nucleus absorbs the adsorptive substance that the nucleus released, eventually gaining an angular momentum. The adsorptive substance that is dispersed as a result of collision is converted into radiated heat.

In a hydrogen atom, the proton charge is equal to the electron; the adsorptive substance generated is equal to the quantity of magnetic moment μ_e, but the proton adsorptive substance is released, generating an electron orbit magnetic moment μ_o and the proton magnetic moment μ_p. The electron with speed v_e is able to release the adsorptive substance, with the generated magnetic moment $\mu_e v_e / c$. A petal is $\mu_e v_e / pc$; its direction is the same as that of the orbit magnetic moment, tangential to the orbit, and participates in generating orbit magnetic moment μ_o, thus

$$\mu_e = \mu_o + \mu_p - \frac{\mu_e v_e}{pc}. \tag{7}$$

If μ_o is Bohr magneton μ_B, then v_e/c is finestructure constant a, the number of petals of electron p is 20, that is, the electron has twenty petals.

$$\mu_e - \mu_p = \mu_B - \frac{a\mu_e}{20}. \tag{8}$$

If the electron absorbs more than one adsorptive substance that the nucleus releases each period, then n is a positive integer

$$n(\mu_e - \mu_p) = \mu_o - \frac{\mu_e v_e}{20c}. \tag{9}$$

The adsorptive substance release and absorption brings about the transfer of the atomic nucleus spin angle momentum to the electron, becoming the electron angle momentum. The electron and atomic nucleus may keep a determinate distance, thus maintaining the stability of the atom.

4. THE MAGNETIC FORCE OF ELECTRON ADSORPTIVE SUBSTANCE

After considering the electron adsorptive substance at a position far from the electron, the magnetic force that the electron motion generates is the same v_e^2/c^2 as the electrostatic force. At a position near the magnetic radius of the electron, the speed of the electron adsorptive substance is equal to the speed of light, with the magnetic force being nearly on the same order of magnitude as the electrostatic force of electron. The adsorptive substance does not have electrostatic force, though it has the same quantity of positive and negative charges and possesses polarity. The polarity rotational frequency and phase of the adsorptive substance tangent to electron, is the same. In a hydrogen atom, after coupling between the magnetic moment of the electron adsorptive substance μ_c and that of the electron motion magnetic moment μ_B, and with the proton magnetic moment μ_p with different rotational frequency and phase the relationship is expressed as follows:

$$\sqrt{2}\left(\mu_c - \mu_B\right) = \mu_p \tag{10}$$

If all frequencies and phases are the same, the coefficient will be equal to 2. When two electrons with the same velocity and direction converge, their adsorptive substances will also have the same rotational frequencies and phases: the magnetic force of motion of electron will correspond to the charge as that of the electrostatic force. Thus, at the magnetic radius of electron, the total magnetic force corresponding to the electron intrinsic magnetic moment is two times the electrostatic force of the electron. At an appropriate direction, the maximum vector sum of the magnetic force is approximately $\sqrt{2}$ times the electrostatic force. This illustrates the case wherein two magnetic moments or magnetic dipole adjoin. When distance r increases, the magnetic force according to r^{-3} likewise decreases. On a regular octahedron, with the edge serving as a boundary that divides the north and south poles, this on a regular octahedron, with the edge serving as a boundary that divides the north and south poles, this arrangement satisfies the two adjoining poles (**Figure 5**), this requires that the number of edges of each vertical be an even number. Other regular polyhedra cannot satisfy this condition.

The sideways radius of the electron adsorptive substance is r_{cm}, and the lengthwise radius is r_a. When the distance of two electrons is greater than $2r_a$, the electrons have the action of both electrostatic force and magnetic moment. When the distance of two electron is near $2r_a$, the magnetic force may be greater than the electrostatic force. When the distance of two electrons is less than $2r_a$, the adsorptive substances are changed, and the electrostatic force is greater than the magnetic force. This property of the electron magnetic force provides a basis for the model of the regular octahedron: if two electrons converge with the same velocity and direction, the magnetic force can be equal to or greater than the electrostatic force.

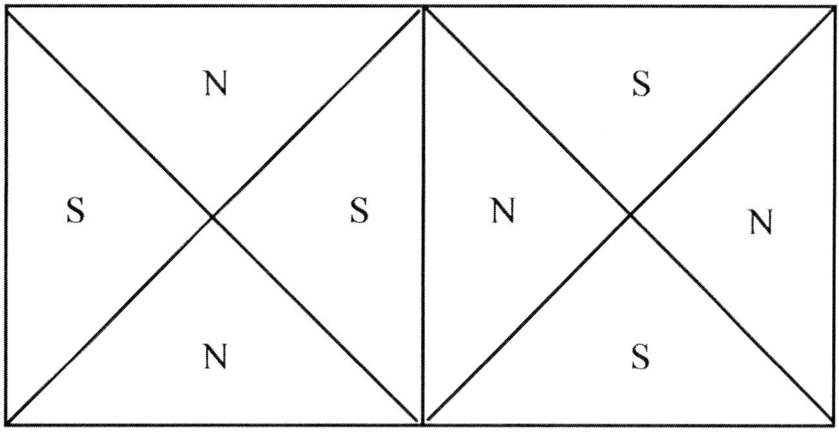

Figure 5. Border upon distribution of north and south poles on regular octahedron.

The magnetic force is two times the electrostatic force, and it also appears in other occasions. The empirical formula of b stability line of atomic nucleus is [2]

$$Z = \frac{A}{1.98 + 0.0155A^{2/3}} \quad (11)$$

Expressing the ratio of the nucleon number A to the proton number Z,

$$\frac{A}{Z} = 1.98 + 0.0155A^{2/3} \approx \left(2 - \frac{1}{50}\right) + \frac{A^{2/3}}{64} \quad (12)$$

The first item explains that at close range, the nuclear force is twice the electrostatic force, the proton number is twice the electron number, 1/50 is caused by the interval of proton and electron. The second item explains that at long distance r, the nuclear force according to r^{-3} decreases, the electrostatic force according to r^{-2} decreases; When the nucleon number A is from 50 to 64, the nuclear force and the electrostatic force at far range are equal, and the atomic nucleus is most stable.

5. THE REGULAR POLYHEDRON CONFIGURATION

The regular octahedron configuration is the only regular polyhedron in which the electron motions converge in the same direction. If allow the electrons converge in reverse direction, then all regular polyhedra satisfy this requirement. If converge in reverse direction, electrons move to a common edge at different times, there are not collision. It is reasonable to imagine that the electronic shell of the inert gases is at the second shell to outermost shell, and every shell is composed of a regular polyhedron. In each electronic shell, circles of electronic motion are faces of the regular polyhedron; electronic number is the faces number of the regular polyhedron. Two electrons of the first shell compose two parallel rings, with the electron moving in the same direction in a cylinder (**Figure 6**). The spins of two electrons are in the same direction, adjoining north and south poles. They can either be the same or in reverse direction of the orbital angular momentum at one time, thus helium has two configurations. The electrons of each shell of the inert gases element composite configuration and the electron motion modes are as follows:

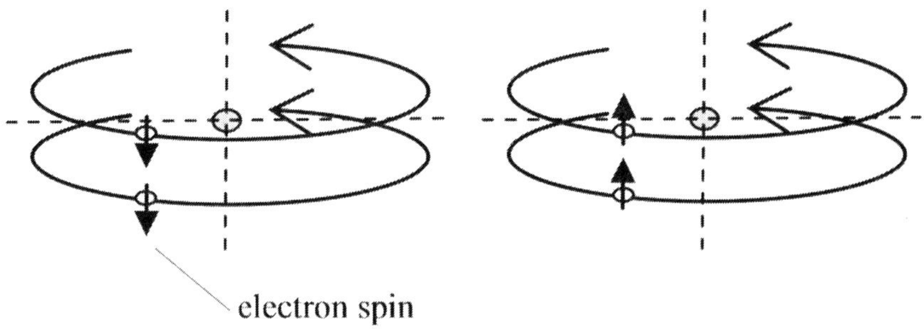

Figure 6. The double ring of same direction and parallel.

FIRST SHELL: cylinder, same direction and parallel;

SECOND SHELL: regular octahedron, converge at same direction;

MIDDLE SHELLS: regular polyhedron, face numbers are 6, 12, 20, allow converge in reverse direction;

OUTERMOST SHELL: regular octahedron, converge at the same direction.

Thus the electron shell configuration of the inert gases element can be denoted by a row of the face number of regular polyhedron, inasmuch as it is in accordance with the electron configuration of periodic table of elements (**Table 1**). 18 is dodecahedron and hexahedron, 32 is dodecahedron and icosahedron.

The orbital magnetic moment of the electron also engenders the adsorptive substance. The speed of the same shell electrons is near equation, the electrons can share the adsorptive substance, via its transfer. The energy of each electron tends to equate, and the adsorptive substance even increases the available distance of the magnetic force between electrons. The spin of atomic nucleus extends out and causes the first shell electrons to form a cylinder. For cylinder and regular octahedron, the electron motions of the parallel plane are in the same direction, forming a dipole, consistent with the spin polarity of the atomic nucleus. For hexahedron, dodecahedron and icosahedron, the electron motions of the parallel plane are in the reverse direction, forming a monopole, such as hexahedron with dodecahedron and hexahedron with icosahedron: these combinations can eliminate polarity. The inner shells of electrons are near the atomic nucleus, receive more adsorptive substances that are released by the nucleus, possess a higher negative charge density, and are at the accumulative position of the adsorptive substances. As such, the angular momentum of inner and outer shells is higher, shaping the regular polyhedron with more face numbers. The angular momentum of the middle shells is lower, shaping the regular polyhedron with lesser face numbers.

Table 1. The regular polyhedra that are composed of inert gases electronic shell.

Inert element	Electron configuration	Regular polyhedron
He	2	2
Ne	2. 8	2. 8
Ar	2. 8. 8	2. 8. 8
Kr	2. 8. 18. 8	2. 8. 12. 6. 8
Xe	2. 8. 18. 18. 8	2. 8. 12. 6. 12. 6. 8
Rn	2. 8. 18. 32. 18. 8	2. 8. 12. 6. 12. 20. 12. 6. 8

Before atomic number 21, the regular polyhedron shell and the electron configuration is same in number. The regular polyhedron electronic shells of elements which atomic number are 1 - 103 see Appendix 2.

6. THE CLUSTER OF INERT GAS ELEMENTS

The cluster of inert gas atoms can be described by models represented by regular polyhedrons or cylinders. For the inert gas atoms Ar, Kr, Xe, clusters form according to the preferred magic numbers [3,4].

An explanation using a regular octahedron model can be described as follows: as atoms approach, two closer regular octahedrons have an edge superposed; consequently, two atoms share an edge. Each atom (one regular octahedron) has twelve edges, so each atom can join with another twelve atoms. If twelve atoms are directly set on the edges of a regular octahedron, the resulting structure shows defective tightness. The orbital adsorptive substances on the edge are actually shared by two atoms and these adsorptive substances can undergo excursion. To obtain a tight structure, in a regular octahedron, using a plane that passes through an edge and cutting a vertex with an appropriate angle (**Figure 7**), twelve edges cut twelve times will result in a regular dodecahedron. If the atoms are then set at its pentagon center, in this way, the resulting structure is the tight, and is in agreement with the descriptions in the literature [4].

A Helium cluster [5] is different from other inert gas elements.

The explanation using the cylinder model is described below: because the helium atom is a cylinder, it can arrange along the direction of a circle and the axis, to form different hexagonal structures, even two layer hexagons (**Figure 8**), the axes of two layers coincide with each other.

7. THE MAGNETIC PROPERTIES OF IRON, COBALT, AND NICKEL

Under the frame of quantum mechanics, the origin of metallic ferromagnetism is an issue that is yet to be resolved [6].

The magnetic properties of iron, cobalt, and nickel can be explained through the regular polyhedron electronic shell found in an atom. The regular polyhedron electronic-

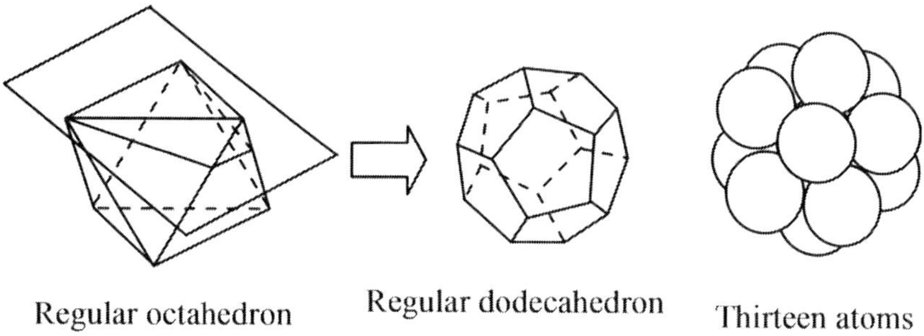

Figure 7. Ar, Kr, Xe cluster.

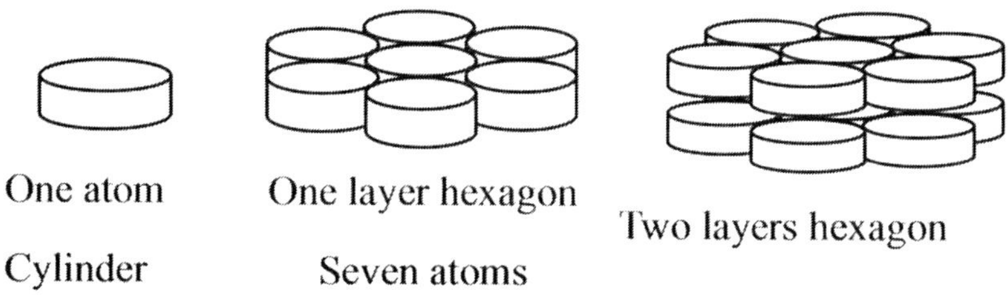

Figure 8. Helium clusters.

shells of iron, cobalt, and nickel are:

26 Fe 2, 8, 12, 3, 1 2, 8, 12, 2, 2

27 Co 2, 8, 12, 3, 2

28 Ni 2, 8, 12, 5, 1 2, 8, 12, 4, 2

All of these have a regular dodecahedron; this is the origin of the magnetic properties of iron, cobalt and nickel. The respective free or valence electrons of the outermost shells are 1, 2, 1 or 2, 2, 2; while those of the hypo-outer shell are 3, 3, 5 or 2, 3, 4. The third outer shells are all regular dodecahedrons. Given that hypoouter shell cannot sufficiently form a whole frame. Thus, the hypo-outer shell and third outer shell form a frame of two shells together; the third outer shell of a regular dodecahedron is thereby fixed.

Meanwhile, the motions of electrons on regular dodecahedrons are divided into three cases:

1) On regular dodecahedron, electrons are separated into two hemispheres according to the directions of the electron circumferential motion (**Figure 9**). At the edges that join the two parts, the directions of the electron motion

are the same; there are ten edges which converge with the same direction. Other edges converge from the opposite direction, thereby counteracting the magnetic effect. Ten edges that converge with the same direction engender an equivalent circular electrical current, producing magnetism in the process. Normally, μ is the magnetic moment of the entire regular dodecahedron. In ten pentagons with a total of ten edges, each edge has two electron paths, and μ_c is the magnetic moment of each electron, expressed as:

$$\mu = \frac{1}{5} \times 10 \times 2\mu_c = 4\mu_c \qquad .(13)$$

2) On the regular dodecahedron of case 1, there is one pentagon each in the center of the two hemispheres; these are marked as A and B in **Figure 10**. As can be seen, one converges with the same direction, while the other con verges from the reverse direction:

$$\mu = 2\mu_c \qquad .(14)$$

3) Two pentagons in the center of the two hemispheres converge from the same direction as the other edges (**Figure 11**):

$$\mu = 0 \qquad (15)$$

If the efficient Bohr magneton numbers of iron, cobalt, and nickel [7] are expressed by $n_B^{Fe} = 2.219$, $n_B^{Co} = 1.715$, $n_B^{Ni} = 0.604$, respectively, then:

$$n_B^{Ni} : n_B^{Co} : n_B^{Fe} \approx 1:3:4 \qquad .(16)$$

The radii of iron, cobalt, and nickel atoms are very propinquity; thus, the magnetic moment μ_c of each electron are deemed as equal in a regular dodecahedron.

All atomic structures of iron are case 1:

$$\mu_{Fe} = 4\mu_c \qquad (17)$$

A half atomic structure of cobalt is case 1, another half atom is case 2:

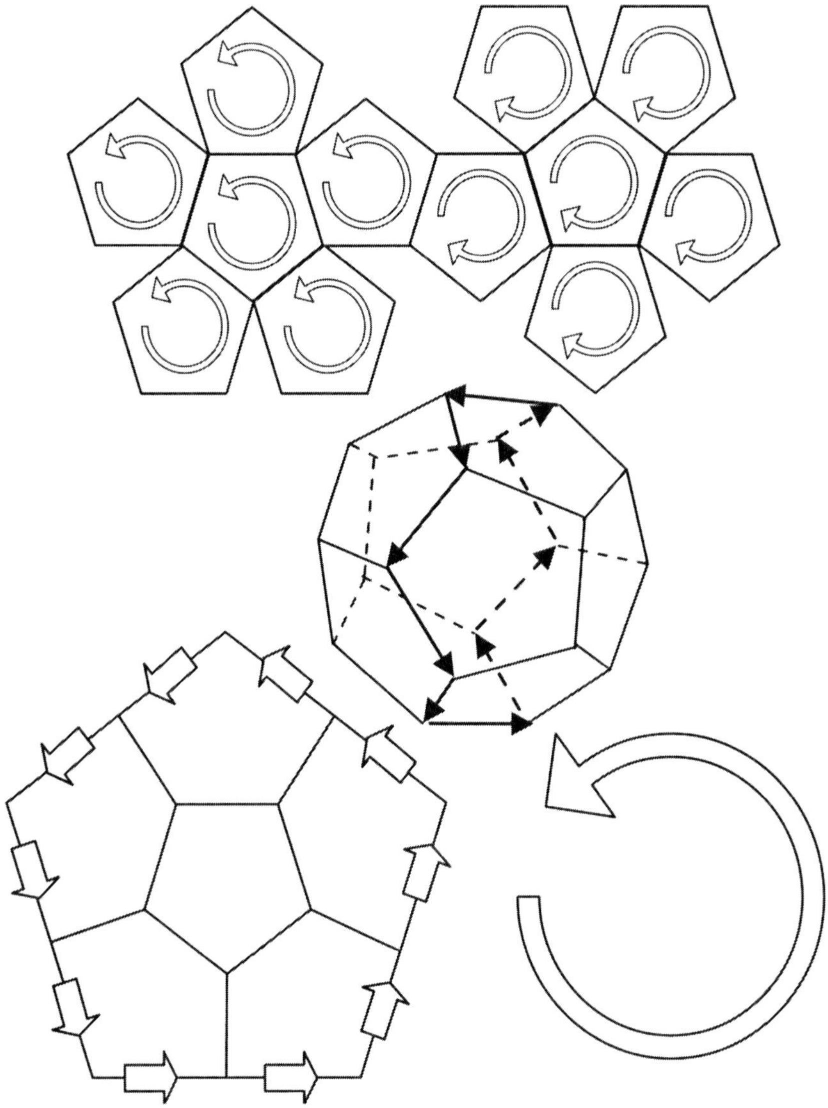

Figure 9. Electron motion and equivalent electrical current in a regular dodecahedron.

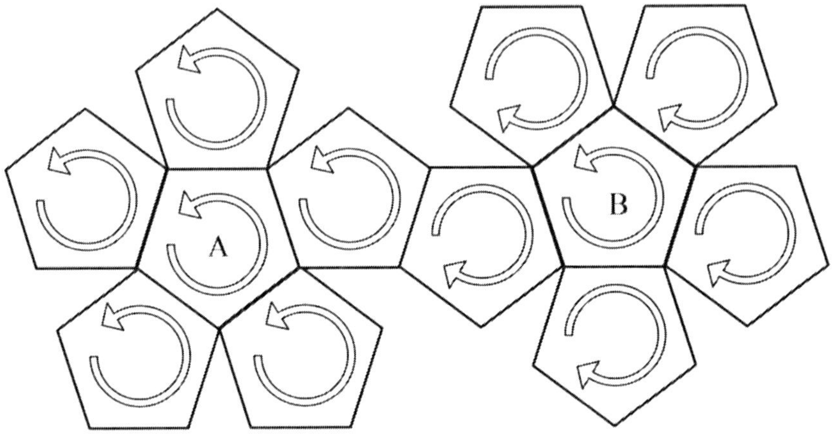

Figure 10. Convergence of the two hemispheres; A, the opposite direction and B, the same direction.

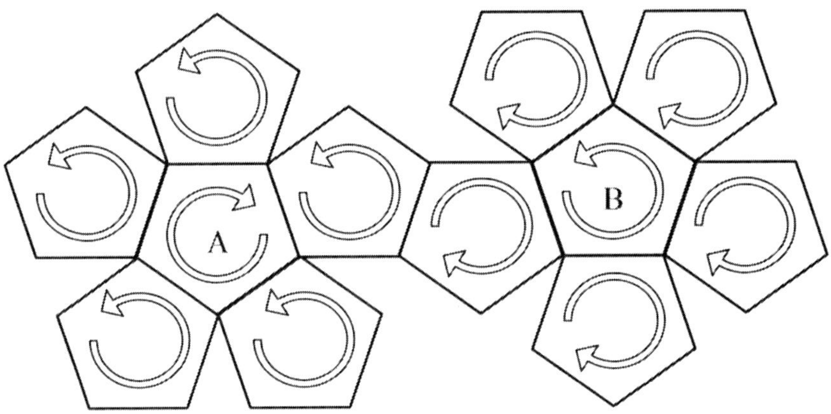

Figure 11. Convergence from the same direction for both A and B.

$$\mu_{\mathrm{Co}} = \frac{4\mu_{\mathrm{e}} + 2\mu_{\mathrm{e}}}{2} = 3\mu_{\mathrm{e}} \qquad .(18)$$

A half atomic structure of nickel is case 2, another half atom is case 3:

$$\mu_{\mathrm{Ni}} = \frac{2\mu_{\mathrm{e}} + 0}{2} = \mu_{\mathrm{e}} \qquad .(19)$$

From Formulas (17)-(19), we gain:

$$\mu_{Ni} : \mu_{Co} : \mu_{Fe} = 1:3:4 \quad (20)$$

Formulas (16) and (20) are concordant.

Therefore the magnetic properties of iron, cobalt, and nickel can be explained by the regular dodecahedron electronic shell of an atom.

The atom radius of iron is 0.126×10^{-9} m, from Formula (2) and $2.219\mu_B = 4\mu_e$, in the regular dodecahedron shell, the electronic velocity is $c/289$.

8. THE THERMODYNAMIC MAGNETON THEORY

The adsorptive substances found in atomic nuclei and electrons are defined as magnetic particles called magnetons.

As a term, "magneton" is commonly used to define or explain a unit of magnetic moment, as in the case of the Bohr and nuclear magnetons, among others. However, in this article, a magneton also denotes a type of particle primarily characterized by a property of magnetism. The magneton is a circle that circumrotates with the speed of light, wherein the radius can change; there are polarities that correspond to the frequency and phase.

The magneton has two different kinds that are separately engendered by either atomic nuclei or electrons; these are called positive and negative magnetons. In an atom, the positive magnetons engendered by the atomic nucleus is released and then arranged into a circulative orbit of the electron, thereby forming an orbit positive magneton (**Figure 12**).

The negative magnetons engendered by an electron have twenty petals and are attracted by electrons. When electrons move, they tend to release negative magnetons. These negative magnetons (1/20) form orbit negative magnetons, while others (19/20) turn into ray radiation. Some atoms do not have steady orbit negative magnetons.

In this text, the orbit positive magneton is called an orbit magneton for short. The circulative orbit and velocity of electrons determine the orbit magneton quantity, which is capable of maintaining the redundant orbit magneton's ability to transfer to another atom or transform into heat radiation even when orbit and velocity change. This means that the orbit magneton transforms into heat radiation when it does not have a containable orbit. After some orbit magnetons transform into heat radiation, an increase in electronic momentum will occur. Thus, the electronic orbit becomes capable of holding added magnetons, after which it can stop the process and then transform the magnetons into heat radiation. The frequency of radiation of the negative magnetons is generally higher than the positive magnetons (**Table 2**).

If the perimeter and density of a magneton are $2\pi r_{\mathrm{m}}$ and, respectively, then its mass could be computed by $2\pi r_{\mathrm{m}} D$. The relationship of the polarity periodicity of magneton N_{mt}, the polarity frequency f_{m}, and radius r_{m} are expressed as:

$$cf_{\mathrm{m}} = 2\pi r_{\mathrm{m}} / N_{\mathrm{mt}} \quad .(21)$$

The polarity frequencies of a positive magneton in all static atomic nuclei are equal.

The magneton radius, periodicity, and the positive magnetons released from the proton are also equal. However, in the case of positive magnetons, which are released from the atomic nucleus (with more than one nucleon), the magneton radius and periodicity are not necessarily equal.

Meanwhile, just as there are attractions between the negative and positive magnetons, similarly, there are also attractions between the negative and orbit magnetons. The positive magnetons that are attracted by an electron first accumulate at two ends of the electron spin axis; from there, they form an impetus magneton. The impetus magneton may be integrated stepwise to the inner electrons, released out along the motion of the negative magneton of the electron occurring after break up. It can then cause the electron to obtain impetus. As many positive magnetons accumulate, these compound into a circlet which becomes the compound positive magneton of the large circle. Here, the polarity periodicity of the compound is the average value before the magnetons compound.

Table 2. Magnetons.

Magneton	Positive magneton	Negative magneton	Orbit magneton	Negative orbit magneton
Polarity	Positive	Negative	Positive	Negative
Location	Proton periphery	Electron periphery	Electron orbit enphery	Within electron orbit circle
Engendered by (transform from)	Atomic nuclei	Electrons	Positive magneton	Negative magneton
Turn into radiation	Heat low freq.	Ray high freq	Heat low freq.	Ray high freq
Magneton radius	$r_{pm} = \lambda_C^{proton} = 2.1031 \times 10^{[illegible]}$ m	$r_{nm} = \lambda_C = 3.8616 \times 10^{[illegible]}$ m	[illegible]	[illegible]
Magneton orbit radius	$r_{[illegible]} = 3.771 \times 10^{[illegible]}$ m	$r_{[illegible]} = 4.475 \times 10^{[illegible]}$ m	[illegible]	[illegible]

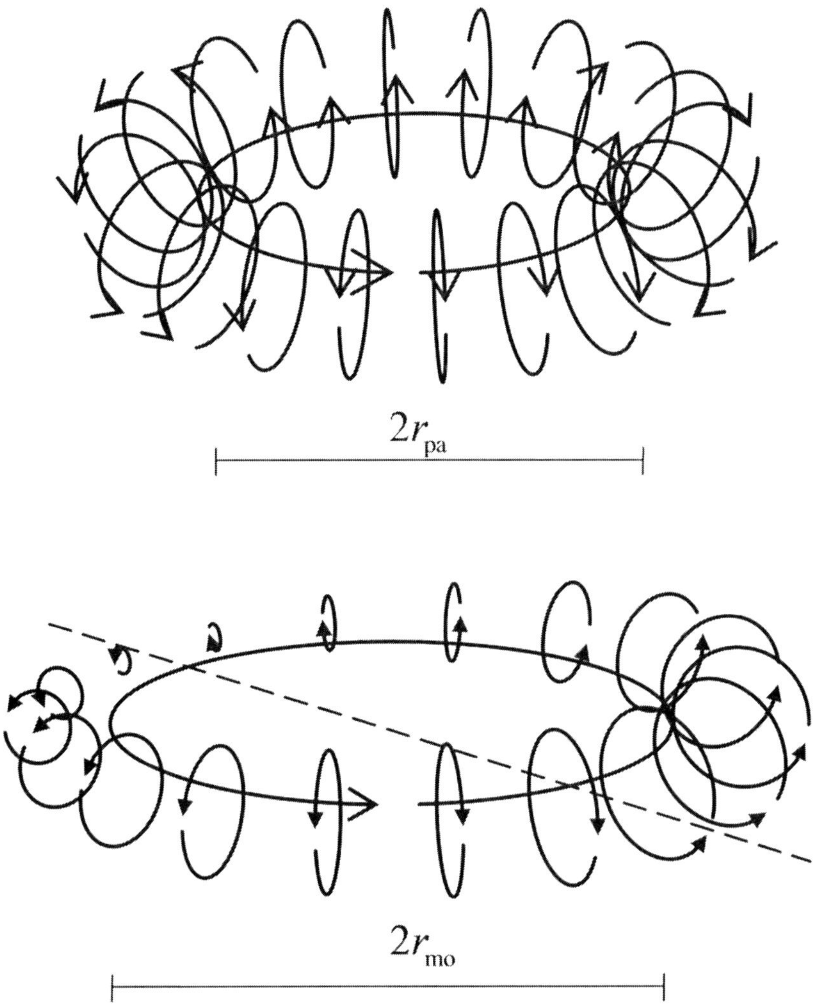

Figure 12. Positive magnetons of proton and orbit positive magneton at electron orbit.

If the radius of the compounded positive magneton is greater than the diameter of the negative magneton of an electron, then the compound positive magneton shall leave two ends of the electron spin axis and transfer in a parallel position to the side of the electron's negative magneton. This combination of compound positive magneton becomes the orbit magneton. The rules of interaction among positive, compound, and orbit magnetons are explained below (**Figure 13**):

1) If two orbit magnetons have equal polarity periodicities and radii, the spin axes are in superposition and the spin direction are the same, then there would be no force action; if the spin axes are in superposition but the spin direction is in reverse, then there would be an at traction. On the other hand, if the circles of two magnetons border upon the same plane and the spin direction is the same, then there would be an attraction; in this case, the linear velocity at adjacent point is in reverse direction.

2) If two orbit magnetons have equal polarity periodicities but unequal radii, their spin axes are in superposition, and if the spin direction is the same, then there would be an attraction; but if the spin direction is in reverse, then there would be a repulsion. If the circles of two magnetons border upon the same plane and the spin direction is in reverse, then there would be an attraction; in this case, the linear velocity at adjacent point is the same direction.

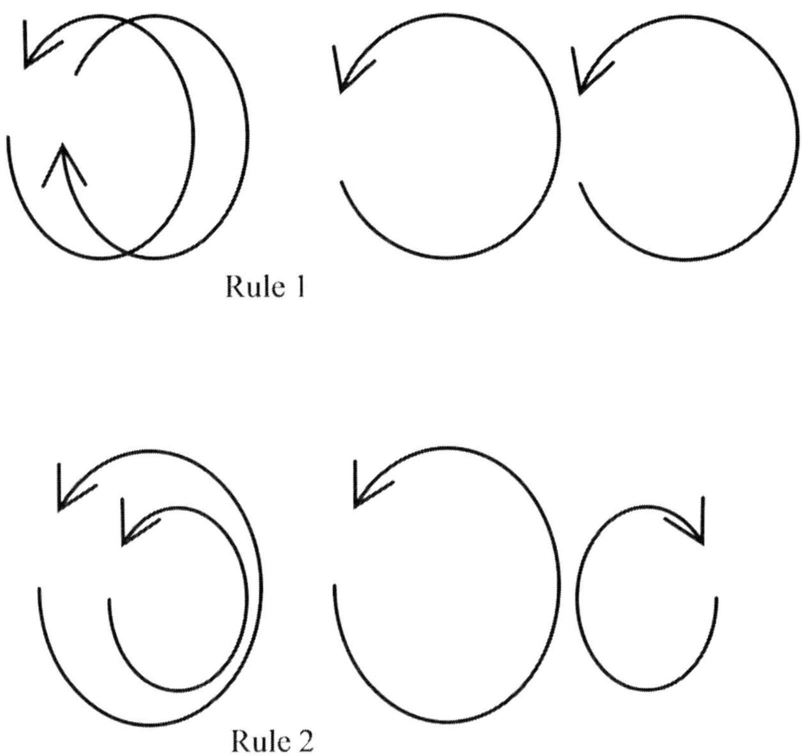

Figure 13. Attraction of orbit magnetons.

3) If two orbit magnetons have unequal polarity periodicities, there would be no force action.

If a multi-orbit magneton forms a ring, the circle of each magneton would be perpendicular to the electron orbit (**Figure 12**). The orbit direction with similar spin directions and arrays would form into a single team or one with a reverse spin direction according to the radii (from large to small). If two magnetons form at the junction and the radii are equal, then the largest or smallest in the team will separate. If the orbit magneton ring on the electron orbit forms 2, 4, 6 and even more teams (two teams at the very least) and the magneton spin directions of the two adjacent teams are in reverse, this will lead to a magnetic moment of zero at the exterior. In **Figure 12**, the broken line separated two teams.

For the electronic shell of regular polyhedron, the orbit magneton distribute in the edge of regular polyhedron.

If the electronic motion does not periodically turn around, then the orbit magnetons would no longer form a ring so that the beginning would connect to the end. Beside the electron, two teams of orbit magneton would form an olive shape (**Figure 14**).

When many atoms or molecules come in contact with each other, some parts of the action between electrons are processed via the orbit magnetons. If the orbit magnetons maintain stable attraction, then the object would be in a solid state. The attraction of the orbit magnetons can decrease the distance between electrons as the electron's negative magneton show no apparent attraction. If the distance between electrons is shortened, this would cause a fraction of the orbit magnetons to be pushed out, leading to slippage between atoms or molecules; if this is the case, then the object is said to be in a liquid state. If most of the orbit magnetons are pushed out and the atoms or molecules are also bounced out, then the object is said to be in a gaseous state.

The attraction between the orbit magnetons is in direct ratio with the amount of magnetons. This is also related to the polarity periodicities and radii of such magnetons.

In the gaseous state, the orbit magnetons that are pushed out when molecules come in contact with each other and bounce out become free positive magnetons. The bounced out molecules can also capture free positive magnetons and may become orbit magnetons. The overall velocity of the free positive magnetons is equal to zero. After the molecule captures the positive magnetons, the speed of the molecule would then be reduced. If molecules have same orbit magnetons, then their speeds are the same.

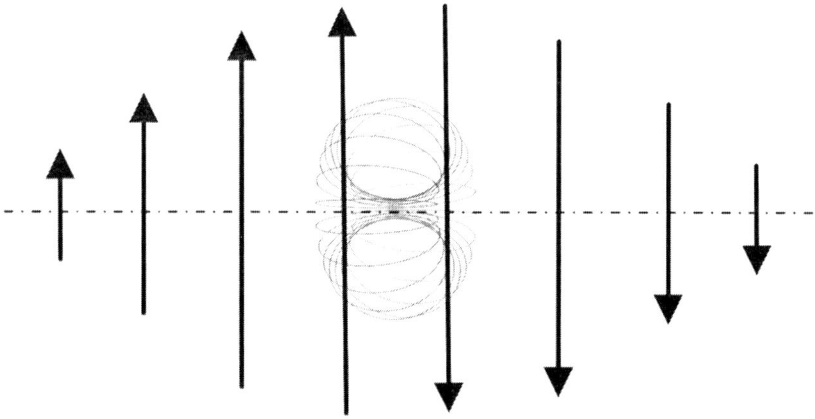

Figure 14. The olive-shaped orbit magneton.

The action of molecules in the positive magnetons is separated into the force of the free positive magnetons f and the force of the captured orbit magnetons F.

If the attraction of molecules in the free positive magnetons is f, the mass of the free positive magnetons is M, and the acceleration is a, then we have:

$$a = \frac{f}{M} \tag{20}$$

If the attraction of molecules in the captured orbit magnetons is F; when captured, the attraction is $\mathrm{d}F$, the mass is $\mathrm{d}M$, and the acceleration is the same as the acceleration of the free positive magnetons a, then:

$$\mathrm{d}F = a\mathrm{d}M = \frac{f}{M}\mathrm{d}M \tag{23}$$

In an equilibrium state, the f of every molecule would be equal. If the molecules with equal $\mathrm{d}M$ are considered as the same kind, then $F(M)$ is the distribution of the attraction according to the same kind of molecular total in M stated as:

$$F(M) = \int a\mathrm{d}M = \int \frac{f}{M}\mathrm{d}M = f\ln M \quad (24)$$

On the basis of the inter-actional rules of orbit magnetons, only the same kind of molecules may attract each other and push out the orbit magneton; in this situation, the system tends to increase as many different kinds of molecules as possible and capture the maximum positive magnetons.

When molecules bounce out, the work of repulsive force equals the negative work of attraction, expressed by the attraction F multiplied by the maximal radius of orbit magneton $r_{\max}$. Meanwhile, the kinetic energy when molecules bounce out is equal to the work of the repulsive force,

$$E_{\mathrm{k}} = -Fr_{\max} . \quad (25)$$

From Formulas (24) and (25), we derive:

$$E_{\mathrm{k}} = -fr_{\max} \ln M \text{, and} \quad (26)$$

$$M = e^{-E_{\mathrm{k}}/fr_{\max}} . \quad (27)$$

This is the relation of the free positive magnetons mass M and the kinetic energy of molecular E_{k}. As compared with the Boltzmann factor $e^{-E_{\mathrm{k}}/kT}$, we can see that the Boltzmann constant k is the attraction of molecules for the free positive magnetons f. If it is constant in the equilibrium state, the temperature T is the maximal radius of orbit magneton $r_{\max}$, expressed as:

$$k = f \text{, and} \quad (28)$$

$$T = r_{\max} . \quad (29)$$

The dimension of kT is work, the dimension of temperature T is length, and the dimension of k is force. From this relationship, we can say that the magneton theory can work well with thermodynamics.

The maximal radius of the impetus magneton is equal to the minimal radius of the orbit magneton $r_{\min}$, and is also equal to the radius of the negative magneton on static electron r_{cm}. The temperature T defined by perfect gas is a macroscopic and statistical quantity of multiunits (molecule); meanwhile, the temperature T of an area near zero K, that is equivalent to: in this area, there are almost no orbit magnetons of the outer shell that may lead to a shift out. Therefore when temperature T nears zero K, the outer shell of most units will have not orbit magnetons; for the few units that have orbit magnetons in the outer shell, the radius of the orbit magneton will take a minimum value as expressed by:

$$T = r_{max} = r_{min} = r_{cm} = \lambda_C. \quad (30)$$

In the outer shell of metal, electrons are far apart from the atomic nucleus, and the number of absorbed positive magnetons could be low such that these would not form into a fixed frame. In such a case, it can rely on the secondary outer shell electrons so that it could be joined together; this is called the hypo-outershell. The outer shell electrons become free electrons in the frame interstice (a free electron has a small quantity of impetus magneton and has an initial kinetic energy E_0). The action between the free electrons is similar to what is occurring in the gaseous state; also, if the free electron has the action of another item from the hypo-outer shell frame via free positive magneton, Formulas (22) and (25) can be changed to:

$$a = \frac{f}{2M}. \quad (31)$$

$$E_k - E_0 = -2Fr_{max}. \quad (32)$$

Using a coefficient i denotinge the scale of two item's action, $0 \le i \le 2$, and from Formulas (24) and (32), we derive:

$$E_k - E_0 = -ifr_{max} \ln M \text{, and} \quad (33)$$

$$M = e^{-(E_k - E_0)/ifr_{max}}. \quad (34)$$

Suppose m_e is the electron mass, v_e is the electron velocity, v_0 is the initial velocity of electron, the momentum of the action between the electron and hypoouter shell frame is $m_e v_e$, and t is the acting time:

$$m_e v_e = Ft = -ft \ln M \text{, and} \quad (35)$$

$$M = e^{-m_c v_c / ft} . \quad (36)$$

From Formulas (34) and (36), we derive:

$$-(E_k - E_0)/ifr_{max} = -m_c v_c / ft , \quad (37)$$

$$\frac{\left(\frac{1}{2}m_c v_c^2 - \frac{1}{2}m_c v_0^2\right)}{ifr_{max}} = \frac{m_c v_c}{ft} \text{, and} \quad (38)$$

$$t = 2ir_{max} \frac{v_c^2 - v_0^2}{v_c} . \quad (39)$$

The electrical resistance of metal mostly comes from the action occurring between the electrons and the hypoouter shell frame. It is derived via the transfer of free positive magnetons, electrical resistivity ρ is namely the acting time t, and the maximal radius of the orbit magneton r_{max} is the temperature T, which is treated i as the probability of all collision. Hence, the collision probability between electrons is $P_c = i/2$,

$$\rho = 4P_c T \frac{v_c^2 - v_0^2}{v_c} . \quad (40)$$

If $v_c < v_0$, or $P_c = 0$, then it is said to be in a superconductive state. The directional motion of the electron can be maintained in the initial kinetic energy E_0. $P_c = 0$ is the non-collision superconductivity, and $v_c < v_0$ the superconductivity when collision speed is less than the initial velocity.

When the temperature T is lower, consider two special cases below:

1) If P_c and v_0 are close to the constants, then

$$\rho \sim T\left(v_c - \frac{A_1}{v_e}\right) . \quad (41)$$

As such, A_1 is approximately constant. From Formula (33) we derive $T \sim v_c^2$, so

$$\rho \sim T\left(\sqrt{T} - \frac{A_1}{\sqrt{T}}\right). \quad (42)$$

2) If v_c and v_0 are close to the constants, the mass of the olive-shaped orbit magnetons would be proportional to the square of the magneton radial; thus, theattraction between electrons would be proportional with the collision probability, $P_c \sim T^2$, leading us to the following:

$$\rho \sim T^3 \quad (43)$$

When the temperature T is higher, then $v_c \gg v_0$. According to the magneton interactional rule, the free electrons attract each other but only if they have the same speed. If there is an increase in free electrons with different speed levels, the range of speed would increase and the collision probability will decrease, $\rho \sim (1/v_c)$, so

$$\rho \sim T. \quad (44)$$

Therefore, the part of electrical resistivity $\rho > 0$ can disport three segments; these are shown in Formulas (42) to (44). For the expression $\rho \sim T^3$ see also [8].

The high critical temperature superconductivity can be seen in non-collision superconductivity. To increase the critical temperature, the following methods are followed:

1) Release the free electrons from the atoms of multiple sorts.

2) The atomic nucleus can release the positive magnetons with different periodicity, these positive magnetons can then form the orbit magnetons of multiple sorts.

3) Increase the speed range of the free electrons.

4) Restrict the motion range of the free electrons.

5) Decrease the density of the free electrons.

In Method 2, the first requirement is an element with a high melting point and the second is an element with low rigidity.

In Formulas (26) to (29), the Boltzmann factor $e^{-E_k/kT}$ is equal to the free positive magnetons mass M. In some liquids, M^{-1} can be viewed as a macroscopic amount referring to viscosity η. In [9,10],

$$\eta = Ae^{Q/RT} \quad (45)$$

where Q is the activation energy, R is the molar gas constant, and A is a constant. If one molecule is transformed to many molecules, then k is transformed to R, and E_k is transformed to Q. In these liquids, the viscosity η is inversely proportional to the free positive magnetons mass M in the area:

$$\eta \sim M^{-1} \quad (46)$$

In [11-13], $\eta = Ae^{B/RT}\left(1+Ce^{D/RT}\right)$, where A, B, C and D are constants, then this is a case when there are two types of free positive magnetons. Fermi distribution is a case when there are two types of attractive particle, Bose distribution is a case when there are two types of repulsive particle.

There is a decrease in free positive magnetons in the process of transformation from liquid to solid state; Its part transform into the orbit magnetons M_o; if their total $M_t = M + M_o$, then E is a constant:

$$\eta = E\left(M_t - M_o\right)^{-1} \quad (47)$$

In the presence of orbit magnetons, similar particles can attract each other and molecules and electrons also behave in this manner. In such a case, the ion is also similar. When the glass of oxide (as SiO_2) change from a liquid to solid state, though free positive magnetons decreasing, there are yet enough orbit magneton, the same ions will attract each other and can separately form an ion group of a similar type. The oxygen ions can then attract each other and form a group of oxygen ions; the silicon ions can attract each other and form a group of silicon ions. There are electrostatic attractions between the groups of oxygen and silicon ions. The structure of a glass is formed when a large amount and space of groups of oxygen ions connect and form a net where a group of silicon ions are embedded.

The chemical reaction is regarded as that occurs the electronic transfer, because only the electron those sort and velocity are same attract each other, the course of electron transfer from a sort of atom to another sort of atom, except electrostatic force require also satisfy two condition: first the electronic velocity is same, second is the mix of magneton: the magneton on electron come from several atomic positive magneton, form the orbit positive magneton on electron after mixed. The magneton mix also may be the change of magneton supply, the source of positive magneton from a sort of atom change to another atom. The magneton mix also may the electronic velocity go to equal. Usually the magneton mix easy in high temperature and pressure, and the electronic kinetic energy easy over the electrostatic force, the chemical reaction easy progress. In the situation of temperature and pressure are not change, the magneton mix becomes mostly factor that influence the chemical reaction.

9. MAGNETON AND ELECTROMAGNETIC INDUCTION

The magneton of electron formerly was circle. When two electrons approaching to a scope of distance, the magneton interact and being distorted, the circle plane occur bend; if the distance of two electrons are not changed anew, the bend of circle keep and are not change; at the farther position there is away the central electron, on the edge, the plane of magneton rotated a angle, the angle between circles become uneven. Formerly the magneton moving along circle with the velocity of light, this the tangent direction, the magneton is tangent with the central electron also, both spins keep be in mesh in certain phase; the circle of magneton swing in other direction when distortion, the advance speed is not speed of light in tangent direction, but is less than speed of light, the electron must moving backward and keep be in mesh. This backward motion of electron is the electromagnetic induction (**Figure 15**).

The motion and spin axis direction of electron incline to superposition when the electromagnetic induction. Two electrons appear the electromagnetic induction; firstly the magnetons of electrons attract each other, the spin axis of electrons incline to same direction, and then the magneton swing, the electron moving backward. Two conditions which the free electron can electromagnetic induction is also two sort swings: the swing of the spin axis of electron and the swing of magneton. The swing of the spin axis cause the spin axis of electrons turning to same direction; the swing of magneton cause the electron that keep be in mesh with magneton moving backward. The electromagnetic induction is the change of the magnetic force causing the swing of magneton of electron, the electron moving, this induction differs from the magnetic force (**Figure 16**).

Two relation of direction are reverse: the relation of the spin axis of electron with the act direction when electromagnetic induction; and the relation of the spin axis with motion direction when thermal motion.

A motion electron comes into magnetic area; firstly the spin axis is turned, the turned place always got behind with interact point, the spin axis incline to same direction with the electron of the back magnet, and then the magneton swing, the electron plus a motion that moving backward along the spin axis, and is the reverse of electron of the back magnet, it is close perpendicular to the motion direction, but is not completely perpendicular to the motion direction. A motion electron is affected by magnet, the force that effecting electron f_m with

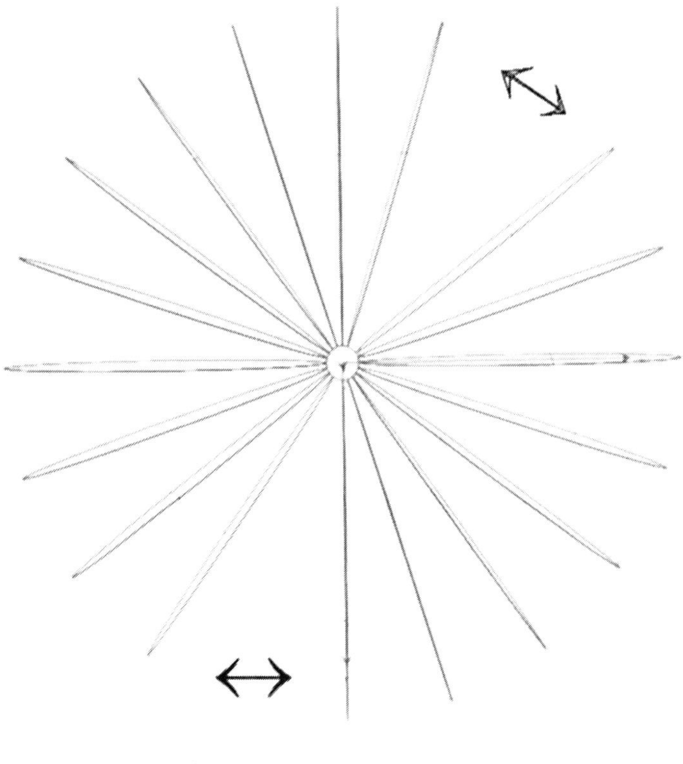

Figure 15. The swing of magneton.

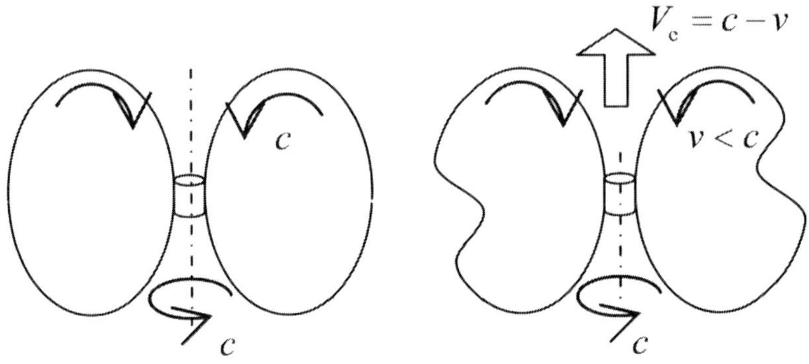

Figure 16. The swing of magnetons causes the electron moving.

the motion direction of electron (or velocity v_e) forms an obtuse angle ¶ > 90° (**Figure 17**).

The magnetic induction satisfies three conditions:

1) The spin axis of electron turning.

2) A revolving ring with velocity of light exists in electron.

3) The swing of revolving ring.

The momentums of instantaneous Lorentz force are not conservation, its action force and effect is perpendicular; the electromagnetism induction satisfies three conditions above are momentums conservation.

The delay and inertia of the electromagnetic induction are greater than the energy of ionization, this explain also three conditions above.

The electromagnetic induction can engender voltage, and engender current if there is a loop. When the area of an electric lead circle changed, each part that lead moving engenders voltage, the directions of these voltages are same; the perimeter of circle changed simultaneity, the perimeter has difference, the voltage to the perimeter difference are in direct ratio, the current is that the perimeter difference multiplies the amount of electron, the perimeter difference to the amount of electron are in direct ratio, the area difference to the square of the perimeter difference are in direct ratio, therefore the current to the area difference are in direct ratio.

10. THE ELECTRONIC SHELL OF CARBON

The electronic shell of the carbolic atom is 2, 4. Four electrons of the outermost shell formed a regular tetrahedron.

In the regular tetrahedron of atom of diamond, all four electrons converge from reverse direction; each atom is connected by superposition of vertex, that is share four vertexes, therefore each atom connect with other four atoms, and is symmetrical (**Figure 18**).

In the regular tetrahedron of atom of graphite, three electrons is converged from reverse direction, these three electrons is converged from same direction with the other one electron, in six edges of the regular tetrahedron, there is three edges of one face is converged from same direction, other three edges is converged from reverse direction. Each atom is connected by superposition of three edges, which are converged from same direction, so each atom is connected with other three atoms, and is in a same plane (**Figure 19**), these atoms can connect into hexagon.

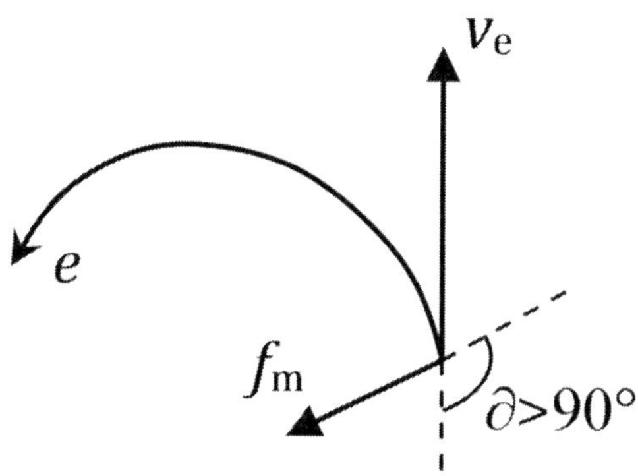

Figure 17. A motion electron is affected by magnet.

The converged electron orbit has two sorts of the same or reverse direction from the rule of magneton attract each other: when converge from same direction, the magneton of same radius attract each other; when converge from reverse direction, the magneton of different radius attract each other. The negative magnetons attract each other when converge from same direction, cause the electron orbit close up, and push out the orbit magneton (**Figure 20**); the negative magnetons do not have attract when converge from reverse direction, electron orbit the do not near mostly, and do not push out the orbit magneton (**Figure 21**). The electron of atom of

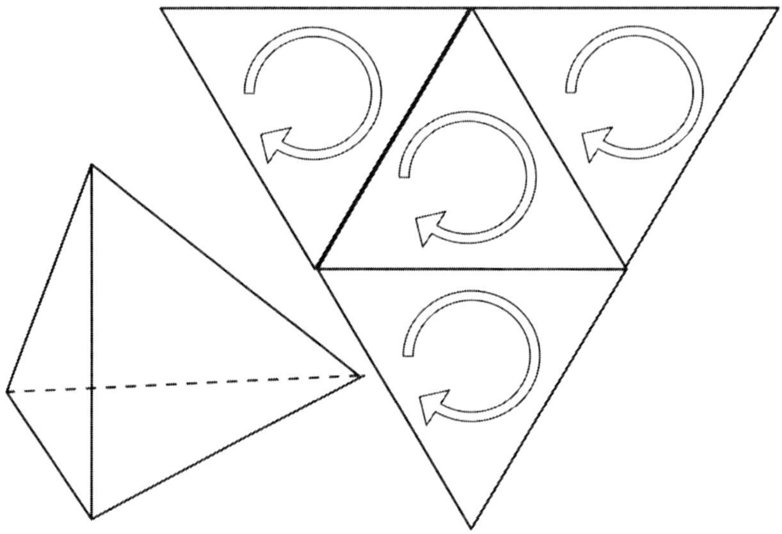

Figure 18. Diamond: four electrons converge from reverse direction.

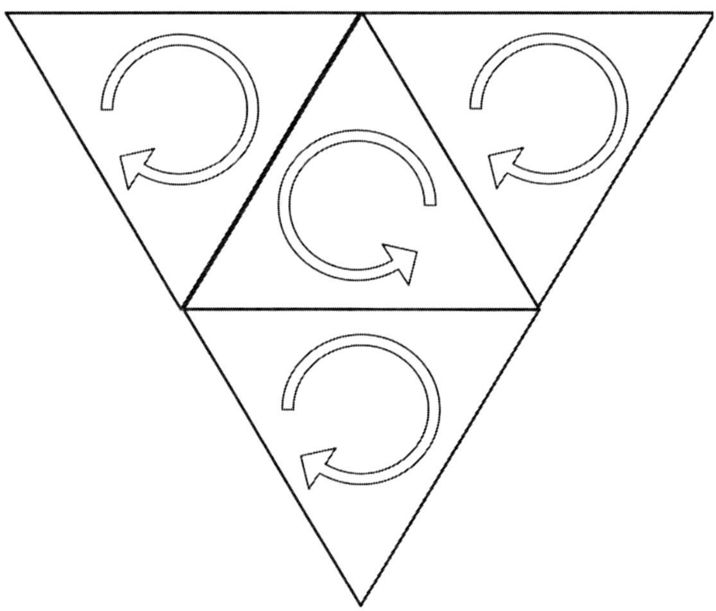

Figure 19. Graphite: three edges of same plane converge from same direction.

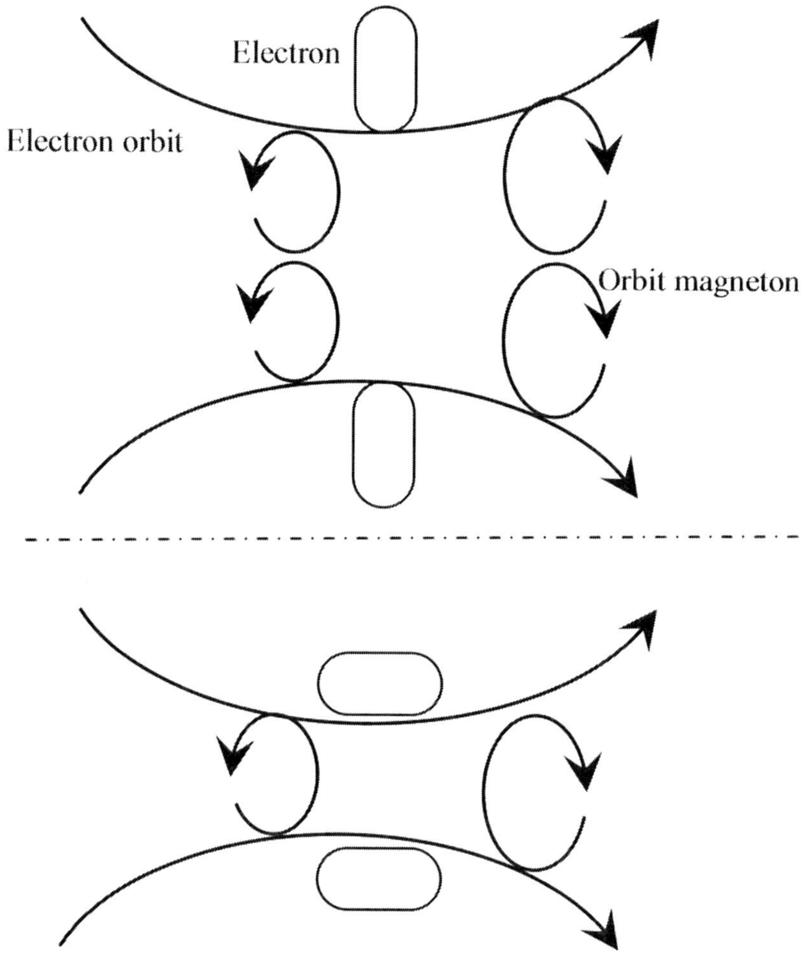

Figure 20. Converge from same direction.

diamond all are converge from reverse direction, electrons of adjacent atom are also converge from reverse direction, therefore do not push out the orbit magneton, have some attraction, and equipoise with electrostatic force of nucleus, the electron orbit is most stable.

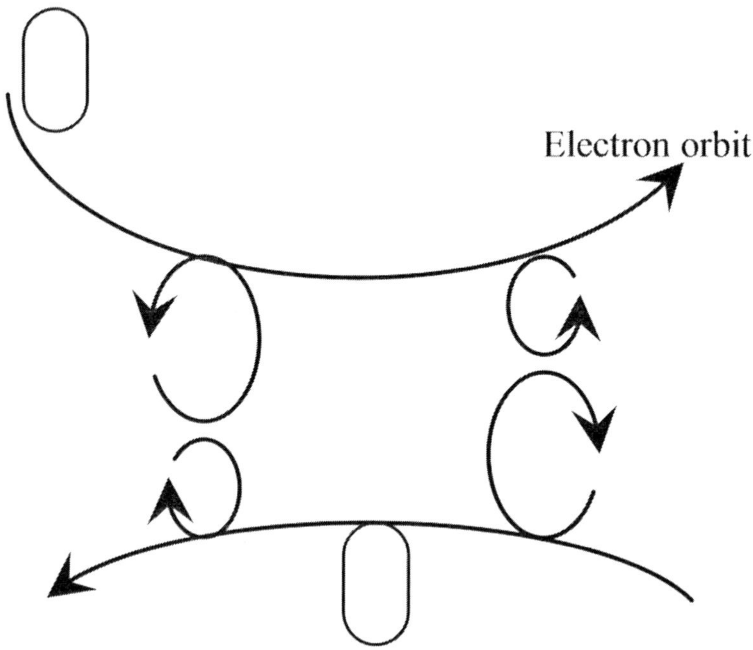

Figure 21. Converge from reverse direction.

11. THE HELICAL LINE MODEL OF THE LIGHT

The structure of the light is a helical line. The axis of the light helix is same from the direction of motion of light. The direction of motion of light is its direction of travel or propagation direction. β is the helix angle of the light helix, $45^\circ \leq \beta \leq 90^\circ$. The helix angle approach 45° if the light density is high; when the light density debases, the helix angle increase, maximum approach 90°, the helix almost become a straight line. The wavelength of light is the screw pitch of the helix. If the helix angle of the light helix is already 45° and the light density continue to increase, then the screw pitch of the helix change into short, that is the wavelength of light change into short (**Figure 22**).

Two planes that are parallel with axis cut symmetrically helix, split helix into three parts, the middle part is a polarized light; if remove the middle part, two remnant parts are also a polarized light.

The light in a magnetic field, the advance have a resistance; the subsequent part of the light helix has an act of pushing for the front part. In the act of resistance and pushing, the front part of the light helix comes into deflection; the deflection of the front part leads the subsequent part and comes into deflection. Therefore the light have a deflection in a magnetic field, the deflection of polarized light can easy be observed.

When a light come close and into an object, have an act first with the magneton of object.

Compton Effect is the result of act of the light and the negative magneton. When light get across the negative magneton, if the axis of the light helix and the spin axis of the negative magneton is superposition, then the helix has a repulsive force, the helix is expanded, the radii of the helix adds the radii of the negativemagneton, that is Compton wavelength λ_C of electron; the screw pitch of the helix adds the perimeter of the negative magneton λ_C, so the wavelength of light adds λ_C.

The orbit positive magneton included two teams of positive magneton with reverse spin, when the light meets the orbit positive magneton, if the axis of the light helix is perpendicular to the circle of the positive magneton, then the light is reflected; if the axis of the light helix is parallel to the circle of the positive magneton, then the light gets across the positive magneton, come into being refraction.

When the velocity of electron change, the radii of the negative magneton of electron decrease, the superfluous substance of the negative magneton become mostly the radiation of light, the few become the orbit negative magneton of electron. Along the straight line if count the frequency of light; along a circle if count the frequency of magneton. A light can have an act with the orbit negative magneton that the frequency is same, and then is absorbed by the atomic nucleus, translate into the positive magneton. The translation about the magneton and the radiation of light or heat express below:

The positive magneton on the atomic nucleus à The orbit (positive) magnetons on the electron orbit à The radiation of heat.

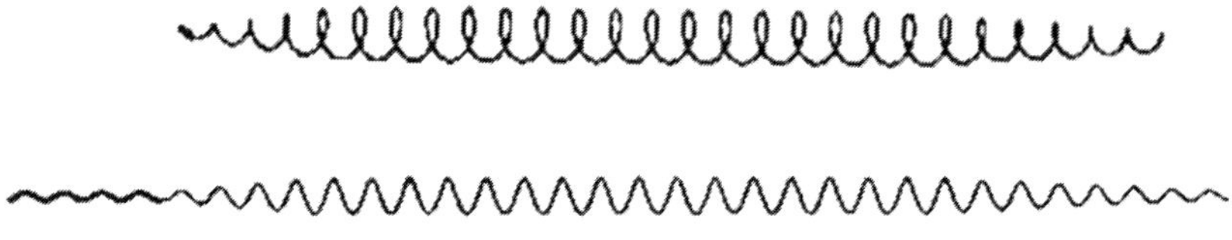

Figure 22. The helical line structure of the light.

The positive magneton on the atomic nucleus à The negative magneton on the electron.

The part of negative magneton on the electron à The radiation of light, the orbit negative magneton.

The radiation of heat, the radiation of light + the orbit negative magneton with same frequency à The positive magneton on the atomic nucleus.

When the magneton that mass is m transform into the radiation of light of half periodic, the velocity of magneton become the velocity of light c, the released kinetic energy is $\frac{1}{2}mc^2$.

The light can separate and form again. The diffraction is that the part of light helix is obstructed. The diameter of the light helix is less a little than aperture, the light passing through immediacy if it is at the center of the aperture; the part is obstructed if the light helix is not at the center of the aperture, and change in the directions. The diffraction of twain gap is that the separate light forms again if these phase is same.

The light form again inside the object can explain the color combination. If the diameter of a light helix is less a several fold than another light, a light helix enter into the another light helix inside, the frequency of the combination light is add up, this is the combination of attraction; if the diameter of a light helix is less a little than another light, the frequency of the combination light is reduced, this is the combination of repulsion.

12. THE MAXIMAL SHELL NUMBERS OF THE ELECTRONIC SHELLS

The negative magnetons on the electron have twenty; the positive magnetons on the atomic nucleus have twenty also, these positive magnetons release out and become the orbit magnetons in the electron orbit, the positive and negative orbit magnetons change to positive magnetons on the nucleus via action; in this cycle, the atomic nucleus release twenty positive magnetons each period. Because the static attraction, the electron will fall on the nucleus if have not the supplement of the positive magnetons, the nucleus change to other nucleus by the action of the electron and nucleus. The number of the positive magnetons that the nucleus released has decided the maximal shell number of the electronic shells, twenty positive magnetons can support the maximal shell number is ten layer, the electronic shells if the shell number exceed ten layer, the electron fall easily on the nucleus, the nucleus is very unstable. If the atomic number is greater than 90, the electronic shells begin exceed ten layer, the most outer shell electrons already cannot form a whole frame, the electrons have no longer subdivide layer. If the atomic number is more, the electron fall easily on the nucleus and bring the nucleus change.

13. CONCLUSION

Zilong Kong Inert gas atoms form a cylinder and a regular polyhedron, forming the same shell electrons. The circle tracks of electrons motion are faces of polyhedron. The electronic number is the faces number of the regular polyhedron. The periodic table of elements is arranged based on a series of regular polyhedrons. The electron has a magnetic outer layer and an electrostatic inner layer. The electron periphery has twenty-petal adsorptive substances, forming a big magnetism radius. If two electrons with the same velocity and direction converge, their adsorptive substances have the same rotational frequency and phase. The magnetic force is $\sqrt{2}$ times the electrostatic force; the existence of adsorptive substance causes the magnetic force greater than the electrostatic force. The magnetic properties of iron, cobalt, and nickel can be explained through the regular dodecahedron electronic shell. The electron orbit converge from reverse direction can explain diamond. The behavior of the adsorptive substances is evolved into the magneton theory, and it can work well with thermodynamic. The free electrons do not have collision in high critical temperature superconductivity. In glass, the same ions will attract each other and can separately form an ion group of a similar type. The structure of the light is a helical line. The maximal shell numbers of the electronic shells are ten.

REFERENCES

1. G. N. Lewis, Journal of the American Chemical Society, Vol. 38, 1916, pp. 762-785.
2. X. T. Lu, H. J. Wang and B. J. Yang, "Atomic Nucleus Physics," Atomic Energy Press, Beijing, 1981, pp. 46-47.
3. O. Echt, K. Sattler and E. Recknagel, Physical Review Letters, Vol. 47, 1981, p. 1121.
4. W. Miehle, O. Kandler, T. Leisner and O. J. Echt, Chemical Physics, Vol. 91, 1989, p. 5940
5. P. W. Stephens and J. G. King, Physical Review Letters, Vol. 51, 1983, pp. 1538-1541.
6. G. S. Tian, "The Origin of Metallic Ferromagnetism," In: 10000 Selected Problems in Sciences, Physics, Science Press, Beijing, 2009, pp. 73-80.
7. H. Stöcker, "Taschenbuch der Physik," Verlag Harri, Deutsch, 1996.

8. A. Chaiken, "Resistivity and specific Heat in Zero Applied Magnetic Field for a Typical Superconductor," 2005. http://en.wikipedia.org/wiki/File:Cvandrhovst.png
9. Y. I. Frenkel, "Kinetic Theory of Liquids," Oxford University Press, Oxford, 1946.
10. M. I. Ojovan, K. P. Travis and R. J. Hand, Journal of Physics: Condensed Matter, Vol. 19, 2007, Article ID: 415107.
11. R. H. Doremus, Journal of Applied Physics, Vol. 92, 2002, pp. 7619-7629.
12. M. B. Volf, "Mathematical Approach to Glass," Elsevier, Amsterdam, 1988.
13. M. I. Ojovan and W. E. Lee, Journal of Applied Physics, Vol. 95, 2004, pp. 3803-3810.

CHAPTER 8

Critical Assessment of Theoretical Calculations of Atomic Structure and Transition Probabilities: An Experimenter's View

Elmar Träbert [1,2]

[1]*AIRUB, Fakultät für Physik und Astronomie, Ruhr-Universität Bochum, 44780 Bochum, Germany; Tel.: +49-234-3223451*

[2]*Physics Division, Physical and Life Sciences, Lawrence Livermore National Laboratory, Livermore, CA 94550, USA*

ABSTRACT

The interpretation of atomic observations by theory and the testing of computational predictions by experiment are interactive processes. It is necessary to gain experience with "the other side" before claims of achievement can be validated and judged. The discussion covers some general problems in the field as well as many specific examples, mostly organized by isoelectronic sequence, of what level of accuracy recently has been reached or which atomic structure or level lifetime problem needs more attention.

Keywords: atomic spectra; lifetimes; fine structure

1. INTRODUCTION

Let me refer to higher authority for guidance in the task prescribed in the title. Burkhard Fricke, Professor of Theoretical Physics at Kassel, some time vice president of his university, recalls advice by his elders as "If theory and experiment disagree, work on theory; if they agree, improve the experiment." Alan Hibbert, professor in the Department of Applied Mathematics and Theoretical Physics at Queens University Belfast and sometime Dean also of the Faculty of Theology, would add a caution to this: "If theory and experiment agree, be aware of the possibility that both may be wrong." Right. Moreover, if one considers another research practicality, one may suffer the consequences "if theory and experiment agree, you won't get funding for further research."

My aim in this article is a discussion of what I as an experimenter expect from theory and atomic structure computations, how I try to assess the quality of the work, and what experiment can or can't do towards a critical validation process of such calculations. After all, experiments and experimenters come in many flavors, too. We all make mistakes, and maybe we can learn from some of them that we have recognized by now. In this context, I dare to digress into anecdotal evidence of various problems in the interplay of atomic structure theory and experiment, into human and technical sources of error and uncertainty, before turning to a more systematic discussion of data on specific isoelectronic sequences. However, I will not present an ultimate recipe for assessing calculated data. Both, experiment and theory, are intertwined in the scientific process, and progress may involve a new perspective, if not always a paradigm shift. More than three decades ago, when I joined Josh Silver's group at Oxford, we wanted to do spectroscopy on two-electron systems with measurements that ought to be good enough to test the smallish QED contributions. We thought QED to be important, and we wanted to challenge theory (and theoreticians). When the agreement with theory turned out only moderately good, we wanted to blame the quality of the QED calculations, but were informed by the theorist A. M. Ermolaev that the theory authors actually believed them to be highly accurate. Those people conceded that the shortcomings lay in the quantum mechanical treatment of the two-electron system—a topic seemingly much less fancy than QED, but nonetheless essential and fundamental. This perspective has been corroborated over and over again since. Thus it is not lofty "theory" that experiment is testing, but computer implementations of many-body quantum mechanics, relativistic wave functions, nuclear magnetic field distributions, multipole expansions of the radiation field, and so on—in a package that combines some of almost everything. The more detail is to be treated, the more unwieldy become the programs, and the fewer cases are being calculated for a given amount of grant money. Any practical atomic structure computation represents a compromise, just as any measurement is a result achieved with limited insight and means. Yet some turn out better than others and need to be recovered from the chaff of computations and whatever unstructured data. Maybe one could develop a feeling for quality?

It is difficult to draw a line at which to begin or stop discussing data and developments, be it on the grounds of progress being made or of data quality. Various good people have written useful surveys of the material, and there also are a number of my own discursive listings of atomic lifetime and spectral data. There is no need to replay such presentations for completeness of content here. I have been guided in the field by some experts such as Lorenzo J. ("Larry") Curtis and Indrek Martinson (for samples of their reviews and systematizations see [1,2,3,4]). Curtis and Martinson have always been extremely polite when discussing other peoples' work, even when finding fault as the result of their own scrutiny. My language is more blunt, especially when I discuss bad science, but it is not intended to exceed the educative. Curtis and Martinson have worked—among other techniques—with beam-foil spectroscopy, from the very early days of the field, and they have developed many of the crucial insights that helped to obtain reliable wavelength and lifetime data. I have come to the same field later and thus had to find my niche by interacting with established groups elsewhere, by working on heavier ions, shorter and longer atomic lifetimes, peculiarities of time-resolved spectroscopy, and so on. In a later period I then learned how to operate ion traps and to exploit a heavy-ion storage ring and electron beam ion traps. My views are, of course, trained and tainted by the experience. My report will largely bypass classical spectroscopy, because much of that has happened before my time, has either been done at national standards laboratories or long since been critically evaluated there. The task put to me is not to recount "how we got there", but to assess the situation now, how to find an orientation and how to possibly move onward. Also, I am not trying to evaluate computational techniques and programs. There is a case (recalled below) in which notionally the same technique has been employed by members of an erstwhile collaboration; one part of the collaboration has proceeded to much better results within a year, while the other collaborators stuck to the poorer version even a year after that. Evidently one should not blame the technique *per se*, but the executors. As a user of computational results, I am rather more interested in whether the results are reliable and useful because they are tied to experimental data or whether they have been obtained *ab initio* and thus represent our theoretical understanding of atomic structure—at the risk of a possibly poorer match with the data.

I am grateful to the editors for letting me combine atomic energy levels and transition rates in the discussion. There is the Ritz variation principle that states how any imperfection in the wave function of the lowest state of a given symmetry will result in a calculated energy higher than in nature/truth. For decades this insight has fired a competition for reaching a lower value for the computed energy of the ground state of the He atom, and the guidance has resulted in excellent achievements. Moreover, in the calculation of level energies theory often separates the wave functions into angular and radial parts. For a long time, orthogonality of the radial wave functions has been considered essential, but nowadays even nonorthogonal operators are employed. There now are some extremely accurate level computations, but why is theory so much less reliable with transition rates?

Firstly, the transition dipole operator *er* for electric dipole (E1) transitions explicitly contains the radial part of the wave function, and this forces the problem of orthogonality to the light, while at the same time there is no variational principle available that defines a boundary condition for guidance.

Secondly, in a single-configuration picture the transitions between fine structure levels or between the ground configuration levels of a multi-electron system (all of the same parity, by definition) are not E1 transitions and these rates do not depend on an operator with radial wave functions. However, the single-configuration picture is simplistic, and deeper investigations by theory as well as accurate experiments show the need for a multi-configuration description and thus for taking care of considerable complexity. Here the transition rate depends on a higher power of the transition energy, and it is difficult to calculate these particular energy differences with a sufficient accuracy.

Thirdly, intercombination transitions are usually modeled as being mediated by spin-orbit coupling and the multiplet mixing of very few levels. The Breit operator, however, ought to be represented by a series expansion, but in practice this is usually truncated after the first term (!), because it is so computationally challenging. Combining that with the need to calculate the influence of all levels with the same parity and total angular momentum, it becomes clear that there may be more problems lurking in dark corners than have been recognized so far. The contributions listed are intertwined; in the discussion below there are cases that are dominated by one or another of these problems. It should come as no surprise that not all of them have been solved yet. It takes comparisons between nature/experiment and theory/computation to find out about the amount of detail that needs to be explored.

1.1. Codes

Among the many codes used in atomic structure computations there are veterans and newcomers, well-trained packages in new disguises, re-packaged well-proven standard parts, and more. Look into the codes and find contributions from many authors, some of them vintage, some newly phrased or added. Debugging new code is tedious and takes a lot of experience and judgement to recognize problems in the product. Some of the codes alluded to below and some of their exemplary developers-cum-users (as far as I know) are:

CIV3 (non-relativistic, with partial relativistic corrections) Configuration Interaction Version 3, developed by A. Hibbert and his group (including many foreign students) at Belfast,

HULLAC Hebrew University Lawrence Livermore Atomic Code, developed by M. Klapisch in Israel and various colleagues as well as many people at LLNL,

MCHF Multi-Configuration Hartree-Fock (non-relativistic, with relativistic corrections, etc.), a classic development over decades by the likes of C. Froese Fischer and her team when she was at Vanderbilt,

MCDF Multi-Configuration Dirac-Fock and MCDHF Multi-Configuration Dirac-Hartree-Fock (relativistic wave functions), with original program versions developed by J. P. Desclaux and P. Indelicato at Paris and by I. P. Grant and his team at Oxford (generic Grant code) and with additional development by C. Froese Fischer and collaborators, among them G. Tachiev, S. Fritzsche and P. Jönsson, (GRASP92, GRASP2K codes and various derivatives),

MCRRPA Multi-Configuration Relativistic Random-Phase Approximation, used by various people, notably by K.-N. Huang at Taipei,

MBPT Many-Body Perturbation Theory and RMBPT Relativistic MBPT, much pushed towards quality by W. R. Johnson, S. A. Blundell, and U. I. Safronova at Notre Dame,

RCI Relativistic Configuration Interaction, practiced and pushed to remarkable accuracy by M. H. Chen and K. T. Cheng at Livermore,

MR-MP Multi-Reference Møller-Plesset, developed by Y. Ishikawa and group (for atomic structure computations: K. Koc, M. J. Vilkas, J. López Encarnación, J. A. Santana, F. Diaz) at San Juan, Puerto Rico,

FAC Flexible Atomic Code, developed for speed, utility, extensive laptop use and collisional-radiative modeling by M. F. Gu when at Stanford, Livermore, Columbia, and Berkeley,

PCFI Partitioned Correlation Function Interaction, a recent appearance aiming at high accuracy in low charge ions, see recent papers by S. Verdebout, P. Rynkun, P. Jönsson, M. Godefroid, G. Gaigalas, and C. Froese Fischer.

There also are updates and improvements of venerable codes such as the Cowan code and its derivatives, HXR, Autostructure, Superstructure, Atom, and so on. Talking about codes, their quality and possible shortcomings, does not befit me as an experimenter who rarely, if ever, runs such a code. Atomic structure theory is a highly developed field. A real (and different) challenge lies in the proper coding for computation, as well as in the instructions for proper use. The results need to be tested by quality experiments (that is where I see my role). In many cases the comparison of computation and measurement will point out a need of further debugging the codes and/or to better experiments.

2. HISTORICAL REMARKS

Here is a nasty remark from me as an experimenter right at the start: One ought to distinguish theory and computation. Not everybody running an atomic structure code qualifies as a theorist. However, the words "doing theory" are frequently used for employing somebody else's programs, and I am not singlehandedly able to abolish that custom. During the Cold War it was said that the Soviet scientists—for lack of computers—did theory, while their Western counterparts ran computers. This statement surmises a development of mathematical formalisms for the explanation of physical phenomena on one side vs. the search for numerical recipes and "mechanical" number crunching on the other. The contrast is certainly overstated, and both sides are needed, but similar biases crop up in other contexts. While the economic situation in the former Eastern bloc countries is still poor enough not to support all good scientists on location there, quite a number have found employment in the West. Thus the Western countries reap the fruits of a different scientific education path, and Eastern-born scientists have joined the number crunching cottage industry, as has the younger Chinese generation.

When atomic structure codes were developed (mostly, but not exclusively, in the West—see the ATOM code, which is based on 1 / Zexpansions), aspiring young scientists from developing countries would go there and learn how to use the programs. Some of the young scientists grasped the underlying physics as well as the need to push out the frontier of knowledge and insight. Others were happy to obtain a copy of the program they had worked with, persuaded their home department of physics to purchase a personal computer for them (much more affordable than the equipment and running costs for an experimental laboratory), and began running the same program on the same problems as before, maybe changing the nuclear charge or the range of principal quantum numbers covered. Yes, I have seen quite a number of manuscripts submitted to learned journals in which the physics context was still at the state of thirty years earlier, most of the references were equally dated, and the results were in agreement with what other people had obtained using the Cowan code thirty years earlier (or using CIV3 not long thereafter). No, this problem is not restricted to theorists, I have also seen it with experimenters. I consider theory (in the present context atomic structure theory) as intellectually demanding. I see the quest for a valid implementation of a workable description of physical phenomena into numerical codes also as demanding (and code debugging has to go on forever in that work as well as in experiments). However, I am much less impressed if what I see is the application of somebody else's program suite to yet another case of the same ilk, the umpteenth paper on Mg-like or Al-like ions (with up to 92 elements plus the transuranes to choose from, one by one), and minute changes in the dressing. I have seen gross misunderstandings of basic quantum physics by a (theory!) coauthor of a long string of computed papers. It would be better (and save many trees and hi-tech resources) to have a code that can be run by nonspecialists on demand, if and when such numbers are actually needed. Hence I see not everything published from atomic structure computations as a product of theory or by a theoretician. In fact, both from experimenters and theoreticians I have seen manuscripts of authors who believed in their computed results (from Western programs, so they must be good?), but showed a profound lack of understanding the principles of atomic structure. Chance coincidences of measured and calculated wavelengths were assumed as proof of identity and classification in the spectra of foil-excited ion beams or of an ECR plasma discharge. The authors clearly disregarded that with non-selective excitation the complete spectrum of each ion charge state must be excited, not just a single line here and there, and that a spectral data analysis without attention to a sensible charge state distribution is moot. Some of that work originated in China (which I would consider as a product of authors low on the learning curve, but on their way up), and some (among other places) from a particular university physics department in the Western United States (where after many such occurrences I doubt that their learning curve was pointing up).

One major topic of my discussions will be atomic lifetime calculations and measurements. However, there is no need to repeat too much of the information discussed in earlier reviews [5,6,7,8,9,10,11,12,13] where also large numbers of references lead to piles of data and to explanations how to collect them. Many significant phenomena can be discussed with four-electron ions, and therefore I am presenting what I want to say in two locations, in the general section here and in a more technical one later.

2.1. Be-Like Ions

Be-like ions with the ground configuration $1s^2 2s^2\ ^1S_0$ may seem not much more complex than a He atom, as long as the $1s^2$ core is not opened. However, there are displaced terms in the $n = 2$ shell (configurations 2s2p and $2p^2$), and hence transitions of a fair variety (E1, spin-forbidden E1, M1, E2, hyperfine induced decays, etc.) can occur. In contrast to He-like ions in which the $\Delta n > 0$resonance transitions are closely related to resonance transitions in H-like ions (which can be computed accurately), the dominant transitions in Be-like ions are $\Delta n = 0$ transitions which are less easily treated by computation. Hence Be-like ions offer a laboratory for testing computational approaches against experimental data some of which challenge technology. For example, the resonance level ($2s2p\ ^1P^o_1$) lifetime is rather short and dips into the picosecond range for ions beyond $Z = 4$. Such lifetimes are routinely measured by beam-foil spectroscopy (see [13]), but a naively obtained apparent lifetime result taken at face value may be severely misleading, if the upper level is not selectively excited, but is suffering from cascade repopulation. This problem and its treatment (by measuring the decay curves of the cascades and performing a correlated analysis, for example in the ANDC scheme [14,15,16,17], or by employing a cascade model fit) have been discussed repeatedly [17,18,19,20,21] and need not be replayed here beyond mentioning their existence (see discussion below). The $2s^2\ ^1S_0$ – $2s2p\ ^3P^o_1$ intercombination decay rate, in contrast, is several orders of magnitude smaller than the resonance decay rate and relates (for low-*Z* ions) to level lifetimes in the second to millisecond range. This is way too long for classical beam-foil spectroscopy, and ion trapping techniques are a necessity, with their own evolutional progress by trial and error, failure and insight. These two extreme cases provide a useful set of stepping stones for the discussion in the present context.

2.1.1. Resonance Transition Rate

The measurement of the resonance transition rate in Be-like ions has been a cornerstone project of beam-foil spectroscopy. When the experimental results did not match those of perceived-to-be-good computations, however, the discrepancy was used by some people to disregard the experimental technique and to claim (even some of the earlier proponents of the technique did so) that beam-foil spectroscopy was generally unable to produce reliable lifetime data. This impression, though formed by a lack of insight, has stuck, not the least because it has been repeated many times in authoritative conference presentations by a few key representatives from NIST and their data base project. On the experimental side, even several decades after the proper procedure of taking cascades into account had been developed, there have been publications of data evidently not involving the cascade corrections that are particularly important in Be-like ions. The cascades are so important in Be-like ions (and correspondingly in Mg- or Zn-like ions), because there are $2p^2\ ^1S_0$ and $^1D_2$ levels in the same shell as the $2s2p\ ^1P_1^o$ resonance level, which are easily populated in the interaction of fast ions passing through a thin foil (solid matter). The $2p^2\ ^1D_2$ level is longer lived than the resonance level by a factor of about 4 to 5 (at low *Z*) and the corresponding cascade is easily recognized and treated. In contrast, the $2p^2\ ^1S_0$ level lifetime differs only by some 20% from that of the resonance level, for many ions along the isoelectronic sequence. This causes a growing-in cascade which, when not recognized, results in a seeming lengthening of the resonance level lifetime by some 30%. Reistad and Martinson [22] and the present author [23] have analyzed the literature data on the members of the isoelectronic sequence and demonstrated the wide scatter of the uncorrected data as well as the isoelectronic consistency of all such lifetime data that had been evaluated with cascade-correction techniques. The isoelectronic trend of the line strength ([24] in a formulation that follows Hylleraas' expansion of basic atomic entities [25,26]) thus established is reliable to within, say, 3% (and incidentally agrees with good quality theory). This is not bad for a system that critics have claimed to prove the unreliability of lifetime measurements by beam-foil spectroscopy.

What the experience actually does prove seems to be the fact that a fully automated analysis of decay curves can have grossly misleading results, because it disregards the information that is available from atomic structure knowledge for the system studied. No engineer would construct a bridge without reviewing the terrain and the ground, but there have been physicists who dreamed of "completely unbiased (by human interference) analyses of multi-exponential decay curves".

Already decades ago Curtis and others have repeatedly demonstrated under which peculiar circumstances naive analyses of such multi-decay curve data may or even must fail, but there is no conceptual obstacle to any such measurement and meaningful evaluation once the underlying atomic structure is recognized and ANDC procedures or related modeling techniques are applied. Of course, the simplest cure for most of these problems would be the selective excitation of a given level. Unfortunately, lasers that produce multiply ionized atoms and also reach an arbitrary level of interest to modify its population are beyond technical reach so far. Let us see in the discussion below where we get with the tools we have.

Because of the aforementioned cascade correction problem and the many early "wrong" lifetime results, beam-foil researchers have mostly become overly cautious (there also is a fraction that persistently overstates the quality of their results) with error estimates for lifetime measurements. This is clearly apparent in the Be sequence. Träbert and Curtis [21] plot the resonance level lifetime (as a scaled line strength, that is, after removing the wavelength as an easily predictable or well-measured entity) and an isoelectronic trend (based on one of Curtis' many systematizations, avoiding atomic structure computations). Figure 1 shows the same situation in a comparison of experimental lifetime data with theory. The scatter of the data around the mean is clearly smaller than the typical individual 10% error bar. This reflects the caution of the experimenters (or maybe only a hand-waving estimate of errors by order of magnitude) as well as the good internal consistency of the data set (once the poor evaluations are discarded). Unfortunately, 3% are an achievement on one hand, but the community would like to have even better numbers to test theory, which in this case has been claimed to be good to ±100 ppm [27].

Computations of neutral atoms are particularly challenging, and, indeed, neutral Be 0 has been declared to be of interest as a benchmark. There are beam-foil lifetime results from early days by Martinson, Gaupp and Curtis [30] and much later ones by Irving *et al.* [31], and they differ by some 4.5%. Since the early data point predates the development of the ANDC cascade correction technique, one would assume that the newer data point benefits from the better insight and must be better. Because the new team included Indrek Martinson and Larry Curtis from the first measurement, one may safely assume that all technical knowledge accumulated in the community was taken care of. However, neither the old nor the newer result differs severely from the theoretical results or the aforementioned isoelectronic trend, but at about 2% the remaining deviation from the latest computations is not satisfactory. This overall reproducibility has been explained by the fact that the aforementioned

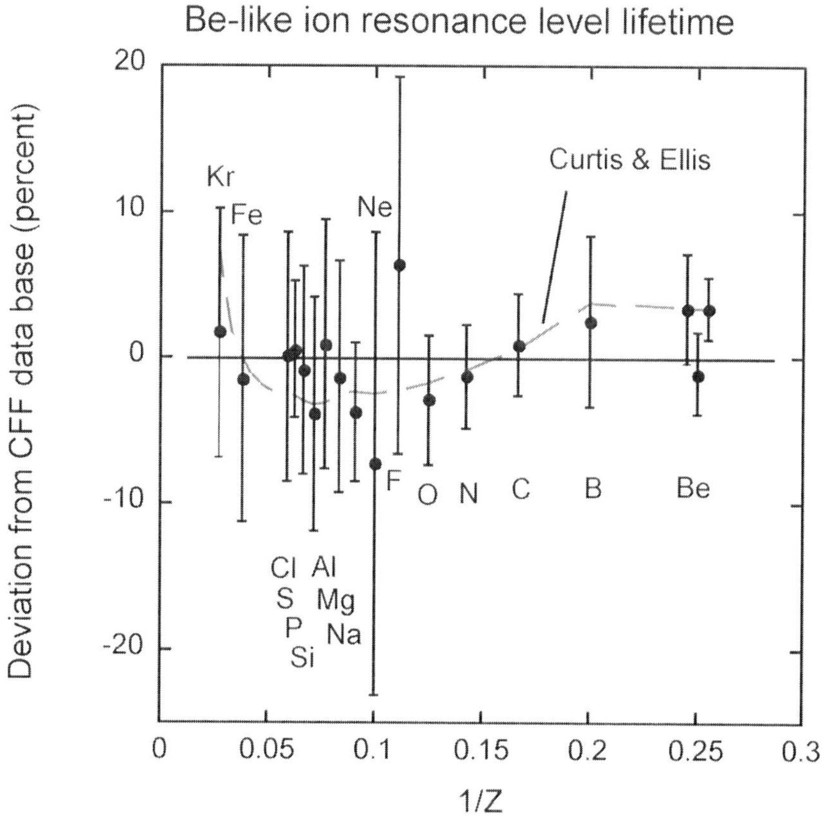

Figure 1. Lifetime data on the 2s2p 1 P 1 o level, the lowest excited singlet level in Be-like ions. The data are shown by their deviation from the lifetime predictions in the Multi-Configuration Hartree-Fock/Multi-Configuration Dirac-Hartree-Fock (MCHF/MCDHF) collection data base provided by C. Froese Fischer and G. Tachiev (the horizontal line) [28]. Also shown (broken line) is the result of a fit to the experimental data [29]. For some elements there have been several measurements, of which only the result of best quality (judged by the author) has been selected. For Be there are three data entries (the most accurate ones), and open questions remain (see text).

strong cascade that plagues the ions in the Be isoelectronic sequence does not matter in the neutral atom, because there the cascading level would lie above the ionization limit and preferentially decay by autoionization. However, it is an open question what fraction of the decay is taken by autoionization, and whether the "surviving" decay branch might still cause a significant shift of the apparent lifetime of the resonance level. To me it is surprising that a system in which the recognized dominant source of systematic error (that particular cascade) is claimed to be absent should yield two lifetime results that are so different, from experiments using the same technique.

Well, I have run lifetime measurements on Na-like ions of various elements, and among those of Ca [32], some results differed significantly from the isoelectronic trend and from theory predictions that were beyond doubt. Although seasoned beam-foil colleagues from various countries tried their individual pet evaluation techniques, in the end we all agreed on the same numbers that differed from prediction—yet we stated in our publication that we recognized the difference and could not explain it. We do not think that theory needs to be modified, because we think the computations on Na-like ions are more reliable than a single measurement (which was repeated, of course, but did not change in result), even as that was done at the technically most suitable accelerator facility (at Bochum). However, an Asian theory colleague then suggested that theory and experiment would agree better on the lifetime, if one assumed that not all radiative decay channels mattered. He tried to sell this idea to various experts who all told him off, and he tried to have a manuscript with such anti-quantum mechanical fallacy published; as far as I know he failed in the attempt, because different journals sometimes turn to the same experts. The other measurements on the same Na isoelectronic sequence do not show such a deviation from the prescribed trend [33,34,35,36]. This leaves at least two explanations; firstly, from a statistical view, there "have to be" outlier results, which we can't always explain. Alternatively, since (negative) Ca ion beams (for a tandem accelerator) are difficult to produce in an ion source, the ion beam current was relatively low and thus a sufficiently high signal rate required observations through wider spectrometer slits which encompass poorer spectral resolution and enhance the likelihood of spectral blends (none has been recognized and identified yet) which might affect the decay curves. So much for the unsolved riddle of Ca X.

What may have also been at play in Be I could be an unconscious need to have a different result from before, because the new effort was more involved. A new evaluational result may therefore have received more credit than a confirmation of the old number. Why this unsubstantiated conjecture? When Schnabel and Kock [37] tried a different experimental technique, laser-induced fluorescence, their lifetime result incidentally corroborated the earlier beam-foil number rather than the newer and purportedly better one. The laser technique, however, is not necessarily guaranteed to yield correct results. A lifetime near 1 ns is short for producing (visible) laser light pulses at a selected wavelength. (Yes, sure, nowadays the attosecond range has been reached, but this is not practical nor sensible for many atomic physics experiments in which valence electrons populate levels that have mean lives in the nanosecond range.) The shape of the laser pulse in the time and wavelength domains needed to be modeled, and also magnetic sublevels played a role via quantum beats (a consequence of coherent excitation of nondegenerate levels). A laser experiment that aims for selective level population therefore is not necessarily as clean as one would hope for. I would not be surprised to learn that eventually experiment might find a result in the middle between those three results mentioned, which would be close to more recent computations (see discussion in [38]), but what is needed is also a smaller experimental uncertainty.

Verdebout *et al.* [38] have used the resonance line of Be I as a test-case of their recently developed PCFI approach. Their theoretical oscillator strength, f = 1.375(1), is close to various other recent theoretical results. So, whatever experimental value is taken for comparison, the theory-experiment mismatch is around 2%–3%, much worse than the theoretical uncertainty estimate. However, isoelectronically smoothed representations of the line strength and similar averaging procedures recounted and discussed by Irving *et al.* [31] pass through the region of overlap of the experimental error bars for Be I. Maybe for now the experimental results should be represented by their weighted average. (The neutral members of some other isoelectronic sequences, especially Ne I, suffer from much worse discrepancies among their experimental lifetime data - see below.)

Evidently there is still room for further experimental work on a strong and fully allowed E1 transition in a four-electron atom. This puzzling situation calls for an experiment that is independent of environmental density (beam-foil is a high-density excitation, low-density observation technique) and that comes close to selective excitation in a low-density environment, perhaps by synchrotron radiation. Beryllium, however, is toxic when in the form of breathable dust, and it would need to be dispersed as atoms (a hollow cathode discharge was used in the Schnabel and Kock work) or clusters, a clearly health-hazardous situation which one tries to avoid (not least when cleaning the apparatus after the experiment or disposing of Be traces safely). It is much easier to call for new and better measurements than to actually do them.

2.1.2. Intercombination Transition Rate

At the risk that the reader might see this report as full of unsubstantiated ranting, let me discuss an example in which both, theory and experiment, show ample blundering (in hindsight). Some learning theory claims that we (only) learn from our own mistakes. Help me show them wrong by learning also from the mistakes of others (at least occasionally)! The $2s^2\ ^1S_0$ - $2s2p\ ^3P_1^o$ intercombination transition in Be-like ions such as C^{2+} (spectrum C III) is of fundamental as well as of astrophysical interest, and the calculation of the transition rate has drawn the interest of more than 30 authors or teams of authors over the years, plus more again on four-electron ions of the neighboring elements in the sequence. In the grand picture (see Figure 2, which is based on an earlier version [6]), the predictions from the 1960s to the 1990s scattered by ±30% around a mean near 100 s^{-1}, earning the transition rate a low accuracy rating in the Wiese tables [39]. The level lifetime τ of a level with an unbranched decay is the inverse of the transition rate A. In 1983, a lifetime measurement at the Harvard-Smithsonian Center for Astrophysics (CfA) employed a radiofrequency (Paul) ion trap and reported a decay rate of 100 ± 20 s^{-1} at a conference [40], followed up a year later with a rate some 25% lower [41,42], and another 9 years later by a rate higher than the first one by some 20% [43]. The error bars of the purportedly improved data left a gap in the middle (where the first report had placed the result). This part of the story might reflect progress with the experiment, and thus would seem laudable. Also, from a laboratory with quite some reputation, one might expect a sound explanation of the technical progress involved, but none was given in the publications. However, I have worked in the same laboratory years later and was told details that are not mentioned in the publications. I was told by bystanders that there was not an original data set later superseded by improved measurements, but that all three publications were based on one single measurement campaign; the different results apparently stemmed from different interpretations or rather from shifting assumptions whether a single or two exponentials should be used in fits to the data, and whether—if using two components—the fast or the slow one should be considered the right one. Such a discussion would have befitted the publications, but it might also have limited the appeal of the work to referees and readers as a source of reliable data (with a 6% error bar in the third installment). In fact, this last data point was adopted in the new issue of the famous Wiese tables [44] at a time when the 'definitive' CfA result was already experimentally and theoretically superseded by much more accurate results—which lay smack in the middle, where the two "improved" CfA-results had left a gap (see below). To me, this is a case of misjudgement (and misinformation) by overly eager experimenters (possibly fueled by Harvard's self-image). Let us now put some blame also on theorists.

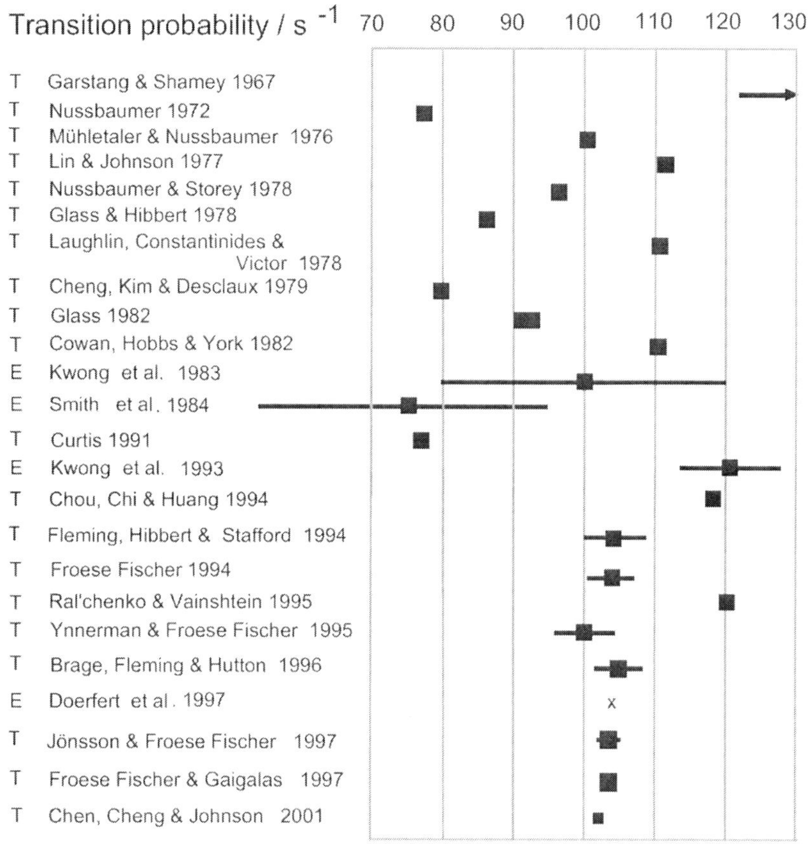

Figure 2. Intercombination transition rate in C III (adopted from Figure 4 in [6]). The timeline combines calculated (T) and measured (E) results and is discussed in the text. The publications in this incomplete listing are identified by authors and year. The uncertainty of the 1997 measurement by Doerfert *et al.* is as small as the center of the cross mark. Note that a number of recent computations also bear error estimates.

In Figure 2, a 1991 data entry by L. J. Curtis [24] follows the 1984 experimental data reported by Smith *et al.* [41,42] at CfA. This is not a calculation *per se*, but Curtis had assumed the CfA data to be reliable and thus used them as an anchor in his semi-empirical analysis of the isoelectronic sequence. Hence the isoelectronic trend he derived (a worthy exercise to bridge the wide gaps left by too few hard data) was flawed at low values of the nuclear charge *Z*, because the experimental reference data were mistaken. The next (and clearly different) CfA data point for the same entity was followed within a year by a theoretical treatment from Taiwan [45] (apparently the authors were unawares of any measurement of the triplet level lifetime), and within another year by another publication from Leningrad [46]; that paper explicitly claimed good "agreement with the experimental results, including the recent ones". When talking to coauthor Vainshtein about this result in the light of later progress in the field, I was told "that this is where experiment was at the time". Evidently the theorists had been satisfied with the outcome of their computations, because it agreed with published experimental data. As an experimentalist I ought to be happy about this trust in experiments, but I am hesitant to rely on computations and experiments that have not been critically evaluated and passed various intrinsic tests, preferably also having been cross-checked against work using other techniques, and so on. Theory claims to be of predictive value; if computations are considered successful because they reproduce experiment (without independent accuracy estimates), then they have no predictive value and should not be considered proper. However, our story continues.

Beginning in 1994, the groups of Alan Hibbert and Charlotte Froese Fischer (and their postdoctoral coworkers), using very different codes, began to place the intercombination transition rate of interest in C III just above 100 s^{-1}, at first with uncertainty estimates of 4% and 3% [47,48,49], respectively, but already stating that their result was incompatible with the aforementioned papers from CfA. In 1996 a new experiment at the Heidelberg heavy-ion storage ring TSR succeeded (published in 1997 [50]). The first case studied with this technique was the very intercombination transition in C III, and the excellent data quality left no options for various ways of evaluation: a single exponential yielded a lifetime (and thus also the transition rate) with an uncertainty of a small fraction of 1%, the most accurate such data point for any multiply charged ion at the time, and in full agreement with the more recent computations. (Curtis [51] lateron has tried to see whether any deviations from an exponential decay law might be recognized in the original data files with their excellent signal-to-background ratio, because that would be of high fundamental interest, but has not found any.) Theory (mostly by the same groups and their associates) since then has dared to specify a theoretical uncertainty of about 1%—again, once one knows the "right" answer, one may feel more confident also about computations that come close. The situation of theory and experiment in Be-like ions then became the topic of a systematic review [27] that included versions of atomic structure packages that are still in use today, and the present manuscript in a way is an offspring of that review, too.

Theory with an uncertainty estimate of 1%, however, does not test an experimental data point of much smaller error bars. Fortunately, Mau Chen, K. T. Cheng and Walter Johnson [52] have mustered the resources of a National Laboratory and applied Relativistic Configuration Interaction (RCI) computations to the problem, with a few hundred thousand configurations. Their computation massively reduced the often astoundingly large discrepancy between results obtained in Babushkin and Coulomb gauges (in nonrelativistic approaches called the length and velocity operator forms), and their error estimate is on the order of half a percent. Their result differs from experiment by slightly more than the combined error bars. In a way, that is a healthy situation. Agreement of experiment and computation is good at a high level in this example, but not perfect. This leaves room for improvement on both sides. However, funding for improved experiments or for even larger computations has not been forthcoming yet. The level of accuracy reached in this atomic lifetime measurement and associated theory carries over to systems which would test QED predictions at the edge of the Standard Model, if only many-body quantum mechanics was developed well enough, as will be discussed later.

2.2. General Problems

The above examples touch on many problems. Within the active time of a given field (fashions come and go also in science) or a given experimental technique, such as beam-foil spectroscopy, there are learning curves of individuals and of the community as well as the development of technical procedures and of equipment. It is important to become aware of laboratories, their attitude, achievements, and possible decline, of people, facilities, funding cycles, alliances and hostilities. This is not limited to experimental work, but also to computations. When computing was still expensive and technically very limited, some shrewd operators (such as Harry Nussbaumer) might find configurations that yielded atomic structure numbers reliably close to reality, from small basis sets that seem ridiculously small in hindsight, while the results obtained by others with similar technical means scattered rather wildly compared to what we know now. The less successful people would not advertise that their means were so limited; it is almost an indicator of self-delusion, if one sees authors praise their own computations without any realistic proof. Nowadays we can expect computations to include convergence tests in which, for example, the number of configurations is systematically varied in order to see whether the result changes significantly. This procedure is expensive in terms of computing time, because it does not increase the number of results, but it is one step of many to estimate the reliability of computations.

Authors with a poor grasp of the concept of reliability seem to prefer to list results with many decimals, most of which are actually meaningless, even as any computer program will reproduce them when starting with the same input. When years ago Pedro Goldman and Gordon Drake had a friendly competition about the accuracy possible in the computation of Bethe logarithms and energy levels in two- and three-electron ions, it was a useful measure to talk about twelfth and thirteenth digits. The two parties challenged each other and tested each other, but always with an eye on the demands of highly accurate experiments. Their computations corroborated experiment at a very high level and then went beyond the experimental data in order to be available when experiment would be improved again. This was excellent work. When nowadays I see manuscripts of computations on core-excited few-electron ions that state ever so many significant digits, I am regularly disappointed, because what matters is not any long number, but usually the difference (term difference) between pairs of them. The really reliable digits disappear in the difference, and the reader is left at a loss when wanting to judge the reliability of the difference.

Of course, this is not a new problem. Dividing one by three leaves you with an infinitely long decimal expression, but of only a single significant digit. Some people write down eight digits from a pocket calculator, or nowadays use double precision computer printouts. This does not enhance the true, inherent accuracy. If an author uses the GRASP code to calculate wavelengths of transitions of ordinary ions in the visible and lists eight significant figures, I know that the author lacks basic physics insight. I have seen tables filling full journal pages with such nonsense, comparing the computations of fine and hyperfine components of lines with poor measurements that yielded (visible) wavelengths with a spectral resolution of maybe a nanometer and a precision that justified no more than 4 significant figures (at best). A journal editor with basic science training should recognize the nonsense, but he/she usually leaves that to a voluntary referee, to whose helpful comments some authors do not concede any value. If rejected by one journal, those authors simply turn to another journal, with the same faulty manuscript.

I have also seen Indian claims of confirming beam-foil lifetime measurements of three decades earlier, with similar error bars stated as in the sound early papers from American colleagues. (In this paragraph I refer, among others, to a specific author who has published various unscientific (in my eyes also fraudulent) claims in Phys. Rev. A, J. Phys. B, Phys. Scr., Z. Physik, Eur. Phys. J., Nucl. Instrum. Meth.—this is not a problem of one journal or one country or continent.) Alas, in none of the papers submitted by the same cluster of authors it was ever proven that the ion accelerator they had used actually did provide ion beams at an energy necessary to produce the high ion charge states in question. In fact, the lead author relied on a French charge distribution model that had been published by its authors for a much higher energy range. The French authors explicitly limited their validation of the model to a certain range (way above the working range of the accelerator used in the experimental work by "those other authors"). Actually, there are valid charge distribution models for the energy range of interest, and they show that the claims by "those other authors" were untenable (I should say fraudulent). Moreover, "those other authors" for almost all of their work used a solid state X-ray detector that yields basically a single spectral line of the X-ray spectrum where with a proper spectrometer one can see dozens of lines from different charge states (which invalidates any interpretation based on that single line). Thirty years earlier spectroscopic measurements had been run in a physically sound way at various laboratories. One might expect that 30 years later people would build on that knowledge and make progress, but "those other authors" instead have elected to turn to fanciful and modern-sounding interpretations of their single-line X-ray data which are physically unsound and in violation of established knowledge. Yet referees and editors have fallen for the competent-sounding manuscripts from a formerly developing country (countries that boast nuclear weapons do not deserve "helpful refereeing" any longer!) and completely overlooked the rotten base. Unpleasant as it is, one has to consider the existence of science colleagues who are not just doing poor science, but who invent meaning, stylize their own work preposterously, operate at least close to fraud. It does not help the field to let their manuscripts pass into print. Well established international physics journals have violated their own policies against serial publications with such non-physical material and have not shown any inclination to shield themselves against such authors in the future after having been warned. Thus neither the status of a journal nor the peer review system provide any warranty for the soundness of published data. If we know our colleagues, we may form an opinion on their work. Newcomers to the field will need some period of apprenticeship with established work groups to gain personal credibility—whatever atomic structure packages they might use. The same holds for experimenters—they are working in a laboratory set up by somebody else, usually beginning with projects thought up by somebody else, and they need to gain and demonstrate experience on their own. Trust and distrust are human factors, but as discussed above, even when judging dry spectroscopic data, some skepticism about authors and institutions is warranted. Historians have turned to institutions such as the Harvard-Smithsonian Center for Astrophysics and revealed its particularly troubled period in the mid-1950s. In other institutions, too, there must have been ups and downs linked to people and funding, and possibly to scientific quality as well.

Although a commonly trusted large institution such as the National Institute of Standards and Technology (NIST) is seemingly beyond reproach, it would be unlikely not to suffer occasionally from flaws in some or other of its sections. After all, they employ people. Maybe such institutions are too easily seen as a brand name and a guarantor of product quality, instead of their true role as an address at which many people work. Still, name recognition may be useful for funding purposes and public relations. However, institutional status and policy can also have drawbacks for science. For example, as part of such an institution, the spectroscopic data base group at NIST feels obliged not to correct even clearly clerical errors in their data base, unless the correction is the result of a fully-fledged research effort of their own with all the in-house refereeing and vetoing prescribed, which is costly and for which the manpower is too scarce. Thus mistakes that have passed by the internal checks and referees into their data compilations can be long-lasting and almost impossible to weed out—as an inadvertent result of their quality-assurance policy. The error rate in the measurements, evaluations, compilations and tabulations by Bill Martin, Victor Kaufman, Jack Sugar, Joseph Reader (to name just a few NIST spectroscopists I had most contact with) and their colleagues is admirably low, but naturally it is not zero.

The sheer size and complexity of any spectroscopic data base such as that built up and maintained at NIST (ASD) have brought about various side effects. No longer are individual entries linked to their sources; the tabulations mix contributions from experiment, derived from a meta-analysis of experimental levels (by the Ritz combination principle), and from computations. A data base aiming at completeness of entries is appreciated, but the data quality is becoming murky. Decades ago a declared goal of several NIST spectroscopy group members was to obtain EUV and soft-X-ray wavelengths with an accuracy of 5 mÅ. By far not all data compiled in the data base have ever come close to that standard. In recent EUV spectroscopy work, a number of NIST ASD spectrum listings have been found to contain much less accurate wavelength data, and there is no simple way anymore to find out what the original error estimate has been. In the very helpful (though dated) Kelly and Palumbo tables [53] the authors state that L. J. Palumbo was responsible for the technical process of porting whatever tabulations into the book format—a formidable task. R. L. Kelly took the blame for all scientific decisions, including the selection of data. That selection, indeed, is where some sources that had stated rather too many digits of the wavelengths were (mis-)judged to be better than others, but actually had been wrong in measurement or line identification. These problems with a group of soft-X-ray publications surfaced when Multi-Reference Møller-Plesset computations were used to interpret high-resolution EUV beam-foil data (see examples and discussion in [54]), and old reference data turned out to be at fault.

By the way, one should also be cautious of new ways of defining error. One group at NIST once introduced a "probable error" different from variance or standard deviation. It made their measurement uncertainties look smaller (when the numbers are expressed in error intervals), but luckily it was not adopted elsewhere. (Although, whenever the community sees the label NIST, people assume that something substantial is being conveyed and try to understand the procedures.) Instead, some eyebrows were raised when the actual NIST X-ray data associated with the new error measures were scrutinized in other laboratories and were found to not support the image of world class reference data. Another nuisance is the claimed 'custom' in various communities to quote one standard deviation (comprising 67% of the data) or two (95% confidence interval). On the odd chance that somebody from a different neighborhood should read a paper, it would be best to state what error measure is being used.

Experiment and theory are both needed at a high quality level before one can begin judging the quality of computations by an intercomparison. It is not a suitable measure of reliability if the output of one computation largely agrees with that done by somebody else (an argument frequently encountered in computational manuscripts from non-Western countries). If it does not agree, it is obvious that there must be problems, but agreement may just be by chance, or by using a similar algorithm with the same shortcomings (perhaps in a program of a different name). When the Grant code was in its infancy, lifetime predictions showed a significant unreliability. Debugging efforts by members of other groups then revealed that certain sets of integrals had only been estimated (by a prominent theory coauthor) in order to obtain some output (not bad for that). When the numerical problems were solved and the integrals properly solved, the results improved markedly. The successor stages of that development have become mainstays of the atomic structure computation business.

Error estimates sometimes also are parts of political plays. There are groups eager to shave a little off an error bar in order to be seen as better than the competition, and there are others who tend towards more conservative estimates as a matter of principle. In the 1970s, there was a competition for the most accurate atomic lifetime measurements on fast beams of atoms. The group of Horst Jürgen Andrä at Berlin was particularly active and was repeatedly credited with the most important work. In early work on Ba II an uncertainty of 0.1% was claimed [55], to which no correspondingly precise theory was available for comparison. The laboratory work therefore was re-oriented towards calculable systems such as the alkali atoms where an intermediate experimental result reached 0.26% (reported at a conference; the final results were later published with error estimates of 0.15% for Li I and 0.18% for Na I [56]). Some hay was made out of the disagreement of any but the poorest computations with the superb measurements. At the same time the group of Schmoranzer at Kaiserslautern obtained data with a slightly larger error bar [57], but their data were internationally almost overlooked, because the Berlin data seemed superior. Using a different technique, laser excitation of atoms in a vapor target, Jörgen Carlsson at Lund measured lifetimes of the same levels (and applied the same measurement techniques also to hyperfine structure studies), but again at slightly larger uncertainties [58,59,60]. In the long run, theory was improved until in the end it was accurate enough and then it contradicted the Berlin measurement. After a few years of uncertainty of whom to believe more, a very different technique (using the weak binding between distant Na atoms [61]) enabled a major improvement. The new data from molecules and from a newer version of the Kaiserslautern fast-beam experiment match recent theory within their 0.2% uncertainty [62,63], and the continuing work at Kaiserslautern has also indicated a likely minor flaw in the early Berlin work. By the way, both, the original Kaiserslautern and Lund results and the clearly improved new ones, are compatible with theory. So Kaiserslautern and Lund were right from the beginning, but lost out on the advertising side.

The Berlin and Kaiserslautern groups used fast atom beams produced by the neutralization of ion beams in a gas, which is associated with a smaller energy loss than ions experience in an interaction with a thin foil. This energy loss (mean and scatter) is one of the disadvantages of beam-foil spectroscopy, because it affects the knowledge of the ion velocity at observation past the foil. In a Swedish lifetime measurement on the He I 1s3p 1 P 1 o level [64], the actual ion beam velocity was measured from quantum beats in neutral He atoms, the beat frequency being provided by highly accurate computations of the $n = 3$ triplet fine structure intervals. The singlet level lifetime result has been published with an uncertainty of 0.27% and still is the most accurate traditional beam-foil lifetime measurement result. Most certainly the 0.27% uncertainty value is the result of a thorough error analysis. However, I have learned [65] that H. J. Andrä, the group leader of Andreas Gaupp, jokingly congratulated Indrek Martinson, the senior scientist in the Swedish group, for having chosen an error bar just a little larger than their own (at the time, on a different atomic system and experiment). Knowing Indrek, who in the 1960s had worked alongside Andrä in Stanley

Bashkin's newly started beam-foil spectroscopy group at Tucson, I would not exclude the possibility that a politically wise conscious adjustment of error bar size may have occurred rather than sheer happenstance. At least the anecdote reflects the competitive spirit of some actors in the field.

3. ISOELECTRONIC SEQUENCES

The recognition of experimental error sources and of error propagation in statistics and data evaluation suggests that we repeat any measurement many times. The finiteness of resources, however, makes us also strive for measurements on other systems. It would be good to correlate the two pools of measurements, for example by studies of ions along isoelectronic sequences. This adds a parameter, the nuclear charge Z or the ion core charge ζ (the spectrum number), to the pool of data. Hylleraas' series expansions of atomic parameters such as level or binding energies, transition rates, line strengths, polarizabilities, and so on, provides simple systematic interconnects [25,26]. Where series expansions and analytic formulas don't reach, one may turn to simple computations as first approximations that give guidance but eventually need to be replaced by expansive work. Examples of such extremely useful tabulations are the MCDF computations of the energies and rates of transitions in ions isoelectronic with first row atoms (Li through F) by Cheng, Kim, and Desclaux [66] and in Al-, Si-, P-, and Cl-like ions all across the periodic table of elements by K.-N. Huang and his coworkers [67,68,69,70] from Taiwan which I have benefited from for decades. (This group uses more up-to-date computational tools nowadays, such as MCRRPA (see [71,72] and much more computing power), but the old tables have not been re-run and re-published.) However, with improvements of spectroscopic tools for beam-foil spectroscopy and with the advent of electron beam ion traps (EBIT), both of which can give access to all charge states of all elements, the Huang tables have reached the limits of their use, because they were computed when computing resources limited such work to rather few configurations and small basis sets. "Modern" data of high spectral resolution need higher quality computational results to compare with. Instead of including examples from those now outdated studies in my examples discussed below, I prefer to expand the scales of the graphs to show what more expansive computations can do nowadays. Similarly, there are series of tabulations of atomic data for such isoelectronic sequences and astrophysical applications (including spectral modeling) contributed, for example, by A. K. Bhatia and G. A. Doschek (Naval Research Laboratory) and various of their colleagues. Those tabulations have had their use and their deserved user community. However, the lowest energy transitions listed in those tabulations are usually the M1 transitions within the ground configuration. Transition rate measurements at heavy ion storage rings and at EBITs have repeatedly shown the tabulations by Bhatia *et al.* to be off by some 30% and more for the rates of these E1-forbidden transitions. It is therefore not appropriate any longer to include those outdated computations in a discussion of the state of the art. At the same time I am skeptical when I see that a computation addresses only a single ion species and not at least a few ions along an isoelectronic sequence. A variation of the parameter Z in principle is simple in atomic structure computations, but might reveal quality deficiencies (of computations or measurements) in a comparison with further experimental data—a chance not to be missed. Moreover, only if a computation reaches a given degree of accuracy before measurement can it boast of significant predictive power.

3.1. H-Like Ions

The Dirac equation is often considered as an exact description of the one-electron atom or ion. However, that would require an infinitely heavy nucleus to avoid recoil effects, which apparently cause problems to the formalism. Hence there is no *exact* description of even the simplest atom. Nevertheless, in the face of a plethora of correction terms and of the series expansions of quantum electrodynamics (QED), the general assumption in the community is that H-like atomic systems can be better calculated than measured. In practice, however, only HI and DI have been measured with extreme accuracy, and for all the ions of the HI isoelectronic sequence, the tools of study are much blunter.

In order to study higher-order expansion terms of QED, it would be interesting to do accurate measurements on the H-like ion 92 U91 +, because there those terms are relatively larger than the measurement accuracy of the same terms in HI. However, the uranium nucleus is far from spherical, and its imprecisely determined shape causes uncertainties in the computations. In contrast, 82 Pb 81 + has a doubly magic nucleus with a somewhat smaller nuclear charge and thus smaller QED contributions, but the expected accuracy of experimental tests of the calculated QED terms is expected to be still higher than for 92 U 91 + .

There are various computations that address single-electron ions of high-Z elements, especially uranium (for a sample, see [73]), and the differences between the results are much smaller than the precision of direct measurements. A major problem for the experiment is the short lifetime of practically all levels of interest, which by the uncertainty principle is linked to a level width and thus to line broadening. The inter-shell ($\Delta n > 0$) E1 transition rates scale as Z^4 as do fine structure splittings and leading QED terms (Lamb shift). The $2s\ ^2S_{1/2}$ level may be long lived in hydrogen ($\tau \approx 1/8$ s), but its two decay modes, 2E1 and M1 decays, scale as Z^8 and Z^6, respectively, resulting in a lifetime prediction of about 9 fs in U^{91+}, with the M1 decay dominating by a factor of 20 over the 2E1 decay branch. This short lifetime is associated with a considerable line broadening.

An alternative experimental approach uses Li-like ions and builds on theory's success in describing three-electron systems. The experimental idea is simple and exploits the fact that here the $1s^2 2s$ level is the ground state and thus not lifetime-broadened. The 2p levels are connected to this ground state by E1 intra-shell decays of comparatively low rate and thus by EUV / soft-X-ray transitions that are not much broadened. Such measurements of $2s$-$2p_{1/2,3/2}$ transition energies have been done at the Livermore electron beam ion trap [74,75,76] and in other experiments. In order to estimate the QED contributions in one-electron ions from the three-electron ion measurements, one needs theory as a bridge (with some not quite developed links) telling of the relative sizes of various QED corrections in high-*Z* ions. In spite of the uncertainties of such a procedure, the result of a measurement on U^{89+} and the extension to U^{91+} by theoretical tools [76] has turned out more accurate as a test of the two-loop Lamb shift contribution to the 2s level energy in a one-electron ion than any of the attempts at a direct measurement so far.

3.2. He-Like Ions

Singly excited two-electron ions are considered to be accurately calculable [79], both for the level structure and the transition rates among the lower levels. At high *Z*, the longest lived level is $1s2p\ ^3P_0^o$ which decays by an intra-shell decay to the $1s2s\ ^3S_1$ level and by hyperfine mixing to the ground state (see the 4-percent beam-foil lifetime measurement on $^{197}_{79}Au^{77+}$ [80]). Multi-photon decay rates increase steeply with the nuclear charge *Z*; a quarter century ago Munger and Gould reported a 6-percent lifetime measurement on $^{238}_{92}U^{90+}$ (even isotope, no hyperfine structure) from Berkeley [81], which recent work at GSI Darmstadt [82] tries to improve on. The easiest detection is by the decay of the $1s2s\ ^3S_1$ level which is very short-lived, but which is then replenished by the stronger decay branch of the $1s2p\ ^3P_0^o$ level. The rate of this decay branch depends on the 2s-2p energy splitting and can thus be exploited for a determination of the $^3P_0^o$ level position. In the highest-*Z* naturally abundant element, uranium, the $2s$-$2p_{3/2}$ transition has also been seen by now [83], but such an observation with its considerable uncertainty is no challenge to theory.

At lower Z, the longest-lived level is the lowest triplet level, $1s2s\ ^3S_1$, with its M1 decay (line 'z' in plasma lingo). The story of how this line was recognized against prediction in solar X-ray spectra [84], how theory struggled for a while to find a valid isoelectronic scaling of the transition rate, how Gordon Drake developed a description based on a relativistic transition operator and extensive nonrelativistic wave functions [85], how C. D. Lin in his Ph. D. thesis introduced a relativistic correction [86] to Drake's results, and how it then took a quarter of a century before a fully relativistic description including a wave function expansion to all orders became available [79], has been told in various places (among others, [6,8,10]) where also most of the experimental sources have been cited. Figure 3 shows a plot of the deviation between experimental data and theory along the isoelectronic sequence. Theory predicts a Z^{10} dependence of the transition rate, which corresponds to a variation by 15 orders of magnitude over the range from neutral He to Xe^{52+} ions, and experiment largely agrees with theory within the experimental uncertainty. It is crucial to minimize the experimental uncertainty on such fundamental measurements. This demand has been met best in the range from B ($Z = 5$) to S ($Z = 16$), with measurements at the TSR heavy-ion storage ring and at the Livermore EBIT [87,88,89,90,91,92]. Here the typical uncertainty is about 0.5%. This is comparable to the difference between Drake's (corrected) nonrelativistic computations and the relativistic computations by Johnson *et al.* [79]. Hence experiment is being challenged to reach an even higher accuracy. This experimental progress is most likely to happen in the range of ion trap experiments, where the accuracy already is highest.

The beam-foil technique used at higher nuclear charges [93,94,95,96] is not easily trained to achieve uncertainties of only, say, two percent (the measurement on Nb ($Z = 41$) [95] is a rare exception). Slow ion beams have been applied to the case of Ar^{16+} which had a history of underestimated systematic errors in the pioneering fast-ion beam measurements at Berkeley. The incorrect lifetime data obtained initially then intrigued various theoreticians and thus yielded a playground for theory development. In the course of that competition a computation seemingly closest to experiment was found to suffer from a sign error, a clear sign that significant errors can be made in either field and need an attentive field of competitors to be spotted and cured. As of now, fast-ion lifetime measurements with serious precautions as well as slow-ion beam measurements based on recoil ions produced in a gas target under fast heavy-ion irradiation [97] have yielded Ar^{16+} $1s2s\ ^3S_1$ level lifetimes near 200 ns that agree with each other and with theory within their experimental error bars of several percent. An improvement may be feasible near Z = 18 with well controlled slow ion beams, perhaps produced from an electron cyclotron resonance ion source (ECRIS) or an electron beam ion source (EBIS).

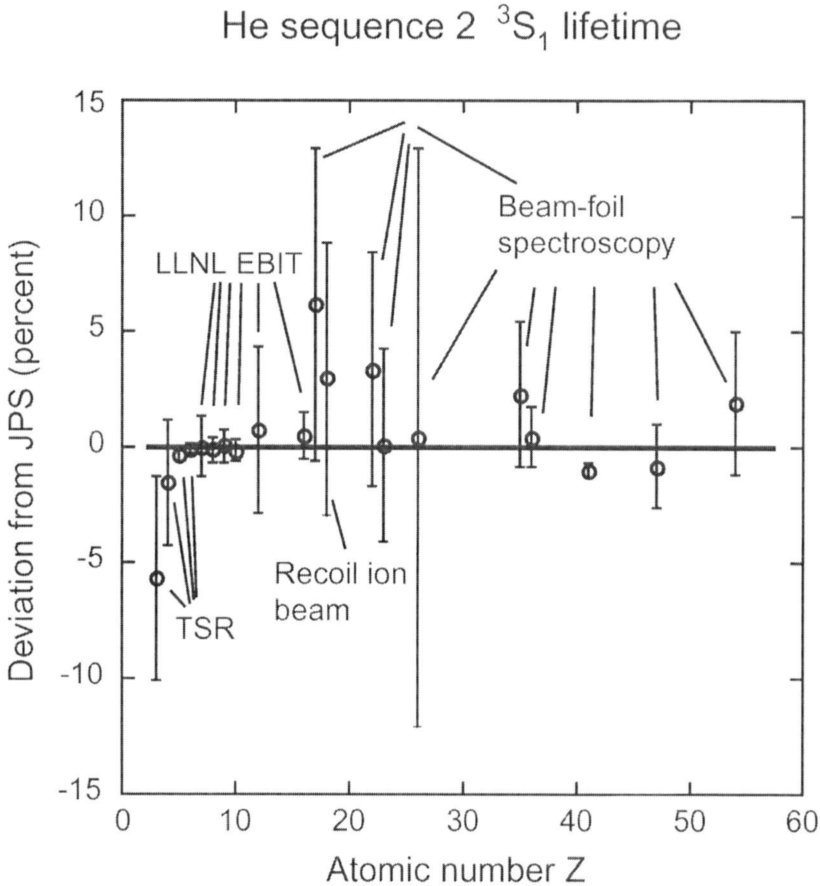

Figure 3. Deviation of the measured M1 decay rate from the lowest triplet level in He-like ions (Z = 3 through 54) from the results of computations by Johnson, Plante, and Sapirstein (JPS) [79]. The experimental data points result from work at the Heidelberg heavy-ion storage ring TSR, the Livermore electron beam ion traps (EBIT), a recoil ion beam experiment at GSI Darmstadt, and from beam-foil measurements in various places (Berkeley, Argonne, Darmstadt). Some older results that have been technically superseded are not shown.

At low Z, the 1s2s $^3S_1$ level lifetime in neutral He is a true challenge because of its magnitude on the order of one hour (see the pioneering measurement by Moos and Woodworth [98]). There is a Li^+ data point with a 1σ-uncertainty of 10% from a measurement at the Heidelberg heavy-ion storage ring TSR that probed the 1s2s $^3S_1$ level by an electron beam and dielectronic recombination. A significant contribution to the error budget is the shortening of the apparent lifetime by ion losses from the ion cloud circulating in the storage ring. In this case the ion loss rate was close in magnitude to the radiative decay rate. A later attempt at TSR has aimed at optically probing (by further laser excitation) the $^3S_1$ level population of Li^+ ions. However, the laser-ion interaction appeared to introduce new problems, and the experiment has remained unfinished [99]. The TSR storage ring has since been closed down and will be moved to the ISOLDE facility at CERN. A new lifetime measurement at a storage ring with longer storage times, perhaps achieved by an even better vacuum enabled by a cryogenic vacuum vessel, seems a worthwhile enterprise in order to test theory in a near-neutral atomic system, which is particularly challenging.

3.3. Li-Like Ions

3.3.1. Levels

Although a bit dated by now, the 1991 publication by Yong-Ki Kim and colleagues on the resonance levels in Li-, Na-, and Cu-like ions [100] still is a most useful benchmark paper, showing excellent agreement of these accurate computations with many—but not all—experimental data. Their computations combine relativistic many-body perturbation theory, Dirack-Fock computations and approximate “screened” QED corrections. Blundell [101] has treated QED in ions of the same sequences in a different way, and experiment indicates that some of his results are superior at very high Z [102,103]. However, his computations address only a selection of elements, and thus a direct comparison with experimental results is possible only in a few cases. This problem aggravates with some of the most detailed computations which often deal only with a single element, or with a few elements with isotopes that offer a chance at studying hyperfine effects (which scale with Z^3). There is a cottage industry of computations of higher-order QED contributions and nuclear magnetic effects to the level energies of Li-like (and H- and B-like) ions of a few promising isotopes. However, this view to the detail appears to blind out the larger picture and isoelectronic trends. If an isotope has not yet been studied for hyperfine effects experimentally, it is much less likely to find theoretical treatment. Of course, it is uneconomic to calculate all isotopes in detail, but it might be useful to identify isotopes that from a theoretical perspective (perhaps taking elemental abundances into account in the selection) appear to be more promising than others.

In the mid-Z range, data have been pursued at tokamaks [104] and by beam-foil spectroscopy in a Giessen-led collaboration working at GSI Darmstadt [105,106,107,108,109], corroborated and almost matched in accuracy by EBIT work at Livermore [102]. EBITs have the advantage of a practically stationary light source where beam-foil spectroscopy suffers from having to correctly account for the Doppler shift of light emitted by a beam of fast ions. The mid-Z range is of interest, because relativistic effects are notable, but not yet dominant. (Appropriate theory has been discussed in the aforementioned references.) With increasing ion atomic number, both, relativistic effects and QED contributions, increase and require fully relativistic computations, not just nonrelativistic computations with relativistic corrections. Nuclear physics effects also change with the atomic number. Very heavy nuclei are usually nonspherical, and the correction of nuclear size effects in QED that are related to actual nuclear structure is difficult, because the nuclear parameters of the heaviest naturally abundant isotopes are not known well enough. Hence QED effects are largest in uranium ($Z = 92$), as has been confirmed by measurement and theory for various highly charged ions [74,75,76], but for a long time the nuclear structure uncertainties posed a limit to accurate computations. Because of this, more accurate tests of QED contributions than with U may be doable with Pb ($Z = 82$), although the absolute magnitude of the QED effects is smaller. Recent development in theory begins to shift this notion, and with improved modeling of the nuclear shape [77,78] now higher-order QED effects may be the main source of uncertainty.

Because the weight of various entities and processes changes in atomic systems and in computations along isoelectronic sequences, one should not expect that a single computational approach easily matches all data. There are wide gaps in the upper half of the Z range of elements, and that implies that the mutual consistency of the experimental data has not yet been assured by sheer statistics. Experimental problems with the ions of any given element possibly still play an unrecognized role which may only be found out when the data points are closing up to each other. If such problems have been overcome, the differences between measured and calculated data points (level energies, wavelengths, lifetimes, *etc.*) should follow smooth trends that can be interpreted for clues to missing contributions to atomic structure computations. As will be demonstrated in the subsequent sections, "we are not there yet". As a general observation, wide-range computations may be in excellent proximity to low-Z data and yet be imperfect elsewhere. Atomic structure effects may appear as missing overall terms or as a change of balance of configuration interaction, relativistic effects, and QED contributions.

In comparisons of measured and calculated values, one would like to show the difference from "truth" which indicates how much further one has to go. Since truth is no physical entity and the correct numbers are not known, one could compare calculated values to measured ones (which bear an error bar), or measurements and computations to one of the latter (which ought to carry an error estimate). In the former presentation, the scatter of measured data is reflected in an opposite scatter of all other data entries. Because computations tend to yield a smooth trend and for the fact that computations can be done for each atomic number while measurements often cannot, the most commonly chosen option is the reference to one computation. This choice does not preclude the possibility of errors and inconsistencies in the trend of the computed numbers.

3.3.2. Lifetimes

Over the years, the $2p_{1/2}$ level lifetime in Li-like ions has been measured from $Z = 3$ up to $Z = 92$, and the $2p_{3/2}$ level lifetime from $Z = 4$ up to $Z = 36$ (Li I was not included here because of quantum beat problems). This situation reflects the absence (largely) of relativistic effects in the $2p_{1/2}$ wave function and the proportionality of the level energy to Z, whereas the fine structure splitting, a relativistic effect (proportional to Z^4) massively modifies the $2p_{3/2}$ level energy. The steeply increased transition energy shortens the $2p_{3/2}$ level lifetime to a few picoseconds in Kr^{33+} [20], whereas the $2p_{1/2}$ level lifetime amounts to much more than that even in U^{89+} ions, namely about 62 ps [110]. These lifetimes in high-Z Li-like ions are in good agreement with the venerable *ab initio* MCDF computations by Cheng, Kim and Desclaux [66]. Theodosiou, Curtis and El Mekki [111] have tried a combination of simpler computations (the numerical Coulomb approximation, which actually predates the MCDF approach by many years) in combination with corrections for experimental level data. Their parametrization effort yields rather similar level lifetimes for practical applications. However, their procedure is not a show of quality of theoretical achievement, but rather a show of what lengths one has to go for in the absence of sufficiently good *ab initio* predictions. The complete lifetime data set has been analyzed again elsewhere (see Figure 1 of [21], where also the references to the experimental data are given). Theoretical interest in the Li sequence has not waned, and only recently a stretch of the sequence has been recalculated for levels, lifetimes and collision strengths [112,113]. By the way, at low Z there are two early beam-foil results for the $2p_{3/2}$ level lifetime in Be^+, neither of which agrees with the mean trend—but the weighted mean of the data does, which suggests that at the time the experimental uncertainties may have been underestimated.

Another *caveat* for the search for 'best' data: The latest publication is not necessarily the most advanced one. For years, L. J. Curtis maintained a web page of his publications that also offered access to a manuscript by Kukla *et al.* in *Phys. Rev. A* style formatting. The manuscript reported on beam-foil observations of Li-like ions of Kr, but it never appeared in the journal. At one stage the manuscript was cited by colleagues who had measured Li-like ions of Ni [114] and seemed to believe that the Kukla work was much younger than a 1995 publication on Li-like Kr [20]. However, I had met Kris Kukla on another project at Argonne when he was active in the field. Thus I had an inkling of the true time frame, and I knew of such (partial) data from Argonne mentioned in a 1999 conference paper [115]. From Curtis I learned that the plan had been to write a conclusive Kr paper on the basis of Kukla's thesis work, and the version offered on the web had (largely) been an unedited chapter of that thesis. I urged to see this material finally brushed up into a proper publication [116], thus adding another data point to the collection, in order to accompany my own earlier lifetime data for Kr^{33+}, albeit from a more limited data evaluation. Still, a second data point (actually, in this case a third) is a good thing for a data base as it indicates the reproducibility.

It is not likely that much more lifetime work will be done on the $2p_{3/2}$ level between Li and Kr, because the data collected so far seem consistent. However, higher accuracy, possibly better than 1%, would be valuable. Selective excitation of trapped ions by VUV or EUV laser light and electronic timing might do for the first few ionization stages. An alternative for more highly ionized species, and with the potential of reaching high accuracy, might employ X-ray laser excitation starting from ions prepared in an EBIT [117]. The lifetime of the other fine structure level, $2p_{1/2}$, could thus be studied in all high-*Z* ions. For this level even the proven technique of beam-foil spectroscopy would do in principle (if practically limited to an accuracy rather above than below 1%), but no accelerator laboratory presently seems interested in such a project.

While the spectra of Li-like ions are dominated by the excitation of the single valence electron outside a closed shell, this is by no means the only option of interest. Double and triple excitation are feasible and cause satellite lines to the diagram lines of H- and He-like resonance line series in the X-ray spectra of many light sources (plus two-electron one-photon satellites). Since a long time the satellites have been exploited as a diagnostic of temperature and density in astrophysical plasmas and in fusion-oriented plasma research. Beam-foil spectra show a profusion of such satellite lines because of the high electron density (and thus a high collision frequency) in the excitation volume (the foil) [118,119,120]. Most multiply excited levels decay rapidly, often by autoionization, a process which in low charge state ions has time scales on the order of ten femtoseconds. However, symmetries sometimes hinder autoionization and/or X-ray decays, and a triply excited level of neutral Li has been observed and its picosecond level lifetime measured [121,122,123], both in good agreement with computations. The wavelength measurement on the decay of the triply excited system incidentally represents one of the most accurate such measurements of beam-foil data in the EUV. For many less special levels the calculated results have been less fitting the anyway not very accurate experimental results [124,125]. This experience cautions against any general claims that three-electron systems are reliably calculated these days. What can be calculated well are singly excited three-electron systems in which the single valence electron outside a closed-shell core dominates.

Years ago a triply excited few-electron ion seemed rather exotic and far from any use. Nowadays synchrotrons, EUV- and X-ray lasers have become such brilliant light sources that with their intense radiation fields they can produce multiply excited ions from neutral matter. However, so many multiply excited levels can be reached (though not necessarily resolved) in the same energy pulse that in the face of such complex spectra so far no highly accurate numbers are needed for modeling and understanding the radiative properties of specific systems or their individual levels.

3.4. Be-Like Ions

Four-electron ions (partly discussed in the introduction), with two electrons in a closed $n = 1$ shell and two valence electrons, add a lot of complexity to that of three-electron ions. Even within the lowest shell ($n = 2$) we now have not only electric dipole (E1) transitions (such as the resonance transition), but also a spin-changing E1 (intercombination) transition, M1 transitions between the 2s2p triplet levels, an E2 transition between the $J = 0$ and $J = 2$ triplet levels, 2p 2 displaced levels, multiplet mixing, magnetic interactions, and so on, and the various interactions scale differently with Z. An exact atomic structure computation would return correct results on all atomic parameters. Existing programs do not achieve that goal. They usually employ approximations and address certain aspects, and we should not be surprised that computations that yield numbers in agreement with experiment for one atomic property do not necessarily serve equally well for the others.

Ulyana and Marianna Safronova, Walter Johnson, and their colleagues have studied various details of Be-like ions along the isoelectronic sequence by RMBPT computations [126,127,128]. Their plethora of results would not fit into any print journal, which is why numerical data often have been tabulated for a few ions only, and trends (especially useful for warning of cross-overs among levels and changes of their labeling) have been demonstrated graphically. It comes as no surprise that a schematic selection of ions (such as every fifth or tenth element) is not likely to provide data on all elements of interest to experimenters, and single-column wide graphics covering data of the whole table of elements that vary by orders of magnitude are not suitable to extract accurate values. The constraints may be imposed by economic limitations, but they also limit the feasibility of comparisons between experiment and computation. For low-lying levels in Be-like ions, among the more recent computations (two decades younger than the venerable MCDF computation by Cheng, Kim and Desclaux [66]) the most dense listing appears to be one of the papers named above [126], making it the most suitable reference for comparisons. Figure 4 shows that the experimental data points are internally consistent up to $Z = 36$, but are very few and less precise above. As to the quality of the computations, the agreement with experiment is best near $Z = 20$ and clearly poorer below and above this range. The three experimental results for high *Z* elements originate from different light sources, for Mo ($Z = 42$) a tokamak plasma [104], and for Xe ($Z = 54$) a beam-foil experiment [108] and an electron beam ion trap [102], and they seem to indicate a different trend than the low-*Z* data do. More experimental data are needed, and other types of computations would be welcome, too. There are, for example, the RCI computations by Chen and Cheng [129] that, however, treat even fewer ions than the Safronovas and Johnson. At low *Z*, these computations are practically coincident with the experimental data (in contrast to Safronova and Johnson); beyond $Z = 30$ they deviate slightly from the perceived trend of deviations between experiment and MBPT results, and at high *Z* the deviation between RCI and MBPT changes sign. Hence there are significant differences between the results obtained with highly developed atomic structure codes, and in this case RCI seems to do better. (Preliminary Livermore EBIT data for Pb ($Z = 82$) fall close to the RCI predictions.)

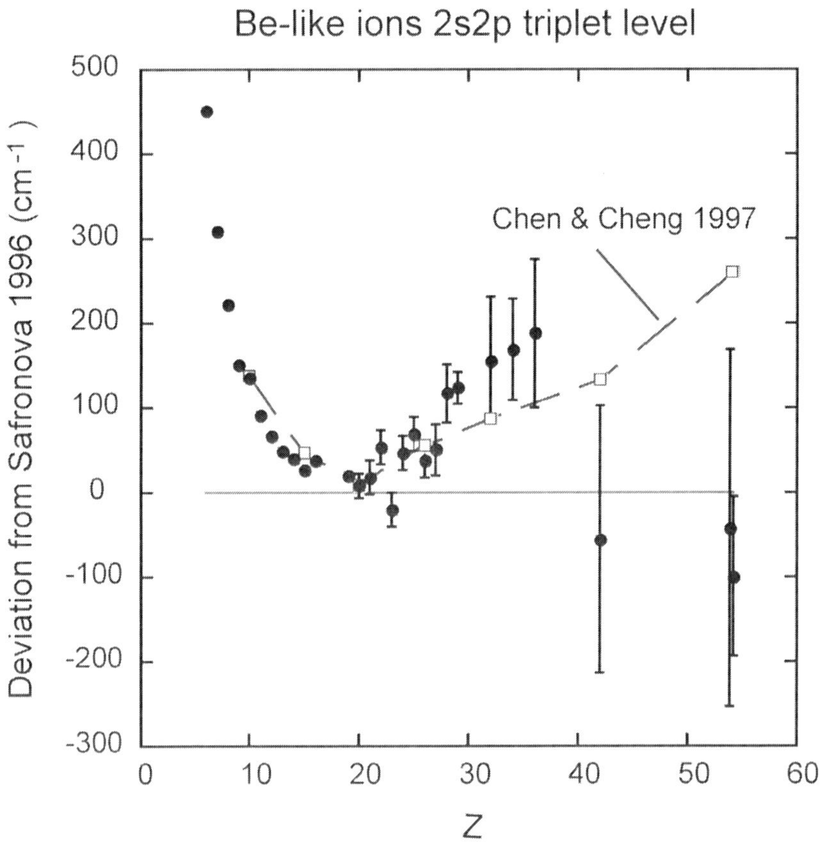

Figure 4. Measurements (full black symbols) and Relativistic Configuration Interaction (RCI) computational results (open blue squares and broken line [129]) of the 2s2p 3 P 1 level in Be-like ions and their deviation from Many-Body Perturbation Theory (MBPT) computations by Safronova *et al.* [126]. Note that at low *Z*, the computation by Chen and Cheng [129] is close to experiment, while the other one is not, and *vice versa* at higher *Z*.

3.4.1. 2s2p Level Lifetimes

For comparisons with experiment it is helpful to find predictions for all elements. Träbert and Curtis [21] have listed much of the experimental lifetime work on ions in this isoelectronic sequence as well as compared the results to simple scaling rules in order to avoid the possible errors of new computations. Because the references are available there, the present discussion focuses on the state of the art computations.

Per Jönsson and Charlotte Froese Fischer have calculated the 2s 2 1 S 0 - 2s2p 1 P 1 o resonance transition rates for many values of *Z*[27]. However, the computations converge better for a more dominant central Coulomb field, that is, for higher *Z*. Computations for low-*Z* ions thus require more elaborate wave functions, with a larger number of configurations, for the same estimated level of accuracy. Considering the much increased computer run time requirement with larger sets of wave functions, the computations then are not formulated identically for different ions. In general, the drop in computing costs is successively permitting the use of larger basis sets for the benefit of computational accuracy. However, not all computations are being updated when the occasion arises, and computations of different quality may be contributing to the same data collection.

Froese Fischer and Tachiev have a web site (presently hosted at NIST) that lists their own MCHF/MCDF/MCDHF results [28]. For Be-like ions, the tables cite MCHF computations of 1999 (for Z = 4 – 12) and MCDHF computations of 2004 (Z = 9 – 30) and 2009 (Z = 10 – 30) . No single set of computations covers all elements. What then can one use as a reference to compare experimental data and the results of other computations with? Irrespective of claims at some time that the computational techniques are so good that the resonance transition rate in Be-like ions may well be known (from computations) with an accuracy on the order of 10 – 4 , the differences among the corresponding tables on this web page amount to several percent. One has a choice if several computations are listed for a given ion, but at low *Z* there are results only from the oldest computations by the team. This does not matter much for the singlet level lifetime, for which experiment has an uncertainty of about 3%. This can be seen in Figure 1 where the experimental data are referred to the Froese Fischer and Tachiev data collection; the figure reveals that a semiempirical analysis of the same data by Curtis and Ellis [29] describes the isoelectronic sequence equally well, but slightly differently at low *Z*. There is an amusing detail for neutral Be (already discussed above): One analysis favors one of the recent experimental results, and the other favors the other.

For the 2s2p 3 P 1 o triplet level lifetime (see Figure 5) the inconsistency of computational results presented on the same web page [28] is notable. The MCDF/MCDHF computations accessible there have used different numbers of configurations along the isoelectronic sequence, and they did not include Xe (Z = 54), the highest *Z* data point long since established by experiment [130]. (The result obtained by Cheng, Kim and Desclaux [66] for Z = 54 was used as an unmarked stand-in in Figure 5, although the isoelectronic trend of those computations is very different and approaches the newer results only in this range.) At low *Z*, results of computations of different vintages are available on the Froese Fischer and Tachiev web page: MCHF for $Z = 5–12$, MCDHF of 2004 for $Z = 9$ to 30, but leaving out $Z = 10$, and MCDHF of 2009 for $Z = 10$ to 30, but leaving out $Z = 14$. When comparing data outside the range of a single computation, one finds that the MCHF and MCDHF computations by the same team of theorists differ by 7% (at low Z); the two sets of MCDHF computations differ from each other by 3% at $Z = 11$ and by about 1.5% at $Z = 30$. Needless to say, it is difficult to trust a computation to, say, 1%, if ever so often new runs by a similar suite of programs result in output that differs by a multiple of the claimed uncertainty from previous results. At low Z, the computations by Brage, Fleming, and Hutton [131] (which, alas, do not treat all elements) agree best with the rather accurate atomic lifetime data from heavy-ion storage ring experiments. For Ne ($Z = 10$) and above, most of the experimental data are rather poor and cannot discriminate between the more modern computations. This calls for further meaningful experiments, for example on Ne, since the present "experimental lifetime" literature entry for Ne is an estimate from plasma observations which is based on theory plus a 50% error estimate, and that does not really imply a meaningful measurement.

3.4.2. Displaced Level Lifetimes

When trying to measure the lifetime of the 2s2p $^1P^o_1$ level in Be-like ions, especially in beam-foil spectroscopy, one has to also obtain information on the lifetimes of the strongest cascades that replenish this level, which originate from the $2p^2$ $^1S_0$ and $2p^2$ $^1D_2$ levels which in turn are replenished by other levels. Also, there is the prominent 2s2p $^3P^o$ – $2p^2$ ^3P line multiplet. (Similar transitions connecting to displaced levels are numerous in the further isoelectronic sequences.) A rather reliable resource on these transitions has been provided by Brian Fawcett [132] who has catalogued wavelengths and transition rates for L-shell ions up to $Z = 28$from a combination of observations and scaled Cowan code (and variants) computations. Within the typical 10% uncertainty of beam-foil spectroscopy, the most widely employed technique for lifetime measurements on such systems, the Fawcett tabulations have been found to be pretty reliable. This compilation is very valuable for the practitioner; I am not aware of any *ab initio* computation with a similarly wide scope and accuracy of the results. Some specific cases will be discussed below.

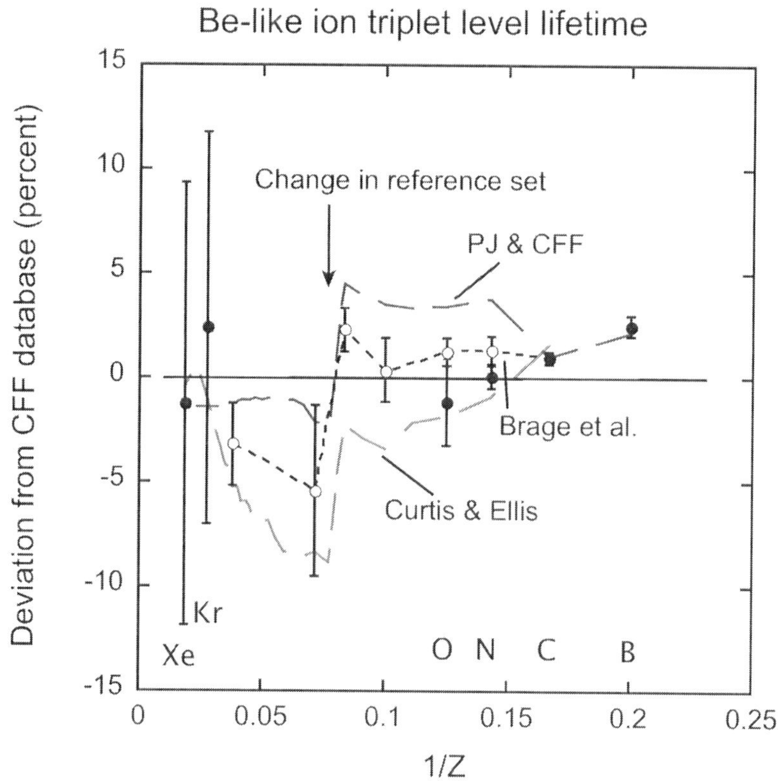

Figure 5. Lifetime of the 2s2p 3 P 1 o level in Be-like ions (intercombination decay) displayed as the relative deviation from the computational web resources (MCHF/MCDHF collection) made available by Froese Fischer and Tachiev [28]. A problem lies in the only partial coverage by any given computation, and therefore one cannot refer all of the available measurements and computations to a single reference data set. Because of this, the trends shown all suffer a step function (by 7%) above Ne (Z = 10), because this is where one of the references ends and the comparison changes to another one. The only continuous reference data set would have been the one derived by a semiempirical averaging procedure applied by Curtis and Ellis [29] (broken red line) to computations and measurements, which thus represents a secondary trend. The broken blue line refers to computations by Jönsson and Froese Fischer [27], the short-dash broken black line connects the computational results reported by Brage *et al.* [131] (open circles, the study has to be lauded for point-by-point estimates of uncertainty). Not all of the experimental data (filled circles) are shown, because for several of them the size of the error bars exceeds the plot boundaries.

Fawcett's computations are basically nonrelativistic (with relativistic corrections). However, there are effects of multiplet mixing that are much more pronounced than fine structure splittings. For example, with increasing nuclear charge *Z*, there increases the probability of a spin-changing decay branch of the $2p^2\,{}^1D_2$ level to the $2s2p\,{}^3P^o_{1,2}$ levels. (A corresponding case, the $3p^2\,{}^1D_2$ level decay, also with in-system and intercombination decay channels, occurs in the Mg isoelectronic sequence [32,133].) One might expect that any atomic structure code operating in intermediate coupling would include this decay branch automatically, but the scarcity of literature data and the scatter of the results of the few available computations tell a different story.

3.4.3. Hyperfine-Induced Decays

Just as the $1s2p\,{}^3P^o_0$ levels in He-like ions with a non-zero nuclear spin can be made to decay by hyperfine interaction (which mixes the level with a hyperfine level of the nearby $1s2p\,{}^3P^o_1$ level), the $2s2p\,{}^3P^o_0$ levels in Be-like ions can be made to decay by hyperfine mixing with a hyperfine level of the nearby $2s2p\,{}^3P^o_1$ level. The intercombination transition rate in Be-like ions is very much lower than that in He-like ions, and so must be the hyperfine-induced decay rate, requiring a combination of highly charged ions and ion trapping techniques. Hence this decay rate has been measured in only a few ions. Theoretical discussions of the situation have been around for several decades, but have recently been modified considerably. Cheng, Chen and Johnson [134] have recently looked into this problem and explained several computational approaches, which is very helpful for the community. The newer computations largely agree with each other, but the "agreement" with experiment is a few percent in one ion and about 20% in another (which represents a massive improvement on previous treatments!). This situation is not yet satisfactory, as has been discussed elsewhere [135], and more measurements are sorely needed. It is too early to make sensible statements on the quality of computations in this case.

3.5. B-Like Ions

With more electrons in the atomic system, the degree of complexity rises. In addition to inter-shell and intra-shell transitions without or with a change of spin, regular and displaced levels (already present in Be-like ions) there now are E1-forbidden transitions within the ground term. Such (E1-) "forbidden" transitions are of wide-spread importance in plasma diagnostics, because their intensity is sensitive to electron density in a much lower range than intercombination (spin-changing) or E1-allowed transitions. In low-charge ions, they serve to understand the extremely dilute planetary nebulae; in moderately charged ions of iron group elements they help to investigate solar and other stellar coronae, and in highly charged heavy ions, where their rate (climbing steeply with *Z*) is no longer low in comparison to E1 transitions, they signal the presence of certain charge states and thus serve as temperature diagnostics. Decades ago, Kaufman and Sugar [136] have compiled a hugely useful collection of wavelength and transition rate data on such E1-forbidden transitions up to Z = 42 (Mo), which in turn has also contributed notably to the NIST data base, combining experimental data with (Cowan code) computations of their own. Isoelectronic lifetime studies by experiment, backed-up by modern computations, however, have revealed a fair number of inconsistencies in those tables. Some of those lifetime measurements (mostly done at a heavy-ion storage ring, some at electrostatic ion traps or at EBITs) are cited in the following sections.

Inspired by Bengt Edlén's early systematizations of level data, a number of computations have produced results for ions in the first third of the periodic table, but far from covering all elements in this range (see, for example, [137,138,139,140,141,142,143,144,145] and further references therein). This patchy coverage renders intercomparisons for extended sections of the isoelectronic sequence difficult or impossible. On the upside, many computations include those of transition rates in addition to levels. Hence measured lifetimes of displaced levels (for Si and P see [146,147], the latest such study may have been the one by Tordoir *et al.* on Na [148]) could be and have been compared to theory, whether semiempirically flavored [132] or calculated *ab initio* [142]. One way out of the page limit of printed journals are electronic depositories. Apparently the authors of the latter work were forced by referee suggestion (or by the journal) to put some of their tables (for three elements) into an electronic depository with the publishers instead which, because of mislabeling, seemed inaccessible 14 years later. (Thanks to A. E. Kramida for pointing out the Erratum with the correct file path!). For anybody interested in the data it is difficult to understand why thus two pages were saved from a print journal, one of which was lost again because of the need for an Erratum.

In B-like ions (and correspondingly in Al-like ions, etc.), several fine structure levels of the displaced configuration 2s2p 2 cross each other along the isoelectronic sequence. The level mixing affects the relative line intensities of 2s-2p transitions. This effect has been systematically studied in beam-foil spectra [149,150]; the observations qualitatively confirm various computational predictions.

A key role in B-like ions is played by the fine structure splitting of the ground term, 1s 2 2s 2 2p 2 P o (Figure 6). The MBPT computations by Safronova *et al.* [138] approximate the experimental level splitting well up to about Z = 36 , and then they gradually, but distinctly, deviate (see [102]). Experimental data at high *Z* are sparse [102]. For spectroscopy there also is a complication: the 2s 22p 2 P o ground state fine structure interval increases with Z 4 , and so does the fine structure splitting of the lowest 2s2p 2 term, 4 P. The fine structure splittings grow faster than the 2s-2p electrostatic level splitting (which increases approximately linearly with *Z*), and therefore at some value of *Z* the lowest quartet level drops below the upper level of the ground term. In addition to the M1 decay that connects the upper with the lower fine structure level of the ground term, there then is the alternate decay path via two intercombination transitions and the lowest quartet level. Incidentally, in Xe the energies of the two intercombination transitions are close to each other and can be obtained from the same spectrum (see [102]). Having to combine results from two spectral lines increases the overall uncertainty, but the measurement takes advantage of a less crowded portion of the EUV spectrum—a measurement on the single-photon decay branch is not likely to provide a more accurate result with present equipment.

3.5.1. E1-Forbidden Transition Rates

The M1 transition rate in the ground term is of some interest in collisional-radiative spectral modeling. Given the vagaries of determining line intensities and line ratios, an accuracy on the order of one or a few percent seems sufficient in practice. Such an accuracy can be expected from a combination of transition energy and line strength. In a first approximation (and a single-configuration picture), the line strength is an exact factor readily determined from Racah algebra. Rather many publications have produced M1 rates for this and related electric-dipole forbidden transitions in a wide variety of ions of astrophysical interest, and the published results have usually agreed to within a few percent with each other and with the above estimate based on transition energy and line strength. However, the actual situation is much more complex. Many people have tried to calculate the fine structure interval (the transition energy), and most computations have fallen short of an accurate result. If one then honestly combines the poor calculated transition energy value with the easily obtained line strength approximation, the transition rate deviates from expectation (see Figure 1 and Figure 2 in [151]). It is deceiving to see most computational papers reporting "good" results, because when the computed energies are poor, a standard escape is the so-called "semi-empirical" correction (recently and more correctly called "rescaling to experimental wavelengths") by which one's own poor energy intervals are replaced by the experimental ones (from well-resolved spectroscopic observations), and then the combination of known energy and geometric factors automatically yields results close to the pack. These results are then (probably) close to the 'true' ones and provide a useful guide for the experimenter, but they have not been consistently obtained from atomic structure theory.

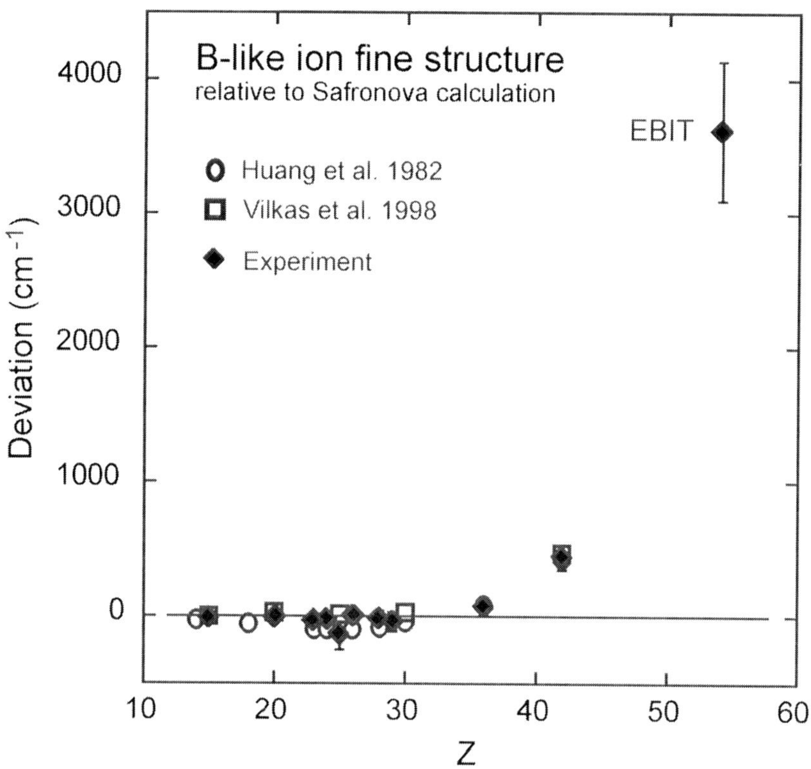

Figure 6. The 2s 2 2p 2 P o J = 1 / 2 - J = 3 / 2 ground state fine structure splitting in B-like ions, shown as the deviation from the computations by Safronova *et al.* [138]. Up to about Z =30 , experiment (filled symbols) and other computations (open symbols) stay close to the trend predicted by Safronova *et al.*, but then they diverge. Unfortunately, in the latter range the accurate computations are extremely sparse. The EBIT data point for Xe (Z = 54) represents the sum of two transition energies, because the lowest quartet level drops below the upper fine structure level of the doublet ground term and thus opens an alternate (two-step) decay path to ground that is more easily observed and resolved spectroscopically than the single-step decay branch.

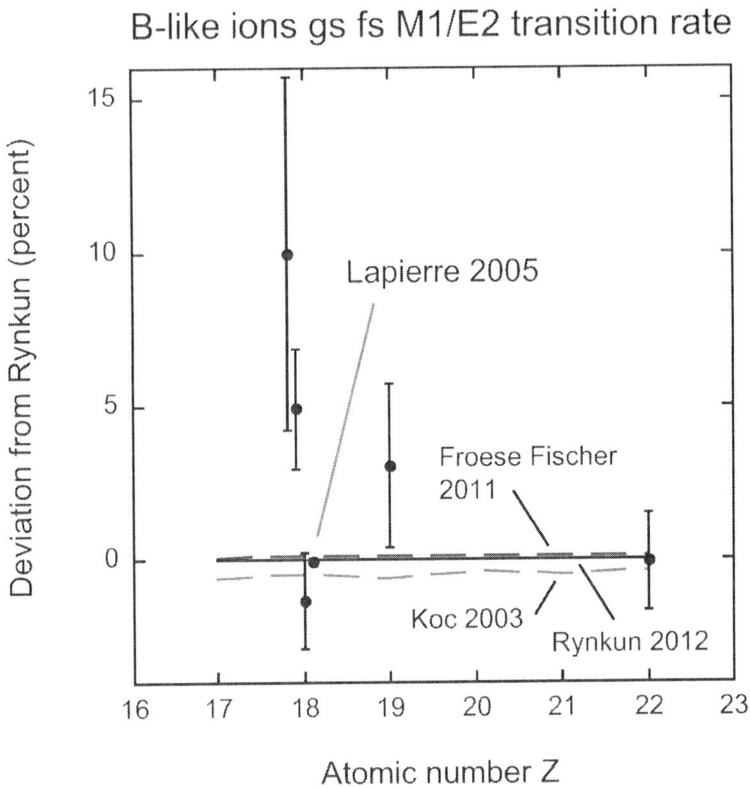

Figure 7. Transition rate of the 2s 2 2p 2 P o J = 1 / 2 - J = 3 / 2 ground state fine structure transition in B-like ions, depicted as the deviation from the predictions by Rynkun *et al.* [144]. Measurements have been done at an electrostatic ion trap [153], electron beam ion traps [154,155,156,157], and at a heavy-ion storage ring [158]. The EBIT measurements have eventually reached high accuracy, as is seen in the data point for Ar that has an uncertainty smaller than the data marker. Three representatives of recent computations are shown (identified by the lead authors), and the best experiment [156,157] agrees with only one of them, the computations by Rynkun *et al.* [144] (full line). It would be good to know whether this computation has taken the anomalous magnetic moment of the electron (EAMM) contribution to the M1 transition moment into account, an effect that amounts to 0.45% and thus to much more than the experimental uncertainty (smaller than the symbol size), but I have not seen it being mentioned.

Progress has been made in computations thanks to cheap computing power, and nowadays more and more computations are arriving at results closer to the experimentally known fine structure intervals. If the transition energies of these E1-forbidden transitions can be calculated, the second factor, the line strength, becomes of interest in more detail, because it reflects the multi-configuration composition of the wave functions. Fortunately, ion trapping techniques (using EBITs or heavy-ion storage rings, including a cross-check for consistency [152], see the original work cited in the reviews [6,7,8,9,10,11,12]) have addressed the millisecond lifetime range and nowadays can measure the transitions rates of many of those E1-forbidden transitions in the ground configurations of multiply charged heavy ions, and they often reach accuracies of better than one percent. One percent—surely computations can do better than that? Skipping all details, the lifetime predictions for levels in the ground configurations often differ by several percent even among more recent computations (Figure 7).

Is it worth worrying about this scatter? Does it matter—since it hardly did matter for decades? Several measurements of E1-forbidden transition rates have reached an accuracy of better than 0.5%, which implies a significant challenge and requires a qualitative step forward for theory. Conceptually established in the 1950s, but largely overlooked and neglected by atomic structure theoreticians, the anomalous magnetic moment of the electron (EAMM)—beyond the (g – 2) QED effect on the level energy—causes an increase of the M1 transition rate by 0.45% [156,157,159]. A lifetime measurement of an accuracy better than this percentage can in principle test a leading QED correction—and there are not many QED contributions as large as this one. One can test an element of the Standard Model by a lifetime measurement that is good to, say, 0.1%, and such measurements are feasible on various atomic systems nowadays. Unfortunately, there is a correspondence to the intended spectroscopic QED measurements on two-electron ions mentioned above: Atomic structure theory (apart from QED) for multi-electron ions is not up to the part (yet). When the Heidelberg EBIT group had analyzed their data and obtained error bars on the order of 0.1% [157] (which as a matter of principle need to be checked by experiments elsewhere), they found the available theoretical predictions of the lifetime (after adding the EAMM correction to all those computations that had not considered it) to scatter within an interval of about 1%, for a relatively simple atomic system consisting of a closed K (1s) shell, a closed 2s subshell, and a single 2p electron. To cut the discussion short, among theorists (by now) there is little doubt about the necessity and amount of the EAMM contribution. However, it cannot presently be tested by experiment, because the much larger contribution by non-QED many-body quantum mechanics is not known with sufficient reliability. Atomic structure theory (many-body quantum mechanics) is not (seriously) in doubt, but its application by means of practical (even large-scale) computations is not good enough yet. Since testing the Standard Model (eventually) by atomic lifetime measurements seems so desirable, a few further examples will be presented below, as well as the interplay with theory in the interpretation of the data and the quest for even higher reliability.

3.6. CNO-Like Ions

In C- (Si-, Ge-) like ions, a noteworthy pair of intercombination transitions connects the $ns^2np^2\ ^3P_{1,2}$ levels of the ground configuration with the lowest quintet level, $nsnp^3\ ^5S^o_2$. Several atomic structure details complicate the computation of the transition rate; the multiplet mixing with J = 2 triplet levels suffers from cancellation effects in the $nsnp^3\ ^3P^o_2$ level decays and requires the inclusion also of mixing with the $nsnp^3\ ^1D^o_2$ level, and there are reasons why the fine structure splittings cannot be employed as a computational bridge as they can in other situations. Hence the triplet-quintet intercombination line has posed a challenge to theory while the lifetime data from early ion trap measurements also scattered notably. Two computations that involved convergence studies (as a function of maximum principal quantum number in the basis set) lastly agreed with each other [160,161] and with the mean of the earlier experimental findings on N^+ ions, but a heavy-ion storage ring experiment [162] came out slightly different just beyond the mutual error bars. As Alan Hibbert jokingly noted at a conference, his second-but-last convergence step (which he had included in his publication) agreed with the new experimental result. By doing worse in procedure he could have reached seemingly perfect agreement with an experiment (then still in the making). (Hibbert's joke was a much better reaction than the anger that was reported to me from a theory colleague who chided me for citing her published second-order approximation computational result—which did not look so good in the end—instead of her unpublished first-order result.) With the given experimental equipment, the same transition could be (and has been) studied in only one other atomic system, the O^{2+} ion [166], but with a larger uncertainty that does not contribute much towards better insight.

Recent computations of C- and N-like ions [163,164,165] have come with extensive bibliographies, which saves me from doubling the effort. The ground configurations of ions with an open n = 2 shell (Li- through F-like) provide prototypes for related systems with an open n = 3 (Na- through Cl-like) or n = 4 shell, and so on. Within the n = 2 shell systems, there also are symmetries that link, for example, the C- and O-like ions which all have 3 P, 1 D, and 1 S levels and similar E1-forbidden transitions between them (though the level sequence within the 3 P ground term is inverted in the O-like ions). This situation is attractive for systematic studies, since it increases the chance that there are ions that within the given performance envelope of a detection system have wavelength and lifetime parameters that are compatible with the apparatus. Moreover, various ion sources yield high output currents for Si, P, S, and Cl (less so for Al and typically much less so for Mg). Various advantages combined at the Heidelberg heavy-ion storage ring TSR permitted high-quality lifetime measurements on C- and O-like ions of Si, P, and S (Figure 8). Lifetimes and wavelengths might have allowed some measurements also on the corresponding Cl ions, but the storage ring was no longer in operation then. N-like ions have been studied as well, but the atomic structure situation is less favorable in these ions, as there are two levels that differ in lifetime by about a factor of two and decay with wavelengths that are not easily discriminated from each other by filters; a combined decay curve then consists of the superposition of both level decays and cannot be reliably disentangled with a small error bar. Measuring three instead of a single element helps with establishing systematic consistency of the data. The comparison of the data set with computations and compilations also indicates that there are sizable mistakes in certain data bases. The measurements have been described in detail [166,167,168] and compared there to many computations, so that we can concentrate on a comparison of the experimental data to only the best computations here (see Figure 8).

3.7. F-Like Ions

The ground configuration in F-like ions resembles the case of the B-like ions, except for the inverted fine structure levels. The E1-forbidden transition in the ground term is relatively prominent in dilute plasmas, and its wavelength has long since been systematized by Bengt Edlén [169]. The J = 1 / 2 level now is the upper level of the ground configuration, and the lower statistical weight (compared to J = 3 / 2 in B-like ions) makes for a systematically weaker M1 transition. Although lifetime measurements have been done (on Ar and K ions at EBITs, and on Sc and Ti ions at the TSR heavy-ion storage ring), they have not reached the same level of statistical reliability as the ones on B-like ions. The latest computations for the lowest three levels and transitions between them appears to have been done by Jönsson *et al.* [170]. Some of the calculated and measured lifetime values for the 2s 2 2p 5 2 P 1 / 2 o level, the upper fine structure level of the ground term, are shown in Figure 9.

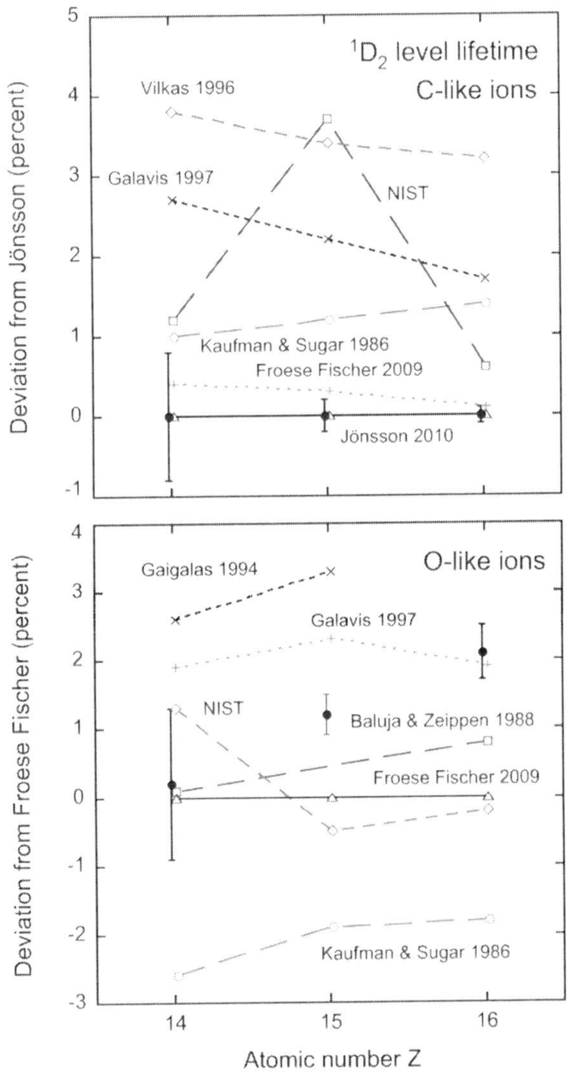

Figure 8. Lifetime of the $2s^2 2p^{2,4}\ {}^1D_2$ level in C- and O-like ions of Si, P, and S. All measurements (full markers with error bars) have been done at the TSR heavy-ion storage ring. The Si measurements could nowadays be redone with much higher statistical reliability, if the storage ring was still available. The data are shown relative to two computations. The computations shown scatter by a few percent around the experimental data, whereas some earlier computations differed by more than 10% (see [168]).

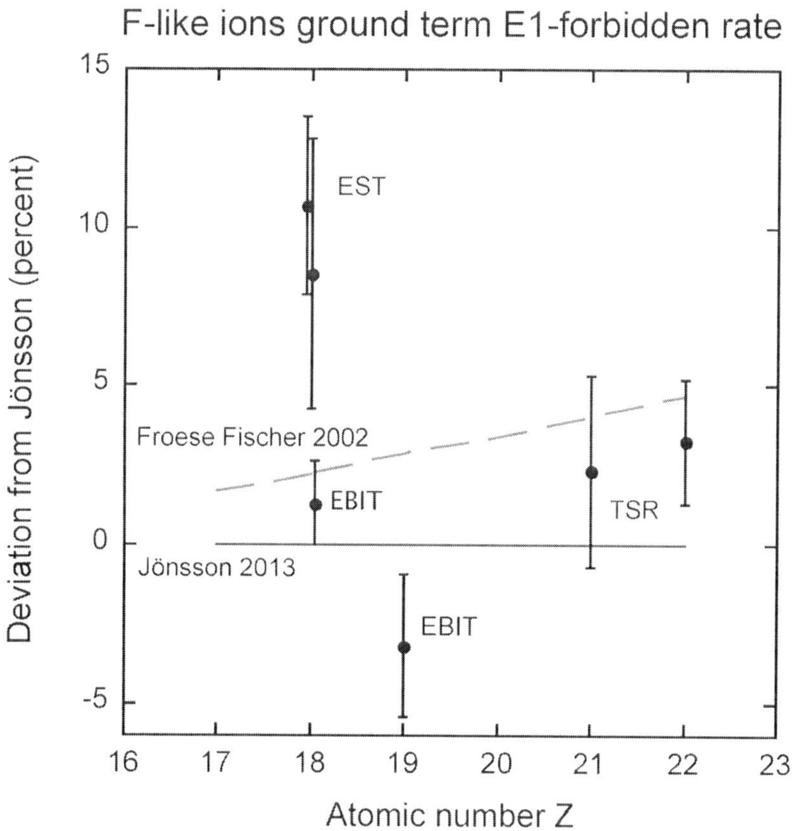

Figure 9. Lifetime of the 2s 2 2p 5 2 P 1 / 2 o level in F-like ions from Cl through Ti. Measurements (full markers with error bars) have been done at an electrostatic ion trap (EST) (Ar), at the Livermore EBIT (Ar, K), and at the TSR heavy-ion storage ring (Sc, Ti). Two of the more recent computations are shown; the experimental errors are still too large to discriminate among these two computations.

3.8. Ne-Like Ions

There are some similarities among the spectra of ions with electron numbers similar to those of the rare gases. A traditional approach to the computation of Ne-like ions, for example, exploits the knowledge of how to treat two-electron systems by approximating Ne-like spectra as arising from the combination of a valence electron and a vacancy in the n = 2 shell. After a plethora of computations by various techniques, complex modern computations no longer depend on such empirical approximations. A recent paper [171] lists many of the earlier studies.

The occurrence of a resonance line and an intercombination transition ($2s^2 2p^6$ – $2s^2 2p^5 3s$) with upper levels $^1P_1^o$ and $^3P_1^o$ has intrigued experimenters and theoreticians and perhaps has led them to assume more simplicity in Ne-like ions than is warranted. Ne-like ions have played a role in the development of certain EUV lasers (often called X-ray lasers), and when those turned out to be more complex and difficult to start than hoped for, atomic structure theory and experiment tried to find out the reasons. The practical problems with those laser experiments have been overcome by now, but Ne-like ions still pose unresolved questions. Nevertheless, there now are plenty of computations to consult, and to follow arguments on whether to maintain a given level designation when levels cross (at high *Z*) according to low-*Z* systematics or whether to change the labels. Let me stick to some seemingly simple phenomena.

3.8.1. Levels

Above the $2s^2 2p^6$ ground state the lowest excitation levels are the 36 levels of the $2s^2 2p^5 3l$ and $2s2p^6 3l$ configurations. Most of these are interconnected by 3s – 3p – 3d E1 transitions (EUV lines for moderately highly charged ions). Selection rules render two of the four 3s levels metastable, leaving one of them with an M2 decay to the ground state and otherwise only M1 transitions in the same subshell. The level lifetimes consequently are very different, and the level populations must be expected to depend strongly on the environmental particle density (collision frequency). Building upon decades of work of others with various light sources, researchers at Lund (relying on laser-produced plasmas) and at Bochum (beam-foil spectroscopy) turned to Ne-like ions of iron group elements; I (as a trainee on the Bochum side) felt relieved when the results largely agreed (where the work overlapped). The process was assisted by beam-foil lifetime work at Bochum, comparing results on the $2s^2 2p^5 3p$ and $2s^2 2p^5 3d$ level lifetimes mostly to Fawcett's tabulations, and by Cowan code computations at Lund. As an example, most of the $2s^2 2p^5 3p$ levels in Ti XIII have a lifetime on the order of 200 ps, while most of the $2s^2 2p^5 3d$ levels in the same system feature lifetimes on the order of 100 ps, experiment on Ti (and correspondingly on S) [172,173] agreeing with the computations.

However, there are two special levels in the $2s^2 2p^5 3p$ manifold, $^1S_0$ and $^3S_1$. The triplet level lies lower than most of the others (so the transition energy towards $2s^2 2p^5 3s$ levels is lower), and hence a longer lifetime than 200 ps should be expected and was found in the Bochum experiment (which corroborated the line identification). A

Russian theory paper nevertheless claimed that the line identification must be wrong, because the measured lifetime disagreed with their own prediction. However, as corroborative evidence they listed a good half dozen other computations as well which together spanned a continuum of lifetime predictions within a factor of about three (experiment fell into the middle of that range). So much for predictive reliability. The $^3S_1$ level is favored in population by its statistical weight (and its lower position) over the $^1S_0$ which lies well above the bulk of the $2s^2 2p^5 3p$ configuration and is predicted to feature a correspondingly shorter level lifetime. Actually, this $J = 0$ level has only two decay channels (to the $2s^2 2p^5 3s\ ^{1,3}P^o_1$ levels), and either of these is expected to be associated with only a weak line because of the low population of the upper level. By collaboration of the Lund and Bochum groups it was tried to identify this line pair in a sequence of elements [174]. However, in all cases at least one of the lines appeared to be blended, preventing any "closure of the loop" which would have served as a strong corroboration of correct identification. Supporting computations left choices among various possible line combinations. Two decades later, Yasuyuki Ishikawa (using his ab initio MR-MP code) and Giulio Del Zanna (using scalable astrophysics modeling codes) tried again, revisited Ne-like ions, found and removed inconsistencies in earlier compilations [175,176] — yet they could not conclusively place the $2s^2 2p^5 3p\ ^1S_0$ levels. This situation is also discussed in the latest paper on such accurate computations of Ne-like ions [171], including further references to earlier ideas about peculiarities of the wave function of the $^1S_0$ level.

Meanwhile, spectra recorded by the *Hinode* spacecraft at wavelengths near 200 Å benefit from much better spectral resolution than has been available decades before in the laboratory or in solar observations from satellites, but in the low-density solar environment the relative level populations of the levels of interest are much lower than in the high-density environment of the ion-foil interaction in beam-foil spectroscopy. In the laboratory, EBIT spectra come closest in (low) electron density to the solar corona, with the additional benefit of a single element contributing (while in the EUV spectrum of the corona, Fe is dominant, but is not the only elemental species). EBIT EUV spectra of Fe, in the range of primary interest from some 170 Å to some 250 Å, feature several lines per 1 Å interval, and a considerable fraction of weak lines presently remains unidentified [177]. For computations to be useful, wavelength predictions ought to be reliable to a fraction of 1 Å. This level has largely been reached (with some caveats due to ongoing development and debugging work) by Ishikawa's MR-MP code when applied to systems with not too many electrons. (However, Fe VIII and Fe

VII feature such a plethora of lines below 170 Å that definitive detailed analyses are not yet in reach there.)

3.8.2. Lifetimes

The rare gases have played (and still play) important roles in the history of spectroscopy. It is amusing to see for how long the Paschen notation on the level structure has remained in use which reflects the process of "digging" into the atom from the outside, in contrast to the "rational" nomenclature of starting with the ground state of an atomic system, which reflects atomic rather than spectral structure. The resonance and intercombination lines of Ne and Ne-like ions have long since been studied for their transition rates (the inverse of the lifetime of their upper levels), also by beam-foil spectroscopy [178]. The neutral rare gas atoms, seemingly easily studied, regularly have yielded several atomic lifetime results that excluded each other by their error bars. At the same time, computations have the largest problems with neutral atoms, because there the central Coulomb field is weakest in comparison to the interactions among electrons. Evidently, the neutral rare gases beyond He pose a particular challenge to experiment and theory. I want to limit my discussion to the ions of the Ne isoelectronic sequence.

By the end of the 1980s, a number of lifetime measurements had addressed the resonance and intercombination transitions in Ne-like ions up to Ar, though only with moderate accuracy. Larry Curtis then developed linearizations of the data in terms of atomic parameters such as the line strength S that would be suitable for systematization and thus for providing better estimates from a sensible smoothing process. One of his tools involved a computation of S for an ion with infinitely high charge ($1 / Z = 0$ in a graphical representation), a reference marker that could be used as an anchor in a plot of low-Z data. The beam-foil group at Lund invited him into a collaborative presentation of their extremely precise results of measurements on Ne-like ions of S and Cl [179]. These experimental results seemed much better than any other lifetime work on the isoelectronic sequence, but the new data points were in gross disagreement with the isoelectronic trend, including the interpretational scheme developed by Curtis (who then dropped the older, less precise data from his analysis [180]). Curtis therefore re-investigated the other published data and asked around for new measurements. He would have liked an explanation for the inconsistent data set on Ne I, too, but that did not happen. Träbert re-analysed old Bochum measurements and extended the local list of results up to S [181]. All data curves showed hefty cascades; at very low Z, the 3d level lifetimes were close to but shorter than that of the 3s triplet level (giving rise to a growing-in cascade), for Al and Si the 3s triplet and several 3d level lifetimes were too close to each other for a meaningful analysis, and only for P and heavier elements the curve tail was dominated by slow cascades. Hence the curve structure could be understood by cascade models, but no high accuracy measurements were possible. Nevertheless, the data trend clearly suggested serious errors (by several standard deviations) in the seemingly precise Lund data on S and Cl. Half a year later, another publication from Lund [173] changed their sulphur lifetime results part of the way to where Curtis' analysis had long since pointed to. The background story of the mistakes is not to be told here, but a take-home lesson is that even highly qualified researchers are not immune against bouts of hubris. For atomic physics, we still do not have highly precise and accurate lifetime data on low-*Z* Ne-like ions, but we now know why: in beam-foil spectroscopy at least, the cascade situation does not permit a better measurement. By the way, the earlier and less precise measurements on elements Cl and beyond that had been assumed to be too poor to be included in analyses any longer when the (seemingly) extremely good-looking data were published have returned to the isoelectronic analysis since, because they were consistent with the overall trend after all. A recent computation [171] also reviews the data base and claims good overall agreement, including the aforementioned 3p and 3d level lifetimes (although in Cl there is some notable mismatch ... with those ominous Lund data).

At a very much higher nuclear charge, $Z = 55$, Ne-like Cs^{45+} ions feature 3s level lifetimes in the femtosecond range. In an electron beam ion trap, the population of the 3p and 3d levels is rather small compared to 3s, which alleviates cascade worries. The ion sample is stationary in such a trap, the ion cloud can be cooled so that the thermal (Doppler) line broadening shrinks below the lifetime broadening, and the instrumental line width can be taken care of by simultaneous studies of longer-lived levels [182]. The lifetime thus measured from the lifetime broadening of the resonance line has come out close to expectation, but more such quality data ought to be recorded for a sensible test of atomic structure theory in the high-*Z* range.

3.8.3. Line Ratio

At first glance, the pair of strongest EUV lines in coronal spectra of various stars (including Sun) might seem too closely related to the aforementioned resonance and intercombination transitions to merit particular attention. The two lines of Fe XVII have been dubbed line 3C ($2p^6\,{}^1S_0 - 2p^5 3d\,{}^1P_1^o$, 15.014 Å) and line 3D ($2p^6\,{}^1S_0 - 2p^5 3d\,{}^3D_1^o$, 15.261 Å), respectively. One might expect that the strongest features in a spectrum ought to be understood first and best. However, the line intensity ratio 3C/3D was regularly calculated as larger than what was seen in low-density plasmas on Earth or in stellar coronae in outer space. An exception was the NIST EBIT; their early result was in the theory ball park, and it has taken years and many corrections to approach the range in which many systematic experimental studies at the LLNL EBIT had been long before—plus all those observers of tokamak plasmas or the stars. Since neither experimenters nor theoreticians found fault with their own ways and results, the systematic discrepancy between experimental and computed line ratios was blamed on a variety of collisional-radiative processes in distant light sources that one could diagnose by the size of the discrepancy. However, Greg Brown's extensive and tedious systematic studies at the LLNL EBIT [183,184] agreed with the stars, but could not suffer from the same collisional effects. Moreover, Ming Feng Gu's computations by his FAC code contradicted those computations that had claimed they could match the observations after all. Work by others eventually suggested that the NIST EBIT problems may have arisen from less well controlled run conditions of their EBIT than the experimenters had believed (causing a blend with a lower charge state ion to matter) and so on.

A digression: Trying to do science, a scientist may well encounter social phenomena that one would hope not to contaminate the community. Among these are camps with followers and entrenchments against others that all believe and claim to be motivated by science. In the present context, I know of Livermore colleagues who encountered at least one referee (but then repeatedly) who wanted to discourage them from applying more experiments (or modeling computations) to this very problem, since—in the strongly expressed view of the referee—a theoretician (let us call him XG) had finished off the problem for good—and anyway the NIST EBIT experimental results agreed with that material. One of the colleagues, seeing his chances for successful grant applications dwindling, because he could no more publish in the key journal, left the field in disgust. The trust this referee had put into one theorist and one institute is, mildly put, surprising. The referee(s) had conveniently overlooked the fact that XG had produced three papers in two years, with different, partly incompatible, sets of results. Which one of those computations was the definite solution to all problems? Why trust computations that change so quickly? How come the NIST EBIT results had to be corrected in the end so that nowadays the "world view" on this topic is where the Livermore group data and computations had been all the time? Referees of a strange ilk and with deep-rooted prejudices can be major obstacles in the process of developing experiment and computations in dialogue.

3.10. Mg-Like Ions

Just as the Na-like ions with their n = 3 valence electron are largely analogous to the Li-like ions with their n = 2 valence shell, a general assumption is that Mg-like ions are not so much more difficult to calculate than Be-like ions. However, Vilkas *et al.* [192] have intercompared four computations reported in the literature up to high values of *Z*, and were surprised by some massively different predictions of the isoelectronic trends of the 3s3p $^{1,3}P^o_1$ level energies (the resonance and intercombination transitions)—see their Figure 1. The authors of those discordant computations are known for good work in many other cases. It has not transpired to me what might explain why several computations went so far astray. However, the case reminds us customers of atomic structure computations to remain cautious also about computations performed by renowned experts. On the other hand, dedicated MBPT, MR-MP, or RCI computations nowadays almost certainly will yield results close to experiment. What went wrong in those computational surveys?

The valence electron in Mg-like ions is ascribed a penetrating orbit; hence the nuclear charge the electron experiences is less well shielded by the inner electrons than otherwise. The result is a shorter level lifetime than in a Be-like ion of similar ion charge. For example, the $2s^2$ - 2s2p intercombination transition in B^+ reveals an upper level lifetime of about 100 ms [193] while the corresponding decay curve of the Al^+ $3s^2$ - 3s3p intercombination transition features a lifetime of about 0.3 ms [193,194]. The latter lifetime is close to the short-lifetime limit of practical work with a heavy-ion storage ring, which unfortunately precludes further such measurements along the isoelectronic sequence. Nevertheless, it is a comforting thought (for experimenters) that two very different experiments (one using a radiofrequency ion trap, the other a heavy-ion storage ring) have yielded practically identical lifetime values—which, by the way, in this case theory is struggling to reproduce.

The measurement situation for the 3s3p $^1P^o_1$ level is very similar to that encountered in the Be isoelectronic sequence, with the same cascade pattern and even a few additional cascades. It is difficult to muster equipment for observing and measuring all the important cascades that play a role after the excitation of a fast ion beam by passing through a solid foil. Quite some time ago the alternative of a population and cascade model has served beam-foil spectroscopy well—with the accordingly somewhat limited accuracy of the results [19]. However, lifetime uncertainties of about 10% are quite respectable after all, and not (yet) easily surpassed by theory (computation) for arbitrary levels.

3.11. Al-Like Ions

Al-like ions have been of interest to spectroscopy in many aspects, the line intensity anomalies resulting from level mixing are an example already mentioned above [150]. In the story of the identification of intercombination transitions in the EUV spectrum of the solar corona [195,196] from beam-foil spectra recorded in the accelerator basement, the 3s 2 3p 2 P o - 3s3p 2 4 P lines have played a key role, because their wavelength and intensity pattern was so characteristic. Mg-like spectra of iron-group elements have a single intercombination transition, which may be bright (and was known from tokamaks), but a single line is difficult to identify. Al-like spectra, however, feature five lines in a reasonably well predictable wavelength and intensity pattern, a fingerprint. With increasing nuclear charge, the fine structure splitting drives the lines apart, but some of them remain close enough to each other to fall into a wavelength range covered by a single position sensitive detector. In this way the lines have been followed in beam-foil spectra up to Z= 79 (Au), and some of the lifetimes have been measured [197]. For the Al isoelectronic sequence, the high-*Z* trend of Huang's MCDF computations was found closer to experiment than similar computations for the other isoelectronic sequences done in the same period. This was likely a chance success, related to a systematic trend: ions with fewer electrons showed one sign of deviation and ions with more electrons deviated to the other side. Reliable *ab initio* computations of those sequences have appeared only some two decades after Huang's original work (see, for example, [192]).

At lower *Z* one might expect that all relatively low-lying levels of Al-like ions are known. Santana *et al.* [198] have calculated (using the MR-MP formalism) the lowest 40 levels of Al-like ions in the iron group (K (Z = 19) through Ge (Z = 32)) and compared the results with the NIST data base. For elements from Ti through Ni, almost all of the levels of the 3s 2 3p, 3s3p 2 , 3s 2 3d, 3p 3 , and 3s3p3d configurations were already known, whereas at a little lower or higher *Z*, the knowledge was and is rather patchy. It is a pleasure to see how closely these *ab initio* computations match measurement, fill gaps, and suggest revisions with almost spectroscopic accuracy. Interestingly, for several ions certain groups of levels differ significantly more (0.5% to 1%) from computation than the majority of others (less than 0.05%), suggesting that the experimental data quality may have been poorer in these cases (there also were apparent misidentifications). It might be possible to close some of the data gaps by going back to the original data recordings and now use the guidance provided by good computations. The quality of the MR-MP computations also shows in the fact that it was possible to calculate (by *ab initio* means) the ground state fine structure in Fe XIV correctly

[199]. A comparable set of computations of Al-like ion levels and transition rates has since been executed by C. Z. Dong's group at Lanzhou, using the GRASP2K code, but I have not seen any published results.

Encouraged by the success of the atomic structure work, MR-MP computations have been used to address the decay channels of those levels the lifetimes of which are largely determined by spin-forbidden and/or E1-forbidden decays, that is the $3s^23p\ ^2P^o_{3/2}$ level, the three $3s3p^2\ ^4P_J$ levels, and the $3s3p3d\ ^4F^o_{9/2}$ level. The latter has long been overlooked as the origin of a cascade with almost the same lifetime that might spoil lifetime measurements of the $3s^23p\ ^2P^o_{3/2}$ level. The likely influence of this cascade is a lengthening of the apparent lifetime, and the effect would be most pronounced in an environment in which the cascade level is more highly populated. Such a pattern is, in fact, apparent in the experimental data [200,201,202,203,204,205], as has been discussed in more detail elsewhere [12]. In Fe XIV the cascade lifetime (calculated by FAC [185] and MR-MP codes) is too close to the primary to be discerned in the decay curve of the $3s^23p\ ^2P^o_{3/2}$ level. However, the cascade lifetime scales differently from the primary decay and the cascade has thus been confirmed in heavier ions nearby [206]. The cascade contribution is smallest in the excellent vacuum of an EBIT, but not really negligible. The most accurate lifetime measurement on Fe XIV [205] is, indeed, on the slow side of theoretical expectation, as would be in the (neglected) presence of the cascade. To repeat a point mentioned above: There also is a need to take the EAMM correction to the M1 transition moment into account, and when one does that, the disagreement of the lifetime measurement result with theory increases and actually becomes quite sizeable (see Figure 11).

Similar 3d cascades plague the lifetime measurements of ground configuration levels also in Si-, P- and S-like ions [206]. However, the 3d levels, especially those of high total angular momentum *J*, have hardly been covered in systematic computations and thus probably are also missing from most collisional-radiative models.

3.12. Si-Like Ions

In Si-like ions, the lowest configurations are $3s^23p^2$, $3s3p^3$, and $3s^23p3d$. The lowest $3s3p^3$ level is a quintet level, $^5S^o_2$, which is of atomic structure interest for reasons similar to the $2p^3\ ^5S^o_2$ level in C-like ions discussed above. At the beginning of the isoelectronic sequence, the two intercombination decays of this level have been known for quite some time. Beam-foil spectroscopy has helped to locate the corresponding lines in the EUV spectra of iron group elements and thus

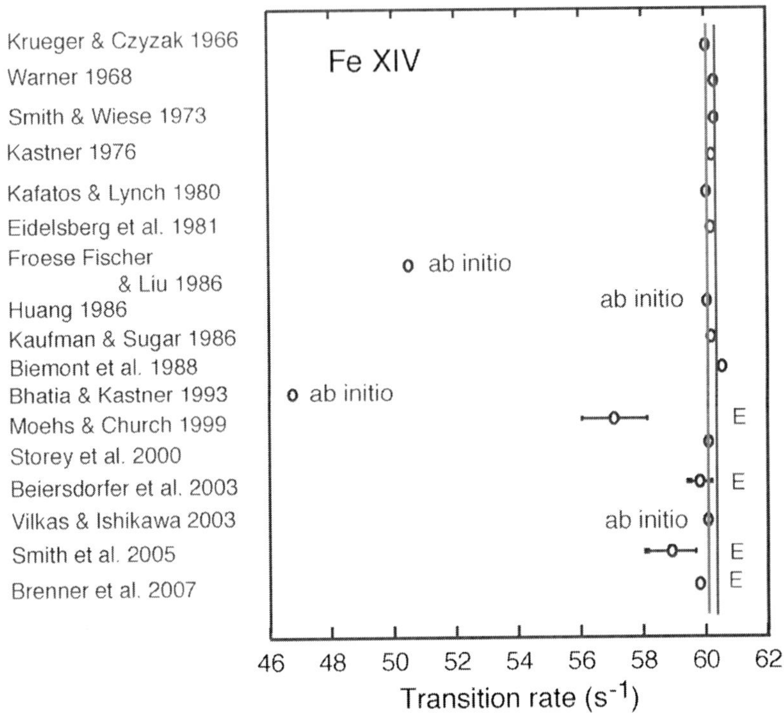

Figure 11. Timeline of the transition rate of the ground state fine structure transition in the Al-like ion (Fe 13 + of iron. The four experimental data points result from two experiments with an electrostatic ion trap and from two that used EBITs. In the latest one [205] the error bar is smaller than the symbol size. The vertical red line (left one of the pair) shows the result of naive theory combining the experimental transition energy and the single-configuration picture line strength. The accompanying dark blue line indicates the same expectation, but corrected for the EAMM contribution from quantum electrodynamics (QED). Note that the best experimental data point does not agree with what presently is seen as a correct theoretical prediction. All of the non-*ab initio* computations fall close to these two lines, because they have been adjusted to the experimental transition energy. Of the four *ab initio* computations only one (MR-MP) [199] coincides with the range of successful experiments by merit (and still misses the EAMM correction). The computational result provided by Huang [67] coincides with experiment as well, but by comparison with other atomic systems one has to conclude that these computations are not accurate enough to call this more than a chance agreement. (Figure adapted from Figure 1 of [198], which includes a list of all the references.)

also in the solar corona [195,196]. However, in Si-like Fe, one of the two lines is blended with a strong He line in the corona, and therefore the accuracy of determining the level position and the lower level fine structure splitting is jeopardized. Elements heavier than Fe ($Z = 26$) are less abundant in the solar corona; consequently the reliability of extrapolations of the atomic structure data suffer. Given that the decays of long-lived levels are not very bright in time-resolved observations such as by beam-foil spectroscopy, and that there are many unidentified lines in the same wavelength range, certain line identifications may be more tentative than one would like. A Swedish-Chinese collaboration assumed the upper hand in the identification process of the intercombination transitions, because they had wavelengths of four elements on a straight-line plot which they saw as proof by isoelectronic trend [207]. However, the newly claimed trend was much steeper than that of the (many more) lower-Z data and also incompatible with any typical atomic structure computational trend. In spectra with many unidentified lines one may occasionally find a sequence of several lines that fit to a common reference by chance, but atomic data analysis would not have gotten to its present state of achievement by chance. Renewed analyses of experimental data [208] as well as extensive MR-MP computations up to Au ($Z = 79$) [209] have since corroborated the orthodox data interpretation.

Si-like ions have another term that has been difficult to handle, $3s^2 3p3d\ ^3F^o$, with three fine structure levels of total angular momentum values $J = 2, 3, 4$. The $J = 2$ level is nothing special and well enough known since a long time. However, the selection rules for radiative transitions render the $J = 3$ level much longer lived than the typical neighbors, and the $J = 4$ level is even metastable, without any E1-allowed decay channel. Biémont's 1986 Cowan code computations [210,211] tried to place the $3s^2 3p3d\ ^3F^o$ term and estimated the fine structure splittings. Guided by these estimates, Träbert identified candidates for the $3s^2 3p^2\ ^2P_2 - 3s^2 3p3d\ ^3F^o_3$ transition in delayed beam-foil spectra, that is, he used the predicted nanosecond lifetime characteristic to enhance the contrast to the surrounding lines which should have lifetimes in the few-dozen to few-hundred picosecond range. The beam-foil observations led to line identifications in spectra of the solar corona that had wavelengths already known with higher accuracy [212]. Again guided by Biémont's calculated fine structure intervals one could then obtain estimates of the $3s^2 3p3d\ ^3F^o_4$ level positions. This level should feature a lifetime in the millisecond range (for many iron group elements), which is much too long for beam-foil spectroscopy, but perhaps measureable in an ion trap. About a decade later, Vilkas and Ishikawa began to apply their MR-MP code to Si-like ions [213,214], and they noted gaps in the NIST data base as well as discrepancies of literature data on the $3s^2 3p3d\ ^3F^o$ term compared to their computations. The latter corroborated Träbert's aforementioned analysis. However, in spite of this progress, nobody has positively identified any of the decay branches of the $^3F^o_4$ level yet. Träbert *et al.* [206] have discussed how the influence of the level decay shows in the decay curve of the $3s^2 3p^2\ ^1D_2$ level, but the decay can take many paths, each of which is elusively weak. The latest dedicated computations on this problem [215] have yielded predictions many of which seemed consistent with weak lines in the solar spectrum at the time, but continuing work with the *Hinode* spacecraft data have led Giulio Del Zanna to state that the comparison with the solar data should no longer be considered conclusive [216,217]. Of course, the solar corona is a low-density environment, and such levels will not be populated very much, leaving all decay channels weak. We thus have a situation in which an experiment at a heavy-ion storage ring clearly demonstrates the cumulative effect of the weak decay channels none of which has individually been observed. (The heavy-ion storage ring also provides a low-density environment, except for the stripper foil in which the high charge state ions are produced. That is the decisive element that boosts the metastable level population so that the contribution is visible and that at the same time precludes a clean measurement of the ground configuration level lifetime.)

Again, as in Fe XIV, the 3d cascade in Fe XIII has about the same lifetime as the 3p level of primary interest. Experimental information can be gained from ions slightly offset along the isoelectronic sequence, because the isoelectronic scaling of the decay paths is different. In such a situation, good transition rate computations are very helpful, even as the measurement does not test the computations with high precision. Unfortunately there are only very few published computations of the 3d level lifetimes. The crossing of the lifetime trends of 3p primary and 3d cascade near Fe is evident in a step function of the experimental lifetime data [206] (relative to any one of the computations). Over all, the scatter of the calculated values for the 3p level lifetimes in Si-like ions is considerable (on the order of ±10%). The MR-MP computations come out best of the *ab initio* ones, but not as close as some scaled ones. The latest computations, by Charlotte Froese Fischer [218], promise progress towards accurate results, but they address only three elements (Fe, Co, Ni).

There also is an interesting statement in that last paper that for me illustrates a difference in world view between experimenters and theoreticians (of course, the quotation is out of context and not intended to be read the way I do): "Experimental lifetimes may be either shorter or longer but generally are within 10% of *ab initio*." For this theoretician, the measure of things appears to be the result of an *ab initio* atomic structure computation, and from there one looks around and probably finds (reasonably well) measured data in the neighborhood. For me as an experimenter, the goal is to do an experiment that comes as close to the (unknown) correct result (nature) as intrinsically possible. Then I can contemplate what a computation yields that mimics a correct description of nature from first principles (*ab initio*), or a computation by whatever means that includes adjustments which "usually yield some sort of agreement with measurement". The goals are to know the correct result (factual knowledge for whatever purpose) and to improve the *ab initio*computation (a physical description of nature) until it matches the correct result, a process that (so we hope) yields (physics) insight. Thus *ab initio* computations *per se* are of limited interest, unless they are intellectually tied to a target. (Under that condition, adjustable computations (such as [219]) may be and have been *helpful*, but only *ab initio* computations are meaningful.)

Recently the venerable non-relativistic CIV3 code has found its way also to Egypt, and it has been applied to two sections of the Si isoelectronic sequence [220,221]. I would be surprised if the range of the latter paper (Z = 30–33) was not served better by relativistic computations. Most of the aforementioned computations have been limited to rather low lying levels in the n = 3 shell. Vilkas and Ishikawa [222] apply their MBPT computations also to n = 4 levels. Much more needs to be done on both sides, experiment and computation.

3.13. P-Like Ions

The $3s^2 3p^3$ ground configuration of P-like ions features a $^4S^o_{3/2}$ ground level, two relatively short lived $^2P^o$ levels and two relatively long lived $^2D^o$ levels. Computations of this system have evolved from semi-empirically adjusted to *ab initio*[223,224, 225,226,227]. The wavelength and lifetime pattern has advantages and drawbacks for the measurement of various ions of the iron group, as might be expected. Surprising to me is that the various computations have problems with the individual levels of the same configuration. Some of the results of the MR-MP computations differ from the trend of the others at low Z, and the results also differ from measurement more than the others (by up to a quarter) for the longest-lived of the ground configuration levels (see [206]), the $^2D^o_{5/2}$ level, while the experiment seems straightforward in this case. This observation raises questions about code debugging, considering that the same code has often produced excellent results in other circumstances. With the P-like ions, the MR-MP computations are among the very few that have been used to address the problem of a few long-lived 3d levels, and there the results seem fine.

3.14. S-Like Ions

Traditional, mostly scaled, computations [228,229,230,231] have approximated the observed spectra of S-like ions, but near-spectroscopic accuracy by *ab initio* computations has only recently been approached [232]. Similar to the corresponding C- and O-like atomic systems, there are structural similarities between Si- and S-like ions. The various spectra of Fe ions with an open n = 3 shell are being re-analyzed by Del Zanna and colleagues [233,234] with the help of adjustable computations and solar spectra (mostly those recorded by the H i n o d e spacecraft) and with the aim to understand the very same solar spectra in detail. As a long-term project, the CHIANTI data base [280,281] is continually upgraded, and meanwhile it has surpassed the quality of the NIST data base [279] in the EUV spectral range. However, the two data bases have been set up with different intentions: The NIST data base aims at recording data for all elements, while CHIANTI aims at understanding the solar spectrum, which implies that elements of different abundance are of different interest to solar research.

As concerns the E1-forbidden decays in the ground configuration of S-like ions, a striking contrast to Si-like ions appears. Instead of a single slow 3d cascade there are six long-lived ones, widely spaced in lifetime [206]. ('Long-lived' meaning the millisecond range when most other levels in the neighborhood feature picosecond lifetimes.) They have been predicted by MR-MP computations, and they are so strong and so numerous that the 3d cascades seriously affect any evaluation of 3p ground configuration decay curves. Any meaningful collisional-radiative model of S-like spectra has to include the 3d levels, but hardly any large-scale computation is available that does.

3.15. Cl-Like Ions

The ground state of Cl-like ions is only one vacancy away from the closed-shell Ar-like system with 18 electrons. Nevertheless, the atomic structure of the excited ions is complex enough to be demanding for computations. Various survey computations have addressed the isoelectronic sequence (see references in [235] and in the specialized publications cited below). Fe X, the Cl-like spectrum of Fe, is rather prominent in the quiet solar corona. Consequently, it has been studied in observational detail [236,237] and by a large-scale *ab initio* MR-MP computations [238].

While Fe X as a topic of a single-spectrum presentation seems well justified by its coronal prominence, there are other studies of individual Cl-like ions which reflect the respective authors' efforts, but have ever so little predictive value. In fact, one of the publications states (on the single atomic lifetime value for which an experimental result is available for that particular ion species) that the calculated result is close to results of this measurement and another computation, and that yet other computations had other (cited) results. There is no explanation, no factual discussion, just the cited rather pointless statement. The results of this computation have been obtained with operators in length and velocity form, and the two results differ by almost a factor of two. The computational uncertainty is not discussed. No, this is not a paper from the dark ages, this is a 2007 publication (not cited here on purpose), using a (good) code (CIV3) as learned by the lead author decades earlier, but it has not been put to work to gain insight. Lest the reader should get me wrong: this paper reported on an apparently good and valid calculation—there are much worse. (By the way, the simply mentioned "other results" scatter by a factor of three or more, as will be discussed below.)

What should be best calculable in Cl-like ions are the lowest levels and the transition rates between them. The fine structure transition in the ground configuration is, of course, technically similar to that in the F-like ions, with a $J = 1/2$ upper level (low statistical weight, and hence a low line intensity). The transition rate is dominantly M1 with a small E2 admixture. In Fe X, this line is the 'red coronal Fe line' which unfortunately lies just outside the good working range of efficient low-noise photomultiplier tubes that have been optimized for the visible range. The level lifetime in Fe X has nevertheless been measured at the Heidelberg EBIT [205]. From measurements of the next heavier elements at the Heidelberg heavy-ion storage ring TSR not only three isoelectronic lifetime results (on Co XI, Ni XII, and Cu XIII) have been obtained, but also—by extrapolation—another one for Fe X [239]. The results (with uncertainties below 1%) are close to theoretical expectation (including the EAMM correction), but - just as for the B-, F-, and Al-like systems discussed above—the non-QED computations are not accurate enough to actually test the complete computational atomic structure package.

There is another test bench in Cl-like ions, and this one is unique so far (in the course of this discussion; there must be many atomic systems with similar features). The lowest excited level above the ground term is the displaced level $3s3p^6\,{}^2S_{1/2}$ with a single electron outside closed subshells and only two decay branches to the two fine structure levels of the ground term $3s^2 3p^5\,{}^2P^o$ where there is a single vacancy difference from a rare gas electron configuration. This arrangement offers a variety of tests, the simplest of which would be the level energy of the 3s3p 6 2 S 1 / 2 level, which is easy to do and only requires the transcription of published tables into work sheets. Let us try something a step more advanced that involves transition rates. The sum of the two transition rates towards the ground state yields the inverse of the level lifetime, and the ratio of the two transition rates is the line ratio. In moderately high charge states, the line ratio is expected to follow a smooth isoelectronic trend, and when using a grazing incidence spectrometer for the observation, no gross discontinuities of the reflectivity or detection efficiency are expected to occur over the narrow wavelength interval of this line pair. A beam-foil study has found agreement among the predicted and measured line ratios within about 10% and agreement of the measured lifetime with some theory at the same level [240]. However, the computations available at the time spanned a factor of two between the various absolute lifetime / transition rate predictions. (The mean of all predictions would have been fine.) Also, the isoelectronic trends of the predictions differed notably. There have been several new computations since then that have confirmed the same picture (with less scatter), but without new lifetime measurements in the Fe group elements (the Bochum experimental set-up has since been forced out of the laboratory), no new insight is expected for those moderately high charge state ions in the near future.

Figure 12 shows the state of affairs of experiment and theory concerning the 3s3p 6 2 S 1 / 2 level lifetime in Cl-like ions. While for moderate charge states the computations merge with the experimental data trend, the various computations diverge badly at low *Z*. Near-neutral ions are more difficult to calculate, but possibly more easily measured. However, the available lifetime data for Ar II differ by much more than their error bars from each other and by large factors from computations available at the time (see plots in a study geared towards low charge states [235]). A single (beam-foil) lifetime value exists also for Ti VI [241], which was followed 26 years later by a single-element computation [242]). Meanwhile a further beam-foil data point for K III has been added by H. P. Garnir at Liège ($\tau = 2.08 \pm 0.2$ ns) [243]. (Unfortunately, the lead experimenter has deceased since.) Most importantly, in a synchrotron experiment Ar gas has been ionized and excited, and the level lifetime in Ar II has been obtained with much improved accuracy [244]. While new computations [245,246] have claimed improved agreement with the beam-foil data, they did not include this most significant data point.

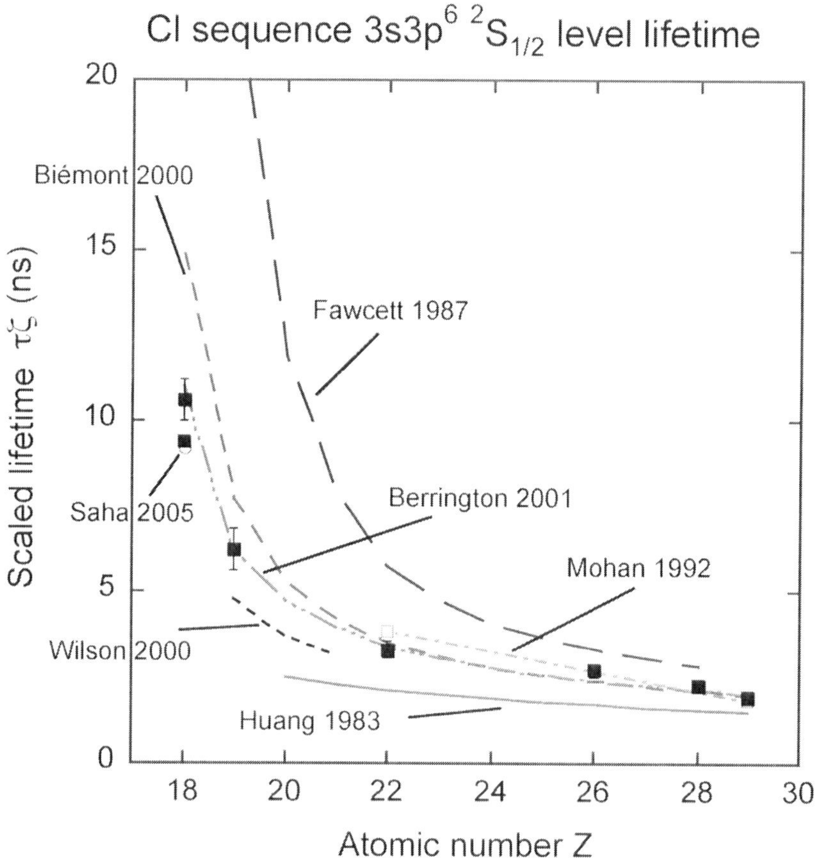

Figure 12. The lifetime τ (scaled by the ion core charge ζ) of the 3s3p 6 2 S 1 / 2 level lifetime in Cl-like ions. Only a selection of experimental data (full squares) and computational results (open symbols for Ar, Ca and Cu, and eye-guiding lines; most of the computations in the literature do not cover low-*Z* systems) are shown. Note how at low *Z* the lifetime predictions diverge by up to an order of magnitude. The computation by Berrington *et al.* comes closest overall to the low-*Z* measurements—except for the measurement on Ar II by Lauer *et al.* [244] (the error bar is smaller than the symbol size) that Berrington did not know of at the time. Saha and Fritzsche, however, matched that one data point well by their computation, but they calculated only this one. The computations by Wilson seemed better than most others at the time—but covered too few elements. Mohan did well in 1992—but why were only two elements computed?

Before too long, theory has corroborated this data point, too, but, alas, the successful computation [247] was done for this one ion only. I am repeating myself: What predictive value is in a computation that follows experiment and covers only the cases that experiment has already treated (here a single ion species)? It is good to know that an advanced atomic structure package has been used and that it has found a result in agreement with a measurement that had not been reproduced by computations before. But why (yes, it costs manpower and money) was the study not extended to a sequence of ions in order to yield an isoelectronic trend and predictions for other ions for which experiment had not yet told a result?

3.16. Ar-Like Ions

The Ar-like spectra do not figure as prominently in tabulations as the other sequences that have been discussed so far, although they are of interest for the interpretation of the cooler parts of the corona [248,249] and have repeatedly been modeled for line intensities (see work by Helen Mason). Spectral modeling usually proceeds with not the most complex type of computation, because one needs to identify candidate lines with peculiar properties under ranges of specific environmental conditions, usually being measured without extreme spectral resolution, other objectives ranking higher. However, there are many lines to consider, which consequently are individually weaker and thus more difficult to measure. Of the three resonance lines of Fe IX, $3p^6\,^1S_0$ - $3p^5 3d\,^1P^o_1$, $^3P^o_1$, and $^3D^o_1$, with very different upper level lifetimes, only two have been seen so far apparently. The strongest of the three resonance lines in Fe IX and Ni XI can be seen in beam-foil spectra [212,250]. The Ar-like spectra appear at relatively low ion beam energies at which the thin carbon foils do not last very long under heavy-ion bombardment, posing a very practical limit to data accumulation. On the other hand, work at the Livermore EBIT has yielded lifetime information on 3d levels in Ar-like ions of Kr (Kr XIX) that have major M2 decay branches [251]. Ar-like spectra are rich in lines, but only very partially known. For the present context, they are still too complex and have not yet been treated well enough.

We now leave the discursive sequence of isoelectronic sequences and jump to cases with many more electrons in total, but with only a few electrons outside of closed shells.

3.17. Cu-Like Ions

The third isoelectronic sequence treated by Kim *et al.* [100] and by Blundell [101] after Li I and Na I is not K I or Cs I (each with a single electron outside a rare-gas core configuration), but Cu I, with a single electron outside of a closed 3d 10 shell (Ni I sequence). At the Livermore EBIT, the 4s 1 / 2 - 4p 3 / 2 resonance line has been followed along the isoelectronic sequence up to Z = 92 , and the data [103,252] have been compared to the veteran computations. For most of the way, the computations are as close to each other as to experiment, but at the highest values of Z, the description by Blundell has worked out better. It is also noteworthy to realize that a fairly small EBIT device can provide Cu-like ions of U, whereas accelerators for beam-foil spectroscopy or lasers for laser-produced plasma work had to push their performance envelope to work in that range. Figure 3 of [103] also shows the systematic errors of the high-*Z*non-EBIT data. If one looks for the computations associated with those other data, one can recognize that some of them had a bias in order to improve agreement with experimental data that by now can be seen as systematically flawed. This should serve as a reminder that ultimately computations with "adjustments" are to be avoided, because they may become misleading.

For the resonance transitions in the Cu I sequence, the deviation between experiment and *ab initio* computation has become smaller than 200 ppm.

3.18. Zn-Like Ions

The same experiments that have yielded excellent agreement of measured and calculated Cu-like ion resonance line energies provide data also on Zn-like ions, but the typical energy deviation has been on the order of 6000 to 8000 ppm, or 40 times more than for the Cu I sequence. Based on the very same spectra and the same calibrations, this wider gap has to be blamed on computations, of which there have been many (see [103,252], plus recent work at Notre Dame, Puerto Rico, or Livermore). Maybe the demonstration of the computational shortcomings has helped to renew interest in such problems on the theory side, and to new attempts and significant progress. From my perspective (possible memory lapses nonwithstanding) the story unfolded about as this:

In 2005, Vilkas and Ishikawa [253] employed their MR-MP second-order perturbation theory algorithms (plus QED corrections) to calculate the structure of high-*Z* Zn-like ions. Their technique was based on Gaussian basis sets [254,255]. The deviations between the results of these atomic structure computations and a number of experiments shrank by one order of magnitude to a band of ±1000 ppm; most Livermore EBIT results and the computations differ by less than 500 ppm. In 2008, Blundell *et al.* [256] published MBPT results that came closer to experiment than most earlier computations, with the exception of Vilkas and Ishikawa's. The deviation was about 500 ppm for W ($Z = 74$) and increased to about 1000 pm for U ($Z = 92$). In 2009, Blundell re-assessed the QED contributions and improved the agreement with experiment to better than 100 ppm [257]. In 2010, Safronova and Safronova, Blundell's coauthors of two years before, presented computations for more ions [258]; where the ion species were the same ones as before, the numbers remained unchanged. Their results and the isoelectronic trend pointed clearly away from the experimental EBIT data, but they matched a set of experimental results from laser-produced plasma work with their larger uncertainties and a systematic error trend (possibly related to the Doppler shift problem in observing a rapidly expanding plasma.). In the same journal issue, Chen and Cheng [259] have presented the results of their RCI computations which (with the exemptions of Os ($Z = 76$) and Pb ($Z = 82$)) agree with all Livermore EBIT data, even slightly better than Blundell's results. The Chen and Cheng results furthermore include elements for which no measurements have been performed yet. At Puerto Rico, the Gaussian basis set has been expanded; as a result, an oscillation of the MR-MP computational results for the resonance transition energy in Zn-like ions around the trends of the Blundell and the Chen and Cheng computations has thus been significantly reduced. The improved results have been published in 2011 [260]; a new EBIT measurement (on Pt ($Z = 78$)) and independent computations for the same ion having been presented already in 2010 [261].

In short, several types of computations of the resonance transition energy in high-*Z* Zn-like ions ($70 \geq Z \leq 92$) have now been brought to the same quality as reached for Cu-like ions decades earlier. Similarly good computations for the *Z* range below are still needed, as well as experimental data on other transitions.

3.19. Ga-Like and Ge-Like Ions

The same MR-MP computations for Pt [262] also cover Ga- and Ge-like ions, corresponding to the Al- and Si-like ion work discussed above. Resonance (4s-4p $3/2$) and intercombination (4s-4p $1/2$) lines, respectively, of ions with one, two, three, or four electrons in the valence shell usually lie in close proximity to each other. This makes it attractive to analyze them together, but it provides/implies a challenge to theory. For the experiment it makes no fundamental difference whether a line arises from a system with a few or many electrons, but for computations, the difference is decisive, and it is a very good test. The experience has regularly been sobering: Li-, Na-, Cu-like spectra can be calculated with high accuracy, and each additional electron brings about a major degradation of the computational accuracy. The Zn-sequence spectra have experienced a major improvement only very recently (as discussed above), but this progress has not reached all atomic structure codes on the market. In fact, the readily observable signature lines in the few-electron spectra are a useful tool to identify by eye and to measure the offset between computations and measurement for the more complex systems [34,192,195,196,263,264,265]. The easily identified line pattern unfortunately disappears along the isoelectronic sequence, because the fine structure intervals grow much faster ($\propto Z^4$) than the gross structure ($\propto Z$). However, the fine structure intervals of the doublet ground term and the lowest quartet term in the systems with three electrons in the valence shell are rather similar, and thus at least certain line pairs or line groups survive for visual recognition. Can we expect theory to treat the many-electron systems equally well in the near future? The individual codes scale differently in the computational demands for more electrons. RCI may do best, but at a high cost. MR-MP does pretty well, as do other MBPT codes if supplemented with a lot of care. There are other codes that may be sufficient for a given purpose.

The intercombination line pair in Ge-like ions is a topic in which experiment still asks theory for assistance. The two lines are not particularly bright, and the wave number difference (the energy difference of the $4s^2 4p^2\ {}^3P\ J = 1$ and $J = 2$ levels) is not known from experiment. In beam-foil spectra, many lines fill the neighborhood, and the line identifications for some of the ions are tentative only. When trying to survey the data base, I have found actual entries of explicitly "estimated" quintet level positions, and by my nearby observations I could tell that the estimates were wrong. The spectra of ions with two, three, or four valence shell electrons appear to provide a useful test of performance quality. I hope that improved computations will become available for a re-analysis of existing spectra so that the frontiers of a reliable data set can be expanded.

4. PM-LIKE, SM-LIKE, EU-LIKE IONS

And now a little recreational digression. A third of a century ago, Curtis and Ellis have come up with a neat idea. Although the neutral atom promethium (Pm) does not have a single valence electron outside of closed shells only (in this case, filled up to $4f^{14}$), high charge state ions in the Pm isoelectronic sequence might feature a 5s-5p pair of resonance line analogous to the alkali atoms [266]. Beam-foil spectra of Au recorded at Bochum [267] showed clusters of very many lines among which it was not possible to pick out the right ones with confidence, but it became obvious that the lines of interest were not as prominent as the resonance lines in the simple alkali spectra. A few years later, Bochum demonstrated that in the adjacent isoelectronic sequences (Sm, Eu, Gd) the intercombination transitions might be filtered out by the time-resolution inherent to beam-foil work, and among many weak lines, candidates were found with level lifetimes of the predicted magnitude [268]. More beam-foil work was tried at RIKEN in Japan, EBIT work (fewer lines) at Berlin, a tokamak observation on U thrown in, and the computations were upgraded. MR-MP computations by Vilkas and Ishikawa [269] indicated that, with the possible exception of the heaviest elements, the 5s-5p line pair would not be easily recognizable—their upper and lower levels were embedded among hundreds of other levels, and no parameter made the lines stand out. While the partial selectivity of the excitation process in EBITs (the spectra are heavily weighted towards ground state transitions) renders most EBIT spectra much simpler than the extremely line-rich beam-foil spectra, this is not good enough to produce the 5s-5p lines prominently. The Berlin EBIT group practically ascertained that they could optimize their EUV spectra of tungsten for specific charge states, but they did not see lines in good agreement with the wavelength predictions for the 5s-5p transitions. This should be no surprise, as has repeatedly been discussed (see [269]). In the latest twist of the story, a Japanese group has calculated that there is another reason, too, in that (at least for Pm-like ions of Bi ($Z = 83$)) there is a long-lived level that acts as a population trap and hinders the population of the signature levels [270].

If one cannot identify the lines in the observed spectra, it is difficult to judge the accuracy of the computations. EBIT spectra with their much fewer lines carry much less potential wavelength information, but what there is, can be mined with higher accuracy (no Doppler shift etc.). However, one has to model theoretical spectra for the excitation conditions in an EBIT. Typical modeling codes are based on HULLAC or FAC computations which are not as accurate in atomic structure terms as are dedicated large-scale atomic structure computations. With hundreds, if not thousands, of levels to consider, it seems highly desirable to develop an interface that feeds the output of accurate structure computations into the data set for the collisional-radiative modeling exercise. This is a tedious task, but it would be helpful for the study of many multi-electron atomic systems.

A much less tedious task, with a much shorter time line, would be to refrain from pretending that the resonance lines of any Pm-like ions are of any likely future use in the diagnostics of fusion reactors. Since decades it has been established that these lines are not prominent (if seen at all, which in most cases remains uncertain). The hot (Doppler broadening) multi-element environment of a fusion plasma won't make the observation and application any easier.

5. W

In the course of development work on controlled fusion, a long-term mantra has been that any material of a nuclear charge higher than Z = 3 must be avoided, because it would cause unbearable radiative losses to the fledgling plasma and thus hinder fusion conditions from being reached. However, the vacuum vessel would likely be built of steel, and this fact motivated data collection on the spectra of Fe (Z = 26) and elements that would likely be used in special alloys, from V (Z = 23) and Cr (Z = 24) to Mo (Z = 42). This purpose is still recognizable in the abundance and distribution of data in prominent data bases. In order to keep contaminations out of the plasma, the vessel surfaces were then coated with light elements such as Be, B, or C. Nowadays the power density of the fusion test beds has become so high that the robustness of the vessel under heat load and neutron bombardment is a primary concern, and now tungsten (W, Z = 74) as the most heat-resistant element is touted as the inevitable material for coating certain surfaces. Tungsten offers the additional advantage that it will not be fully ionized in the center of the hot plasma. Therefore spectral observations of W contaminations can be made in all locations of the plasma, and the spectra will tell about the temperature (by charge state distribution) and other plasma parameters without any need to inject additional contaminants. This set of ideas has initiated an avalanche of studies of the various spectra of W and of charge exchange (CX) processes. NIST Gaithersburg has undertaken efforts (first by Joseph Reader and now by Alexander Kramida) to catalogue the available data [271] as well as pursuing some computations and measurements of their own (Yuri Ralchenko). The International Atomic Energy Agency (IAEA) has supported seminars at which people interested in studying W supposedly coordinate their efforts. In his time at the Livermore EBIT group, Joel Clementson has started a tungsten project ("WOLFRAM") with experiments at various plasma machines, and among other things he has demonstrated high-resolution spectra of the famous E2 and M3 lines in Ni-like W (first recognized years ago in EBIT spectra of Th and U [272]) which are discussed for ITER diagnostics [273,274,275,276,277]. Besides the observers/experimenters, theorists have risen to the challenges of multi-electron system atomic structure. For example, Marius Vilkas at

UPR has used the MR-MP code to calculate levels and transition rates of W ions with charges 7+ (Ho-like) to 15+ (Pr-like), but the results were too rich to be published in print. Jon Grumer at Lund is working on the next higher charge states using MCDF codes. The Safronovas have addressed Ne-, Na-, Mg-, Cu-, Ag-, Pd-, Er-, Yb-, Y-, Pm-like spectra of W (and probably even more), and so on. (I am not going to add a catalog of my own.) There are spectroscopic details of some individual studies that ought to be taken up by data bases. However, besides some lighthouse features there are thousands of lines in measured spectra and computations, and nobody has the resources to actually combine the two reservoirs in detail. Although it would be good to extract the information once and for all and have it available for easy retrieval, this is not likely to happen. The fusion plasma community most likely will need collisional-radiative modeling of some quality and fidelity to accurate data, but without the explicit atomic structure background detail. This may be intellectually disappointing to spectroscopists who try their best to extract the detail, but it reflects the limitations of funding. Tungsten spectra have very many lines, but there will not be enough funding available, and certainly not in proportion to the number of spectral lines seen or calculated.

6. DISCUSSION

The task put to me was to discuss how an experimenter judges computations and their results. Theory (computation) and experiment, however, are entangled entities of atomic physics which one cannot readily separate. Beam-foil spectroscopy has thrived, because the ion-foil interaction effectively populates all levels. Unfortunately, the spectra of the excited ions then often are so full of lines (many from the decays of core-excited levels) that any attempt at analysis is overburdened. The technique works beautifully on few-electron ions and on transitions with outstandingly bright lines or with decays slow enough compared to others that (easily achieved) time discrimination helps. A bane of beam-foil spectroscopy is the need for Doppler shift correction of observed wavelengths, which would be relatively easy if enough reference lines were already known. Laser-produced plasmas also suffer from Doppler shifts (in an expanding plasma) which have tainted some measurements that were hoped to determine wavelength standards. Electron beam ion traps have no Doppler shift problem, since the trapped ion cloud is rather stationary. However, the light source operates as a low-density environment in which most ions will be in their ground state or metastable states, and the spectra are dominated by ground state transitions, in a stark contrast to beam-foil spectroscopy (where a high total angular momentum lends a large statistical weight and thus offers plenty of phase space to population processes). Fewer lines in the spectrum avoid many problems of line blends and position determination, but one often needs collisional-radiative modeling computations to ascertain which lines should be seen and which ones should not appear. The difference can be drastic (see and compare [34,192,197,278]). The difference also appears in astrophysical data. For example, the NIST online data base ASD [279] carries the legacy of line intensities obtained by whatever measurement or estimate over the decades, from the days of visual estimates to modern photoelectric detectors. In contrast, the CHIANTI data base [280,281] aims at providing accurate wavelength data of all elements and charge states of interest in the solar corona, but the line intensity data are derived from modelling for specific representative electron densities and temperatures. This wavelength data base combines observations and scaled computations, as the user has to be aware of.

Highly charged heavy ions are much more easily produced and excited in an EBIT (which fits into a moderate-size laboratory) than in a heavy-ion accelerator that has to accelerate ions to so much higher energies than the electron beam in an EBIT—because of the electron-nucleon mass ratio—for reaching the same relative velocity between projectile and target. There are various active collaborations trying to combine experimental data and computational results on the same atomic systems. Such co-operative efforts can be very fruitful, providing a link between observations of nature and theoretical concepts, between atomic structure computations and reality. The toolboxes on either side are not yet exhausted. A major tool, long time in the making, is the selective excitation of multiply charged ions. Some two decades ago the Oxford EBIT group has started a project on laser excitation. Meanwhile the Heidelberg EBIT group has laser-excited E1-forbidden transitions of trapped highly charged ions [282] and has moved an EBIT to the FLASH facility at Hamburg to effect selective excitation of highly charged ions by EUV light from a free electron laser [117]. As already mentioned, this line of work has then been extended to excitation by X-ray light at the LCLS facility at Stanford [187], with results in a clear discordance with theoretical expectation. Much more is possible. Experimenters just like to test (and challenge) theoretical computations of atomic structure and transition probabilities by accurate measurements!

ACKNOWLEDGMENTS

Over the years, the German research council DFG has been very supportive to my work. I have worked in various places, most extensively at the Bochum accelerator laboratory and with the Livermore EBIT group, and I have benefited greatly also from further experimental collaborations and from most helpful theoreticians. Since this report is (partly) an opinion piece, I refrain from naming any colleagues in particular—they are practically all cited below, with some part of other of their work. In a citation list at least twice as long, I might have come somewhat closer to a fair representation of all their contributions to my present topic.

CONFLICTS OF INTEREST

The author declares no conflict of interest.

REFERENCES

1. Curtis, L.J.; Martinson, I. Recent Progress in Atomic-Structure Investigations with Fast-Ion Beams. Comments Atom. Molec. Phys. 1980, 10, 1–8.
2. Martinson, I.; Ellis, D.G. Recent studies of intercombiantion lines in ions. Comments Atom. Mol. Phys. 1985, 16, 21–33.
3. Ellis, D.G.; Martinson, I.; Träbert, E. Intercombination transitions in multiply ionized atoms. Comments Atom. Mol. Phys. 1989, 22, 241–262.
4. Curtis, L.J.; Martinson, I. Lifetime Measurements in Multiply Ionized Atoms. Comments Atom. Mol. Phys. 1990, 24, 213–233.
5. Curtis, L.J. Precision Oscillator Strength and Lifetime Measurements. In Atomic, Molecular, and Optical Physics Reference Book; Drake, G.W.F., Ed.; AIP Press: New York, NY, USA, 1996; pp. 206–212.
6. Träbert, E. Atomic lifetime measurements with ion traps of many sizes. Phys. Scr. 2000, 61, 257–286.
7. Träbert, E. Lifetime measurements of highly charged ions. Phys. Scr. T 2002, 100, 88–97.
8. Träbert, E. Precise measurements of long atomic lifetimes using stored ion beams and ion traps. Can. J. Phys. 2002, 80, 1481–1503.
9. Mannervik, S. Experimental lifetime studies of metastable levels. Phys. Scr. T 2003, 105, 67–75.
10. Träbert, E. Atomic lifetime measurements using electron beam ion traps. Can. J. Phys. 2008, 86, 73–97.
11. Träbert, E. Problems with accurate atomic lifetime measurements of multiply charged ions. Phys. Scr. 2009, 79, 068101.
12. Träbert, E. In pursuit of high precision atomic lifetime measurements of multiply charged ions. J. Phys. B: At. Mol. Opt. Phys.2010, 43, 074034.

13. Träbert, E. Radiative-lifetime measurements on highly-charged ions. In Accelerator-based Atomic Physics Techniques and Applications; Shafroth, S.M., Austin, J.C., Eds.; American Institute of Physics: Washington, DC, USA, 1997; pp. 567–607.
14. Curtis, L.J. A diagrammatic mnemonic for computation of cascading level populations. Am. J. Phys. 1968, 36, 1123–1128.
15. Curtis, L.J.; Berry, H.G.; Bromander, J. A meanlife measurement of the 3d 2D resonance doublet in Si II by a technique which exactly accounts for cascading. Phys. Lett. A 1971, 34, 169–170.
16. Engström, L. CANDY, a computer program to perform an ANDC analysis of cascade correlated decay curves. Nucl. Instrum. Meth. 1982, 202, 369–376.
17. Engström, L. Studies of atomic decay properties using fast ion beam spectroscopy. Phys. Scr. 1989, 40, 17–24.
18. Träbert, E.; Heckmann, P.H. Beam-foil lifetimes of Be-like ions of elements from Mg to S. Phys. Scr. 1980, 22, 489–492.
19. Träbert, E.; Johnson, B.M.; Jones, K.W.; Gregory, D.C.; Kruse, T.H. Beam-foil lifetimes of n = 3 levels in Fe XIII to Fe XVI. II. Analysis of decay curves. Phys. Lett. A 1982, 87, 336–340.
20. Träbert, E.; Doerfert, J.; Granzow, J.; Büttner, R.; Brauckhoff, J.; Nicolai, M.; Schartner, K.-H.; Folkmann, F.; Mokler, P.H. Lifetimes of n = 2 levels in foil-excited Kr32+ and Kr33+ ions. Phys. Lett. A 1995, 202, 91–98.
21. Träbert, E.; Curtis, L.J. Isoelectronic trends of line strength data in the Li and Be isoelectronic sequences. Phys. Scr. 2006, 74, C42–C50.
22. Reistad, N.; Martinson, I. Accurate transition probabilities in ions obtained by isoelectronic smoothing of line strengths. Phys. Rev. A 1986, 34, 2632–2637.
23. Träbert, E. Isoelectronic smoothing of oscillator strengths in the beryllium sequence. Z. Phys. D 1988, 9, 143–145.
24. Curtis, L.J. Isoelectronic smoothing of line strengths in intermediate coupling. Phys. Scr. 1991, 43, 137–143.
25. Hylleraas, E.A. Neue Berechnung der Energie des Heliums im Grundzustande, sowie des tiefsten Terms von Ortho-Helium. Z. Phys. 1929, 54, 347–366.
26. Hylleraas, E.A. Über den Grundterm der Zweielektronenprobleme von H-, He, Li+, Be++ usw. Z. Phys. D 1930, 65, 209–225.
27. Jönsson, P.; Froese Fischer, C.; Träbert, E. On the status and perspectives of MCDF computations and measurements of transition data in the Be isoelectronic sequence. J. Phys. B: At. Mol. Opt. Phys. 1998, 31, 3497–3511.
28. Froese Fischer, C. The MCHF/MCDHF collection. 2004. Available online: nlte.nist.gov/MCHF/ (accessed on 15 Nov 2013).
29. Curtis, L.J.; Ellis, D.G. Predictive systematization of line strengths for the 2s2 - 2s2p resonance and intercombination transitions in the Be isoelectronic sequence. J. Phys. B: At. Mol. Opt. Phys. 1996, 29, 645–654.
30. Martinson, I.; Gaupp, A.; Curtis, L.J. Comments on the Be I 2s2 1S - 2s2p 1P transition probability. J. Phys. B: At. Mol. Phys.1974, 7, L463–L466.
31. Irving, R.E.; Henderson, M.; Curtis, L.J.; Martinson, I.; Bengtsson, P. Accurate transition probabilities for the 2s2 1S – 2s2p 1P transition in Be I and B II. Can. J. Phys. 1999, 77, 137–143.
32. Träbert, E.; Pinnington, E.H.; Kernahan, J.A.; Doerfert, J.; Granzow, J.; Heckmann, P.H.; Hutton, R. Beam-foil study of the lifetimes of n = 3 levels in Na-like Ca X, Mg-like Ca IX and Si-like Ca VII. J. Phys. B: At. Mol. Opt. Phys. 1996, 29, 2647–2659.
33. Hutton, R.; Engström, L.; Träbert, E. Observation of a discrepancy between experimentally determined atomic lifetimes and relativistic predictions for highly ionized members of the Na I isoelectronic sequence. Phys. Rev. Lett. 1988, 60, 2469–2472.
34. Träbert, E.; Doerfert, J.; Granzow, J.; Büttner, R.; Staude, U.; Schartner, K.H.; Rymuza, P.; Mokler, P.H.; Engström, L.; Hutton, R. Experimental n = 3 level lifetimes of Na-like heavy ions in the realm of massive relativistic effects. Phys. Lett. A 1994, 188, 355–360.

35. Kink, I.; Hutton, R.; Nyström, B.; Martinson, I.; Ishii, K.; Ando, K.; Kambara, T.; Nakai, Y.; Kojima, T.M.; Awaya, Y. Lifetime of the 3p 2P3/2 level in Na-like Kr25+. Phys. Rev. A 1997, 55, 3229–3232.
36. Vasilyev, A.; Jasper, E.; Berry, H.G.; Livingston, A.E.; Curtis, L.J.; Cheng, S.; Dunford, R.W. Lifetime of the 3p 2P3/2 level in sodiumlike bromine (Br XXV). Phys. Rev. A 1998, 58, 732–735.
37. Schnabel, R.; Kock, M. f-value measurement of the Be I resonance line using a nonlinear time-resolved laser-induced-fluorescence technique. Phys. Rev. A 2000, 61, 062506.
38. Verdebout, S.; Rynkun, P.; Jönsson, P.; Gaigalas, G.; Froese Fischer, C.; Godefroid, M. A partitioned correlation function interaction approach for describing electron correlation in atoms. J. Phys. B: At. Mol. Opt. Phys. 2013, 46, 085003.
39. Wiese, W.L.; Smith, M.W.; Glennon, B.M. NSRDS-NBS4—Atomic Transition Probabilities; Government Printing Office: Washington, DC, USA, 1966; Volume I Hydrogen Through Neon, pp. 1–169.
40. Kwong, H.S.; Smith, P.L.; Parkinson, W.H.; Knight, R.D. Transition probability of the C III 190.9 nm intersystem line. Bull. Am. Phys. Soc. 1983, 28, 781.
41. Smith, P.L.; Kwong, H.S.; Johnson, B.C.; Parkinson, W.H. Measurements of A-values for C III (1909 Å) and O III (1660, 1666 Å) intersystem lines used in solar transition zone electron density determinations. Bull. Am. Astron. Soc. 1983, 15, 703.
42. Smith, P.L.; Johnson, B.C.; Kwong, H.S.; Parkinson, W.H.; Knight, R.D. Measurements of transition probabilities for spin-changing lines of atomic ions used in diagnostics of astrophysical plasmas. Phys. Scr. T 1984, 8, 88–94.
43. Kwong, V.H.S.; Fang, Z.; Gibbons, T.T.; Parkinson, W.H.; Smith, P.L. Measurement of the transition probability of the C III 190.9 nanometer intersystem line. Astrophys. J. 1993, 411, 431–437.
44. Wiese, W.L.; Fuhr, J.R.; Deters, T.M. Atomic Transition Probabilities of Carbon, Nitrogen, and Oxygen: A Critical Data Compilation; Series: Journal of Physical and Chemical Reference Data, Monograph, No. 7; American Chemical Society: Washington, DC, USA, 1996; pp. 1–522.
45. Chou, H.-S.; Chi, H.-S.; Huang, K.-N. Core polarization effects on the intercombination and resonance transitions in Be-like ions. Chin. J. Phys. 1994, 32, 261–268.
46. Ral'chenko, Y.V.; Vainshtein, L.A. Intercombination transitions in Be-like ions. Phys. Rev. A 1995, 52, 2449–2452.
47. Fleming, J.; Hibbert, A.; Stafford, R.P. The 1909 Å intercombination line in C III. Phys. Scr. 1994, 49, 316–322.
48. Froese Fischer, C. Allowed transitions and intercombination lines in C III and C II. Phys. Scr. 1994, 49, 323–330.
49. Ynnerman, A.; Froese Fischer, C. Multiconfigurational-Dirac-Fock computation of the 2s2 1S0 – 2s2p 3P1 spin-forbidden transition for the Be-like isoelectronic sequence. Phys. Rev. A 1995, 51, 2020–2030.
50. Doerfert, J.; Träbert, E.; Wolf, A.; Schwalm, D.; Uwira, O. Precision measurement of the electric dipole intercombination rate in C2+. Phys. Rev. Lett. 1997, 78, 4355–4358.
51. Curtis, L.J. Private Communication.
52. Chen, M.H.; Cheng, K.T.; Johnson, W.R. Large-scale relativistic configuration-interaction computation of the 2s2 1S0 – 2s2p 3P1 intercombination transition in C III. Phys. Rev. A 2001, 64, 042507.
53. Kelly, R.L.; Palumbo, L.J. Atomic and Ionic Emission Lines Below 2000 Å, Hydrogen through Krypton; Naval Research Laboratory: Washington, DC, USA, 1973; pp. 1–1001.
54. Vilkas, M.J.; Ishikawa, Y.; Träbert, E. Relativistic many-body perturbation computations of boron-like silicon, Si X. Phys. Scr.2005, 72, 181–199.
55. Andrä, H.J.; Gaupp, A.; Wittmann, W. New method for precision lifetime measurements by laser excitation of fast-moving atoms. Phys. Rev. Lett. 1973, 31, 501–504.

56. Gaupp, A.; Kuske, P.; Andrä, H.J. Accurate lifetime measurements of the lowest 2P1/2 states in neutral lithium and sodium. Phys. Rev. A 1982, 26, 3351–3359.
57. Schulze-Hagenest, D. Ph.D. Thesis, University of Kaiserslautern, Kaiserslautern, Germany, 1979. cited in [56].
58. Carlsson, J.; Sturesson, L. Accurate time-resolved laser spectroscopy on lithium atoms. Z. Phys. D 1989, 14, 281–287.
59. Carlsson, J. Laser spectroscopic studies of lifetimes in neutral atoms. Phys. Scr. 1989, 39, 442–446.
60. Carlsson, J.; Jönsson, P.; Sturesson, L.; Froese Fischer, C. Multi-configuration Hartree-Fock calculations and time-resolved laser spectroscopic studies of hyperfine constants in sodium. Phys. Scr. 1992, 46, 394–398.
61. McAlexander, W.I.; Abraham, E.R.I.; Ritchie, N.W.M.; Williams, C.J.; Stoof, H.T.C.; Hulet, R.G. Precise atomic radiative lifetime via photoassociative spectroscopy of ultracold lithium. Phys. Rev. A 1995, 51, R871–R874.
62. McAlexander, W.I.; Abraham, E.R.I.; Hulet, R.G. Radiative lifetime of the 2P state of lithium. Phys. Rev. A 1996, 54, R5–R8.
63. Volz, U.; Majerus, M.; Liebel, H.; Schmitt, A.; Schmoranzer, H. Precision lifetime measurements on Na I 3p 2P1/2 and 3p 2P3/2by beam-gas-laser spectroscopy. Phys. REv. Lett. 1996, 76, 2862–2865.
64. Astner, G.; Curtis, L.J.; Liljeby, L.; Mannervik, S.; Martinson, I. A high precision beam-foil meanlife measurement of the 1s3p 1P level in He I. Z. Phys. A 1976, 279, 1–6.
65. Martinson, I. Private Communication.
66. Cheng, K.T.; Kim, Y.-K.; Desclaux, J.P. Electric dipole, quadrupole, and magnetic dipole transition probabilities of ions isoelectronic to the first-row atoms, Li through F. At. Data Nucl. Data Tables 1979, 24, 111–189.
67. Huang, K.-N. Energy-level scheme and transition probabilities of Al-like ions. At. Data Nucl. Data Tables 1986, 34, 1–77.
68. Huang, K.-N. Energy-level scheme and transition probabilities of Si-like ions. At. Data Nucl. Data Tables 1985, 32, 503–566.
69. Huang, K.-N. Energy-level scheme and transition probabilities of P-like ions. At. Data Nucl. Data Tables 1984, 30, 313–421.
70. Huang, K.-N.; Kim, Y.-K.; Cheng, K.T.; Desclaux, J.P. Energy-level scheme and transition probabilities of Cl-like ions. At. Data Nucl. Data Tables 1983, 28, 355–377.
71. Huang, K.-N.; Johnson, W.R. Resonance transitions of Mg-and Zn-like ions from multiconfiguration relativistic random-phase approximation. Nucl. Instrum. Meth. Phys. Res. B 1985, 9, 502–504.
72. Chou, H.-S.; Chi, H.-C.; Huang, K.-N. Relativistic excitation energies and oscillator strengths including core-polarization effects for the intercombination and resonance transitions in Mg-like ions. J. Phys. B: At. Mol. Opt. Phys. 1993, 26, 4079–4089.
73. Johnson, W.R.; Soff, B. The Lamb shift in hydrogen-like atoms, $1 \leq Z \leq 110$. At. Data Nucl. Data Tables 1985, 33, 405–446.
74. Beiersdorfer, P.; Osterheld, A.; Scofield, J.; Crespo López-Urrutia, J.; Widmann, K. Measurement of QED and hyperfine splitting in the 2s1/2-2p3/2 X-ray transition in Li-like 209Bi80+. Phys. Rev. Lett. 1998, 80, 3022–3025.
75. Beiersdorfer, P.; Knapp, D.; Marrs, R.E.; Elliott, S.R.; Chen, M.H. Structure and Lamb shift of 2s1/2 - 2p3/2 levels in lithiumlike U89+ through neonlike U82+. Phys. Rev. Lett. 1993, 71, 3939–3942.
76. Beiersdorfer, P.; Chen, H.; Thorn, D.B.; Träbert, E. Measurement of the two-loop Lamb shift in lithiumlike U89+. Phys. Rev. Lett. 2005, 95, 233003.
77. Kozhedub, Y.S.; Andreev, O.V.; Shabaev, V.M.; Tupitsyn, I.I.; Brandau, C.; Kozhuharov, C.; Plunien, G.; Stöhlker, T. Nuclear deformation effect on the binding energies in heavy ions. Phys. Rev. A 2008, 77, 032501.
78. Shabaev, V.M.; Andreev, O.V.; Bondarev, A.I.; Glazov, D.A.; Kozhedub, Y.S.; Maiorova, A.V.; Plunien, G.; Tupitsyn, I.I.; Volotka, A.V. Quantum electrodynamics effects in heavy ions and atoms. AIP Conf. Proc. 2011, 1344, 60–69.

79. Johnson, W.R.; Plante, D.R.; Sapirstein, J. Relativistic Computations of Transition Amplitudes in the Helium Isoelectronic Sequence. In Advances of Atomic, Molecular, and Optical Physics; Bederson, B., Walther, H., Eds.; Academic Press: San Diego, CA, USA, 1995; Volume 35, pp. 255–329.
80. Toleikis, S.; Manil, B.; Berdermann, E.; Beyer, H.F.; Bosch, F.; Czanta, M.; Dunford, R.W.; Gumberidze, A.; Indelicato, P.; Kozhuharov, C.; et al. Lifetime of the 2 3P0 state of He-like 197Au. Phys. Rev. A 2004, 69, 022507.
81. Munger, C.T.; Gould, H. Lamb shift in heliumlike uranium (U90+). Phys. Rev. Lett. 1986, 57, 2927–2930.
82. Reuschl, R.; Gassner, T.; Spillmann, U.; Bräuning-Demian, A.; Ananyeva, A.; Beyer, H.; Blumenhagen, K.-H.; Chen, W.; Hagmann, S.; Hegewald, M.; et al. Lifetime measurement of the 2 3P0 state in He-like uranium. Phys. Scr. T 2013, 156, 014024.
83. Trassinelli, M.; Kumar, A.; Beyer, H.F.; Indelicato, P.; Märtin, R.; Reuschl, R.; Kozhedub, Y.S.; Brandau, C.; Bräuning, H.; Geyer, S.; et al. Observation of the 2p3/2 → 2s1/2 intra-shell transition in He-like uranium. Europhys. Lett. 2009, 87, 63001.
84. Gabriel, A.H.; Jordan, C. Long wavelength satellites to the He-like ion resonance lines in the laboratory and the sun. Nature1969, 221, 947–949.
85. Drake, G.W.F. Theory of relativistic magnetic dipole transitions: Lifetime of the metastable 2 3S state of the heliumlike ions. Phys. Rev. A 1971, 3, 908–915.
86. Lin, C.D. Ph.D. Thesis, Columbia University, New York, NY, USA, 1975.
87. Schmidt, H.T.; Forck, P.; Grieser, M.; Habs, D.; Kenntner, J.; Miersch, G.; Repnow, R.; Schramm, U.; Schüssler, T.; Schwalm, D.; et al. High-precision measurement of the magnetic-dipole decay rate of metastable heliumlike carbon ions in a storage ring. Phys. Rev. Lett. 1994, 72, 1616–1619.
88. Saghiri, A.A.; Linkemann, J.; Schmitt, M.; Schwalm, D.; Wolf, A.; Bartsch, T.; Hoffknecht, A.; Müller, A.; Graham, W.G.; Price, A.D.; et al. Dielectronic recombination of ground-state and metastable Li+ ions. Phys. Rev. A 1999, 60, R3350–R3353.
89. Träbert, E.; Gwinner, G.; Knystautas, E.J.; Wolf, A. Heavy-ion storage-ring quest for atomic lifetimes in Li+ and Be2+. Can. J. Phys. 2003, 81, 941–952.
90. Crespo López-Urrutia, J.R.; Beiersdorfer, P.; Savin, D.W.; Widmann, K. Precision measurement of the lifetime of the 1s2s 3S1metastable level in heliumlike O6+. Phys. Rev. A 1998, 58, 238–241.
91. Träbert, E.; Beiersdorfer, P.; Brown, G.V.; Smith, A.J.; Utter, S.B.; Gu, M.F.; Savin, D.W. Improved electron-beam ion-trap lifetime measurement of the 1s2s 3S1 level in Ne8+. Phys. Rev. A 1999, 60, 2034–2038.
92. Crespo López-Urrutia, J.R.; Beiersdorfer, P.; Widmann, K. Lifetime of the 1s2s 3S1 metastable level in He-like S14+ measured with an electron beam ion trap. Phys. Rev. A 2006, 74, 012507.
93. Dunford, R.W.; Church, D.A.; Liu, C.J.; Berry, H.G.; Raphaelian, M.L.A.; Hass, M.; Curtis, L.J. Lifetime of the 2 3S1 state of heliumlike 79Br33+. Phys. Rev. A 1990, 41, 4109–4111.
94. Birkett, B.B.; Briand, J.-P.; Charles, P.; Dietrich, D.D.; Finlayson, K.; Indelicato, P.; Liesen, D.; Simionovici, A. Hyperfine quenching and measurement of the 2 3P0 - 2 3P1 fine-structure splitting in heliumlike silver (Ag45+). Phys. Rev. A 1993, 47, R2454–R2457.
95. Simionovici, A.; Birkett, B.B.; Marrus, R.; Charles, P.; Indelicato, P.; Dietrich, D.D.; Finlayson, K. Cascade-free Doppler-tuned precision measurement of the 2 3S1 state in He-like niobium (93Nb39+). Phys. Rev. A 1994, 49, 3553–3556.
96. Marrus, R.; Charles, P.; Indelicato, P.; de Billy, L.; Tazi, C.; Briand, J.-P.; Simionovici, A.; Dietrich, D.D.; Bosch, F.; Liesen, D. Lifetime of the 2 3S1 state of heliumlike xenon (Xe52+). Phys. Rev. A 1989, 39, 3725–3727.
97. Hubricht, G.G.; Träbert, E. The argon 2 3S1 lifetime from a measurement on recoil ions. Z. Phys. D 1987, 7, 243–250.
98. Moos, H.W.; Woodworth, J.R. Observation of the forbidden 2 3S1 → 1 1S0 spontaneous emission line from helium and measurement of the transition rate. Phys. Rev. Lett. 1973, 30, 775–778.

99. Saathoff, G.; Reinhardt, S.; Karpuk, S.; Träbert, E. Quest for higher precision on the Li II M1 transition rate. 2004. unpublished.
100. Kim, Y.-K.; Baik, D.H.; Indelicato, P.; Desclaux, J.P. Resonance transition energies of Li-, Na-, and Cu-like ions. Phys. Rev. A1991, 44, 148–166.
101. Blundell, S.A. computations of the screened self-energy and vacuum polarization in Li-like, Na-like, and Cu-like ions. Phys. Rev. A 1993, 47, 1790–1803.
102. Träbert, E.; Beiersdorfer, P.; Lepson, J.K.; Chen, H. Extreme ultraviolet spectra of highly charged Xe ions. Phys. Rev. A 2003, 68, 042501.
103. Träbert, E.; Beiersdorfer, P.; Chen, H. Wavelengths of the 4s1/2-4p3/2 resonance lines in Cu- and Zn-like heavy ions. Phys. Rev. A 2004, 70, 032506.
104. Denne, B.; Magyar, G.; Jacquinot, J. Berylliumlike Mo XXXIX and lithiumlike Mo XL observed in the Joint European Torus. Phys. Rev. A 1989, 40, 3702–3705.
105. Büttner, R.; Staude, U.; Nicolai, M.; Braukhoff, J.; Schartner, K.-H.; Folkmann, F.; Mokler, P.H. Measurement of the 2s 2S1/2 - 2p 2P3/2 wavelength in Li-like nickel. Nucl. Instrum. Meth. Phys. Res. B 1995, 98, 41–44.
106. Staude, U.; Bosselmann, Ph.; Büttner, R.; Horn, D.; Schartner, K.-H.; Folkmann, F.; Livingston, A.E.; Ludziejewski, T.; Mokler, P.H. Measurements of 2s 2S1/2 – 2p 2P3/2,1/2 transition energies in lithiumlike heavy ions: Experiments and results for Ni25+and Zn27+. Phys. Rev. A 1998, 58, 3516–3523.
107. Bosselmann, Ph.; Staude, U.; Horn, D.; Schartner, K.-H.; Folkmann, F.; Livingston, A.E.; Mokler, P.H. Measurements of 2s 2S1/2 - 2p 2P1/2,3/2 transition energies in lithiumlike heavy ions. II. Experimental results for Ag44+ and discussion along the isoelectronic series. Phys. Rev. A 1999, 59, 1874–1883.
108. Feili, D.; Bosselmann, Ph.; Schartner, K.-H.; Folkmann, F.; Livingston, A.E.; Träbert, E.; Ma, X.; Mokler, P.H. Measurements of 2s 2S1/2 - 2p 2P1/2,3/2 transition energies in lithiumlike heavy ions. III. Experimental results for Sn47+ and Xe51+. Phys. Rev. A 2000, 62, 022501.
109. Feili, D.; Bosselmann, Ph.; Schartner, K.-H.; Mokler, P.H.; Ma, X.; Livingston, A.E.; Folkmann, F.; Träbert, E. Access to two-photon QED contributions via 2s 2S1/2 - 2p 2P1/2 transitions in heavy Li-like ions. Phys. Scr. T 2001, 92, 300–302.
110. Schweppe, J.; Belkacem, A.; Blumenfeld, L.; Claytor, N.; Feinberg, B.; Gould, H.; Kostroun, V.E.; Levy, L.; Misawa, S.; Mowat, J.R.; et al. Measurement of the Lamb shift in lithiumlike uranium (U89+). Phys. Rev. Lett. 1991, 66, 1434–1437.
111. Theodosiou, C.E.; Curtis, L.J.; El Mekki, M. computations of 2p Lifetimes in the Li Sequence. Phys. Rev. A 1991, 44, 7144–7157.
112. Aggarwal, K.M.; Keenan, F.P. Energy levels, radiative rates, and electron impact excitation rates for transitions in Li-like ions with 12 ≤ Z ≤ 20. At. Data Nucl. Data Tables 2013, 99, 156–248.
113. Aggarwal, K.M.; Keenan, F.P. Energy levels, radiative rates, and electron impact excitation rates for transitions in Li-like ions with 21 ≤ Z ≤ 28. At. Data Nucl. Data Tables 2013, 98, 1096–1108.
114. Zou, Y.; Hutton, R.; Feili, D.; Neacsu, C.; Ma, X.; Schartner, K.-H.; Mokler, P.H. Precision lifetime measurement of the 2p 2P3/2 level for Li-like Ni by beam-foil spectroscopy. Nucl. Instrum. Meth. Phys. Res. B 2005, 235, 192–196.
115. Jasper, E.; Vasilyev, A.; Kukla, K.W.; Vogel Vogt, C.M.; Livingston, A.E.; Berry, H.G.; Cheng, S.; Curtis, L.J.; Dunford, R.W. Lifetime measurements for allowed and forbidden transitions. Phys. Scr. T 1999, 80, 466–468.
116. Kukla, K.W.; Livingston, A.E.; Vogel Vogt, C.M.; Berry, H.G.; Dunford, R.W.; Curtis, L.J.; Cheng, S. Extreme-ultraviolet wavelength and lifetime measurements in highly-ionized krypton. Can. J. Phys. 2005, 83, 1127–1139.
117. Epp, S.W.; Crespo López-Urrutia, J.R.; Brenner, G.; Mäckel, V.; Mokler, P.H.; Treusch, R.; Kuhlmann, M.; Yurkov, M.V.; Feldhaus, J.; Schneider, J.R.; et al. Soft X-ray laser spectroscopy on trapped highly charged ions at FLASH. Phys. Rev. Lett.2007, 98, 183001.

118. Träbert, E.; Armour, I.A.; Bashkin, S.; Jelley, N.A.; O'Brien, R.; Silver, J.D. The x-ray spectra of H-like, He-like and Li-like silicon ions after foil excitation. J. Phys. B 1979, 12, 1665–1676.
119. Armour, I.A.; Fawcett, B.C.; Silver, J.D.; Träbert, E. X-ray spectra and satellite classification of foil-excited Mg and Al. J. Phys. B 1980, 13, 2701–2709.
120. Träbert, E.; Fawcett, B.C.; Silver, J.D. One- and two-electron one-photon satellite transitions in the x-ray spectrum of foil-excited F and Si. J. Phys. B 1982, 15, 3587–3597.
121. Mannervik, S.; Cederquist, H.; Träbert, E. On the EUV spectrum of Li I. Phys. Scr. 1986, 34, 143–145.
122. Mannervik, S.; Short, R.T.; Sonnek, D.; Träbert, E.; Möller, G.; Lodwig, V.; Heckmann, P.H.; Blanke, J.H.; Brand, K. Bound triply excited states in neutral lithium. Phys. Rev. A 1989, 39, 3964–3968.
123. Träbert, E.; Mannervik, S. Satellite lines from core-excited states in neutral lithium. Phys. Rev. A 1989, 40, 1667–1669.
124. Mannervik, S.; Cederquist, H.; Kisielinski, M. Determination of autoionization rates by means of optical emission spectroscopy. Phys. Scr. 1984, T8, 107–111.
125. Mannervik, S. Optical studies of multiply excited states. Phys. Scr. 1980, 40, 28–52.
126. Safronova, M.S.; Johnson, W.R.; Safronova, U. I Relativistic many-body computations of the energies of n = 2 states for the berylliumlike isoelectronic sequence. Phys. Rev. A 1996, 53, 4036–4053.
127. Safronova, U.I.; Johnson, W.R.; Safronova, M.S.; Derevianko, A. Relativistic many-body computations of transition probabilities for the 2l1 2l2 [LSJ] – 2l3 2l4 [L'S'J'] lines in Be-like ions. Phys. Scr. 1999, 59, 286–295.
128. Safronova, U.I.; Johnson, W.R.; Derevianko, A. Relativistic many-body computations of magnetic dipole transitions in Be-like ions. Phys. Scr. 1999, 60, 46–53.
129. Chen, M.H.; Cheng, K.T. Energy levels of the ground state and the 2s2p n(J = 1) excited states of berylliumlike ions: A large-scale, relativistic configuration-interaction computation. Phys. Rev. A 1997, 55, 166–174.
130. Möller, G.; Träbert, E.; Lodwig, V.; Wagner, C.; Heckmann, P.H.; Blanke, J.H.; Livingston, A.E.; Mokler, P.H. Experimental transition probability for the E1 intercombination transition in Be-like Xe50+. Z. Phys. D 1989, 11, 333–334.
131. Brage, T.; Fleming, J.; Hutton, R. A review of intercombination lines in beryllium-like ions. Mol. Phys. 2000, 98, 1057–1065.
132. Fawcett, B.C. Wavelengths and classifications of emission lines due to 2s22pn-2s2pn+1 and 2s2pn-2pn+1 transitions, Z ≤ 28. At. Data Nucl. Data Tables 1975, 16, 135–164.
133. Hutton, R.; Engström, L.; Träbert, E. Experimental oscillator strengths for the resonance lines in Na- and Mg-like iron. Nucl. Instrum. Meth. Phys. Res. B 1988, 31, 294–299.
134. Cheng, K.T.; Chen, M.H.; Johnson, W.R. Hyperfine quenching of the 2s2p 3P0 state of berylliumlike ions. Phys. Rev. A 2008, 77, 052504.
135. Träbert, E. E1-forbidden transition rates in ions of astrophysical interest. In Proceedings of the ASOS11 Conference, Mons, Belgium, 4–9 August 2013. Phys. Scr. T 2014, submitted.
136. Kaufman, V.; Sugar, J. Forbidden lines in ns2 npk ground configurations and ns np excited configurations of beryllium through molybdenum atoms and ions. J. Phys. Chem. Ref. Data 1986, 15, 321–427.
137. Huang, K.-N.; Kim, Y.-K.; Cheng, K.T.; Desclaux, J.P. Correlation and relativistic effects in spin-orbit splitting. Phys. Rev. Lett.1982, 48, 1245–1248.
138. Safronova, M.S.; Johnson, W.R.; Safronova, U.I. Relativistic many-body computations of energies of n = 2 states for boronlike ions. Phys. Rev. A 1996, 54, 2850–2862.
139. Merkelis, G.; Vilkas, M.J.; Gaigalas, G.; Kisielius, R. MBPT computation of energy spectra and E1 transition probabilities for boron isoelectronic sequence. Phys. Scr. 1995, 51, 233–251.

140. Vilkas, M.J.; Ishikawa, Y.; Koc, K. Second-order multiconfigurational Dirac-Fock computations on boronlike ions. Int. J. Quantum Chem. 1998, 70, 813–823.
141. Li, J.-G.; Jönsson, P.; Dong, C.-Z.; Gaigalas, G. Two-electron-one-photon M1 and E2 transitions between the states of the 2p3and 2s22p odd configurations for B-like ions with 18 ≤ Z ≤ 92. J. Phys. B: At. Mol. Opt. Phys. 2010, 43, 035005.
142. Safronova, U.I.; Johnson, W.R.; Livingston, A.E. Relativistic many-body computations of electric-dipole transitions between n = 2 states in B-like ions. Phys. Rev. A 1999, 60, 996–1004, erratum 2000, 61, 039901.
143. Jönsson, P.; Li, J.-G.; Gaigalas, G.; Dong, C.-Z. Hyperfine structures, isotope shifts, and transition rates of C II, N III, and O IV from relativistic configuration interaction computations. At. Data Nucl. Data Tables 2010, 96, 271–298.
144. Rynkun, P.; Jönsson, P.; Gaigalas, G.; Froese Fischer, C. Energies and E1, M1, E2, M2 transition rates for states of the 2s22p, 2s2p2, and 2p3 configurations in boron-like ions between N III and Zn XXVI. At. Data Nucl. Data Tables 2012, 98, 481–556.
145. Jönsson, P.; Ekman, J.; Gustafsson, S.; Hartman, H.; Karlsson, L.B.; du Rietz, R.; Gaigalas, G.; Godefroid, M.; Froese Fischer, C. Energy levels and transition rates for the boron isoelectronic sequence: Si X, Ti XVIII – Cu XXV. Astron. Astrophys. 2013, 559, A100.
146. Träbert, E.; Heckmann, P.H.; Schlagheck, W.; Buttlar, H.V. Beam-foil lifetime studies of highly ionized silicon. Phys. Scr.1980, 21, 27–34.
147. Träbert, E.; Heckmann, P.H. EUV spectrum and lifetimes of foil-excited phosphorus. Phys. Scr. 1980, 21, 35–39.
148. Tordoir, X.; Biémont, E.; Garnir, H.P.; Dumont, P.-D.; Träbert, E. Atomic lifetimes and transition probabilities in boron-like (Na VII) and beryllium-like (Na VIII) sodium ions. Eur. Phys. J. D 1999, 6, 1–7.
149. Doerfert, J.S.; Träbert, E. Relative intensities of 2s22p 2P – 2s2p2 2S, 2P transitions in B-like ions of oxygen through chlorine. Phys. Scr. 1993, 47, 524–530.
150. Engström, L.; Kirm, M.; Bengtsson, P.; Maniak, S.T.; Curtis, L.J.; Träbert, E.; Doerfert, J.S.; Granzow, J. Extended analysis of intensity anomalies in the Al I isoelectronic sequence. Phys. Scr. 1995, 52, 516–521.
151. Träbert, E. On the transition rates of the FeX and Fe XIV coronal lines. Astron. Astrophys. 2004, 415, L39–L42.
152. Träbert, E.; Beiersdorfer, P.; Gwinner, G.; Pinnington, E.H.; Wolf, A. M1 transition rate in Cl12+ from an electron-beam ion trap and heavy-ion storage ring. Phys. Rev. A 2002, 66, 052507.
153. Moehs, D.P.; Church, D.A. Magnetic dipole transition rates from measured lfetimes of levels of Be-like and B-like argon ions. Phys. Rev. A 1998, 58, 1111–1114.
154. Serpa, F.G.; Gillaspy, J.D.; Träbert, E. Lifetime measurements in the ground configuration of Ar13+ and Kr22+ using an electron beam ion trap. J. Phys. B: At. Mol. Opt. Phys. 1998, 31, 3345–3352.
155. Träbert, E.; Beiersdorfer, P.; Utter, S.B.; Brown, G.V.; Chen, H.; Harris, C.L.; Neill, P.A.; Savin, D.W.; Smith, A.J. Experimental M1 tansition rates of coronal lines from Ar X, Ar XIV, and Ar XV. Astrophys. J. 2000, 541, 506–511.
156. Lapierre, A.; Jentschura, U.D.; Crespo López-Urrutia, J.R.; Braun, J.; Brenner, G.; Bruhns, H.; Fischer, D.; González Martínez, A.J.; Harman, Z.; Johnson, W.R.; et al. Relativistic electron correlation, quantum electrodynamics, and the lifetime of the 1s22s22p 2P 3/2 o level in boronlike argon. Phys. Rev. Lett. 2005, 95, 183001.
157. Lapierre, A.; Crespo López-Urrutia, J.R.; Braun, J.; Brenner, G.; Bruhns, H.; Fischer, D.; González-Martínez, A.J.; Mironov, V.; Osborne, C.J.; Sikler, G.; et al. Lifetime measurement of the Ar XIV 1s22s22p 2P 3/2 o metastable level at the Heidelberg electron-beam ion trap. Phys. Rev. A 2006, 73, 052507.
158. Träbert, E.; Gwinner, G.; Wolf, A.; Tordoir, X.; Calamai, A.G. Magnetic dipole transition rates in B-like and F-like titanium ions measured at a heavy-ion storage ring. Phys. Lett. A 1999, 264, 311–317.

159. Nemouchi, M.; Godefroid, M.R. Irreducible tensor fomathrm of the relativistic corrections to the M1 transition operator. J. Phys. B: At. Mol. Opt. Phys. 2009, 42, 175002.
160. Hibbert, A. Estimation of inaccuracies in oscillator strength computations. Phys. Scr. T 1996, 65, 104–109.
161. Brage, T.; Hibbert, A.; Leckrone, D.S. Transition rates of the intercombination UV0.01 multiplet in N II. Astrophys. J. 1997, 478, 423–429.
162. Träbert, E.; Wolf, A.; Pinnington, E.H.; Linkemann, J.; Knystautas, E.J.; Curtis, A.; Bhattacharya, N.; Berry, H.G. Measurement of the N1 2s2p3 5S 2 o level lifetime using a heavy-ion storage ring. Phys. Rev. A 1998, 58, 4449–4452.
163. Jönsson, P.; Bieroń, J. Relativistic configuration interaction computations of energy levels, isotope shifts, hyperfine structures, and transition rates in the 2s22p2 - 2s2p3 transition array for the carbon-like sequence. J. Phys. B: At. Mol. Opt. Phys. 2010, 43, 074023.
164. Jönsson, P.; Rynkun, P.; Gaigalas, G. Energies, E1, M1, E2 transition rates, hyperfine structures, and Landé gJ factors for states of the 2s22p2, 2s2p3, and 2p4 configurations in carbon-like ions between F IV and Ni XXIII. At. Data Nucl. Data Tables 2011, 97, 648–691.
165. Rynkun, P.; Jönsson, P.; Gaigalas, G.; Froese Fischer, C. Energies and E1, M1, E2, and M2 transition rates for states of the 2s22p3, 2s2p4, and 2p5 configurations in nitrogen-like ions between F III and Kr XXX. At. Data Nucl. Data Tables 2013, 100, 315–402.
166. Träbert, E.; Calamai, A.G.; Gillaspy, J.D.; Gwinner, G.; Tordoir, X.; Wolf, A. Intercombination and forbidden transition rates in C- and N-like ions (O2+, F3+, and S9+) measured at a heavy-ion storage ring. Phys. Rev. A 2000, 62, 022507.
167. Träbert, E.; Wolf, A.; Pinnington, E.H.; Linkemann, J.; Knystautas, E.J.; Curtis, A.; Bhattacharya, N.; Berry, H.G. Heavy-ion storage ring measurement of forbidden transition rates between ground-configuration levels in Si6+ and Si8+ ions. Can. J. Phys.1998, 5876, 899–906.
168. Träbert, E.; Grieser, M.; Hoffmann, J.; Krantz, C.; Repnow, R.; Wolf, A. Heavy-ion storage-ring-lifetime measurement of metastable levels in the C-, N-, and O-like ions of Si, P, and S. Phys. Rev. A 2012, 85, 042508.
169. Edlén, B. Comparison of theoretical and experimental level values of the n = 2 complex of ions isoelectronic with Li, Be, O and F. Phys. Scr. 1983, 28, 51–67.
170. Jönsson, P.; Alkauskas, A.; Gaigalas, G. Energies and E1, M1, E2 transition rates for states of the image and image configurations in fluorine-like ions between Si VI and W LXVI. At. Data Nucl. Data Tables 2013, 99, 431–446.
171. Jönsson, P.; Bengtsson, P.; Ekman, J.; Gustafsson, S.; Karlsson, L.B.; Gaigalas, G.; Froese Fischer, C.; Kato, D.; Murakami, I.; Sakaue, H.A.; et al. Relativistic CI computations of spectroscopic data for the 2p6 and 2p53l configurations in Ne-like ions between Mg III and Kr XXVII. At. Data Nucl. Data Tables 2014, 100, 1–154.
172. Träbert, E. Beam-foil lifetime study of Neon-like Ti12+. Z. Phys. D 1986, 1, 283–286.
173. Kirm, M.; Bengtsson, P.; Engström, L. Experimental investigation of atomic lifetimes for the 2p53l levels in Ne-like sulphur. Phys. Scr. 1996, 54, 167–173.
174. Träbert, E.; Jupén, C. Identification of 2s22p53p 1S0 decays in the spectra of Ne-like ions. Phys. Scr. 1987, 36, 586–590.
175. Ishikawa, Y.; López Encarnaciön, J.M.; Träbert, E. N = 3–3 transitions of Ne-like ions in the iron group, especially Ca10+ and Ti12+. Phys. Scr. 2009, 79, 025301.
176. DelZanna, G.; Ishikawa, Y. Benchmarking atomic data for astrophysics: Fe XVII EUV lines. Astron. Astrophys. 2009, 508, 1517–1526.
177. Beiersdorfer, P.; Träbert, E.; Lepson, J.K.; Brickhouse, N.S.; Golub, L. High-resolution laboratory measurements of coronal lines in the 198-218 Å region. Astrophys. J. Suppl. Ser. 2014. In print.
178. Schectman, R.M.; Curtis, L.J.; Chojnacki, D.A. Redetermination of the meanlife of the 2p level in Ne I using cascade analysis. J. Opt. Soc. Am. 1973, 63, 99–100.

179. Westerlind, M.; Engström, L.; Bengtsson, P.; Curtis, L.J. Experimental lifetimes of the 2p5 3s 1P1 and 3P1 levels in Ne-like sulfur and chlorine. Phys. Rev. A 1992, 45, 6198–6201.
180. Curtis, L.J.; Maniak, S.T.; Ghrist, R.W.; Irving, R.E.; Ellis, D.G.; Henderson, M.; Kacher, M.H.; Träbert, E.; Granzow, J.; Bengtsson, P.; Engström, L. Measurements and data-based predictions for $\Delta n = 1$ resonance and intercombination transitions in the Be and Ne sequences. Phys. Rev. A 1995, 51, 4575–4582.
181. Träbert, E. Beam-foil Lifetimes of 2p53s 1,3P 1 o levels in Ne-like ions of Na through S. Phys. Scr. 1996, 53, 167–173.
182. Beiersdorfer, P.; Osterheld, A.L.; Decaux, V.; Widmann, K. Observation of Lifetime-Limited X-Ray Linewidths in Cold Highly Charged Ions. Phys. Rev. Lett. 1996, 77, 5353–5356.
183. Brown, G.V.; Beiersdorfer, P.; Widmann, K. Systematic measurement of the relative electron-impact excitation cross section of the 3d→ 2p 1P1 resonance and 3D1 intercombination lines in mid-Z neonlike ions. Phys. Rev. A 2001, 63, 032719.
184. Brown, G.V.; Beiersdorfer, P.; Liedahl, D.A.; Widmann, K.; Kahn, S.M.; Clothiaux, E.J. Laboratory measurements and identifications of the Fe XVIII-XXIV L-shell X-ray line emission. Astrophys. J. Suppl. Ser. 2002, 140, 589–607.
185. Gu, M.F. The flexibe atomic code. Can. J. Phys. 2008, 86, 675–689.
186. Brown, G.V.; Beiersdorfer, P. Reply to a comment. Phys. Rev. Lett. 2012, 108, 139302.
187. Bernitt, S.; Brown, G.V.; Rudolph, J.K.; Steinbrügge, R.; Graf, A.; Leutenegger, M.; Epp, S.W.; Eberle, S.; Kubicek, K.; Mäckel, V.; et al. An unexpectedly low oscillator strength as the origin of the Fe XVII emission problem. Nature 2012, 492, 225–228.
188. Cowan, T.E.; Bennett, C.L.; Dietrich, D.D.; Bixler, J.V.; Hailey, C.J.; Henderson, J.R.; Knapp, D.A.; Levine, M.A.; Marrs, R.E.; Schneider, M.B. Precision measurement of the 3s1/2-3p3/2 transition energy in Na-like platinum ions. Phys. Rev. Lett. 1991, 66, 1150–1153.
189. Chen, M.H.; Cheng, K.T.; Beiersdorfer, P.; Sapirstein, J. Transition energies of the 3s-3p3/2 resonance lines in sodiumlike to phosphoruslike uranium. Phys. Rev. A 2003, 68, 022507.
190. Gillaspy, J.D.; Draganić, I.N.; Ralchenko, Yu.; Reader, J.; Tan, J.N.; Pomeroy, J.M.; Brewer, S.M. Measurement of the D-line doublet in high-Z highly charged sodiumlike ions. Phys. Rev. A 2009, 80, 010501.
191. Gillaspy, J.D.; Osin, D.; Ralchenko, Yu.; Reader, J.; Blundell, S.A. Transition energies of the D lines in Na-like ions. Phys. Rev. A 2013, 87, 062503.
192. Vilkas, M.J.; Ishikawa, Y.; Träbert, E. Relativistic multireference many-body perturbation theory computations on Au64+ - Au69+ ions. Eur. Phys. J. D 2006, 41, 77–93.
193. Träbert, E.; Wolf, A.; Linkemann, J.; Tordoir, X. Optical measurement of the B+ and Al+ intercombination and Sc12+ forbidden transition rates at a heavy-ion storage ring. J. Phys. B: At. Mol. Opt. Phys. 1999, 32, 537–552.
194. Johnson, B.C.; Smith, P.L.; Parkinson, W.H. Transition probability of the Al II λ 2669 intersystem line. Astrophys. J. 1986, 308, 1013–1017.
195. Träbert, E.; Hutton, R.; Martinson, I. Identification of intercombination transitions in Fe XIV and Fe XIII in the spectra of foil-excited ions and solar flares. Mon. Not. R. Astron. Soc. 1987, 227, 27p–31p.
196. Träbert, E.; Heckmann, P.H.; Hutton, R.; Martinson, I. Intercombination lines in delayed beam-foil spectra. J. Opt. Soc. Am. B1988, 5, 2173–2182.
197. Träbert, E.; Staude, U.; Bosselmann, P.; Schartner, K.H.; Mokler, P.H.; Tordoir, X. Intercombination transition rates in Al-like Au66+ ions. Eur. Phys. J. D 1998, 2, 117–123.
198. Santana, J.A.; Ishikawa, Y.; Träbert, E. Multireference Møller-Plesset perturbation theory results on levels and transition rates in Al-like ions of iron group elements. Phys. Scr. 2009, 79, 065301.

199. Vilkas, M.J.; Ishikawa, Y. Relativistic multireference many-body perturbation-theory computations on the multiple openshell states in siliconlike Ar and aluminumlike Fe ions. Phys. Rev. A 2003, 68, 012503.
200. Moehs, D.P.; Bhatti, M.I.; Church, D.A. Measurements and computations of metastable level lifetimes in FeX, FeXI, FeXII, FeXIII, and Fe XIV. Phys. Rev. A 2001, 63, 032515.
201. Träbert, E.; Gwinner, G.; Wolf, A.; Knystautas, E.J.; Garnir, H.-P.; Tordoir, X. M1/E2 transition rates in Fe X through Fe XIII measured at a heavy-ion storage ring. J. Phys. B: At. Mol. Opt. Phys. 2002, 35, 671–689.
202. Träbert, E.; Calamai, A.G.; Gwinner, G.; Knystautas, E.J.; Pinnington, E.H.; Wolf, A. M1/E2/M2 decay rates in Fe VII, Fe IX, Fe X and Fe XIII measured using a heavy-ion storage ring. J. Phys. B: At. Mol. Opt. Phys. 2003, 36, 1129–1141.
203. Beiersdorfer, P.; Träbert, E.; Pinnington, E.H. Experimental transition rate of the green coronal line of Fe XIV. Astrophys. J.2003, 587, 836–840.
204. Smith, S.J.; Chutjian, A.; Lozano, J.A. Measurement of metastable lifetimes for transitions in Fe9+, Fe10+, and Fe13+. Phys. Rev. A 2005, 72, 062504.
205. Brenner, G.; Crespo López-Urrutia, J.R.; Harman, Z.; Mokler, P.H.; Ullrich, J. Lifetime determination of the Fe XIV 3s23p 2P3/2 o metastable level. Phys. Rev. A 2007, 75, 032504.
206. Träbert, E.; Hoffmann, J.; Krantz, C.; Wolf, A.; Ishikawa, Y.; Santana, J.A. Atomic lifetime measurements on forbidden transitions of Al-, Si-, P- and S-like ions at a heavy-ion storage ring. J. Phys. B: At. Mol. Opt. Phys. 2009, 42, 025002.
207. Bengtsson, P.; Ando, K.; Kambara, T.; Awaya, Y.; Hutton, R. Intercombination lines in highly charged Al- and Si-like ions. Phys. Scr. T 1997, 73, 81–82.
208. Träbert, E. Contradictory observations in beam-foil EUV spectroscopy at high ion energies. Phys. Scr. 1999, 59, 443–445.
209. Ishikawa, Y.; Vilkas, M.J. Relativistic multireference Møller-Plesset perturbation theory computations on the term energy and lifetime of the 5S 2 o state in siliconlike ions with Z = 28–79. Phys. Scr. 2002, 65, 219–226.
210. Biémont, E. Energy-level scheme and oscillator strengths for the 3s-3p and 3p-3d transitions in silicon sequence for elements vanadium through nickel. Phys. Scr. 1986, 33, 324–335.
211. Biémont, E. 3s23p2 - 3s3p3, 3s23p2 - 3s23p3d, and 3s23p2 - 3s23p4s transitions in the silicon sequence for elements potassium through titanium. J. Opt. Soc. Am. B 1986, 3, 163–176.
212. Träbert, E. Solar EUV line identifications from delayed beam-foil spectra. Mon. Not. R. Astron. Soc. 1998, 297, 399–404.
213. Ishikawa, Y.; Vilkas, M.J. Relativistic multireference many-body perturbation-theory computations of the magnetic-dipole and electric-quadrupole transition probabilities of ions in the silicon isoelectronic sequence. Phys. Rev. A 2001, 63, 042506.
214. Vilkas, M.J.; Ishikawa, Y. High-accuracy computations of term energies and lifetimes of silicon-like ions with nuclear charges Z= 24–30. J. Phys. B: At. Mol. Opt. Phys. 2004, 37, 1803–1816.
215. Träbert, E.; Ishikawa, Y.; Santana, J.A.; Del Zanna, G. The 3s23p3d 3Fo term in the Si-like spectrum of Fe (Fe XIII). Can. J. Phys. 2011, 89, 403–412.
216. Del Zanna, G. Benchmarking atomic data for astrophysics: Fe XIII. Astron. Astrophys. 2011, 533, A12.
217. Del Zanna, G.; Storey, P.J. Atomic data for astrophysics: Fe XIII soft X-ray lines. Astron. Astrophys. 2012, 543, A144.
218. Froese Fischer, C. Towards accurate transition data for 3p2, 3p3, 3p4 levels of Fe, Co and Ni ions. J. Phys. B: At. Mol. Opt. Phys. 2010, 43, 074020.
219. Mendoza, C.; Zeippen, C.J. Transition probabilities for forbidden lines in the 3p2 configuration—II. Mon. Not. R. Astron. Soc.1982, 199, 1025–1032.

220. Abou El-Maaref, A.; Uosif, M.A.M.; Allam, S.H.; El-Sherbini, Th.M. Energy levels, oscillator strengths and transition probabilities for Si-like P II, S III, Cl IV, Ar V and K VI. At. Data Nucl. Data Tables 2012, 98, 589–615.
221. Abou El-Maaref, A.; Allam, S.H.; El-Sherbini, Th.M. Energy levels, oscillator strengths and radiative rates for Si-like Zn XVII, Ga XVIII, Ge XIX, ans As XX. At. Data Nucl. Data Tables 2014, 100, 155–182.
222. Vilkas, M.J.; Ishikawa, Y. Relativistic many-body perturbation computations on extreme ultraviolet and soft-x-ray transition energies in siliconlike iron. Phys. Rev. A 2004, 69, 062503.
223. Mendoza, C.; Zeippen, C.J. Transition probabilities for forbidden lines in the 3p3 configuration. Mon. Not. R. Astron. Soc. 1982, 198, 127–139.
224. Biémont, E.; Hansen, J.E. Radiative transition rates in the ground configuration of the phosphorus sequence from argon to ruthenium. Phys. Scr. 1985, 31, 509–518.
225. Biémont, E.; Palmeri, P.; Quinet, P.; Träbert, E.; Zeippen, C.J. Level lifetimes in 3p and 3d configurations of Fe XII. Eur. Phys. J. D 2002, 20, 37–44.
226. Vilkas, M.J.; Ishikawa, Y. Relativistic multireference many-body perturbation theory calculations on ions of the phosphorus isoelectronic sequence. J. Phys. B: At. Mol. Opt. Phys. 2004, 37, 4763–4778.
227. Del Zanna, G.; Mason, H.E. Benchmarking atomic data for astrophysics: Fe XII. Astron. Astrophys. 2005, 433, 731–744.
228. Mendoza, C.; Zeippen, C.J. Transition probabilities for forbidden lines in the 3p4 configuration—III. Mon. Not. R. Astron. Soc.1983, 202, 981–986.
229. Biémont, E.; Hansen, J.E. Forbidden transitions in 3p4 and 4p4 configurations. Phys. Scr. 1986, 34, 116–130.
230. Saloman, E.B.; Kim, Y.-K. Energy levels and transition probabilities in the ground-state configuration of sulfur-like ions. At. Data Nucl. Data Tables 1989, 41, 339–356.
231. Chou, H.-S.; Chang, J.-Y.; Chang, Y.-H.; Huang, K.-N. Energy-level scheme and transition probabilities of S-like ions. At. Data Nucl. Data Tables 1996, 62, 77–145.
232. Ishikawa, Y.; Vilkas, M.J. Relativistic many-body computations of excited-state energies and transition wavelengths for six-valence-electron sulfurlike ions. Phys. Rev. A 2008, 78, 042501.
233. Del Zanna, G. Benchmarking atomic data for astrophysics: Fe XI. Astron. Astrophys. 2010, 514, A41.
234. Del Zanna, G.; Storey, P.J. Atomic data for astrophysics: Fe XI soft X-ray lines. Astron. Astrophys. 2013, 549, A42.
235. Biémont, E.; Träbert, E. Transition rates of the resonance line doublet in the Cl I sequence, Ar II-Ge XVI. J. Phys. B: At. Mol. Opt. Phys. 2000, 33, 2939–2946.
236. Del Zanna, G.; Berrington, K.A.; Mason, H.E. Benchmarking atomic data for astrophysics: Fe X. Astron. Astrophys. 2004, 422, 731–749.
237. Del Zanna, G.; Berrington, K.A.; Mason, H.E. Atomic data for astrophysics: Fe X soft X-ray lines. Astron. Astrophys. 2012, 541, A90.
238. Ishikawa, Y.; SAntana, J.A.; Träbert, E. Relativistic multireference many-body perturbation theory for open-shell ions with multiple valence shell electrons: the transition rates and lifetimes of the excited levels in chlorinelike Fe X. J. Phys. B: At. Mol. Opt. Phys. 2010, 43, 074022.
239. Träbert, E.; Saathoff, G.; Wolf, A. M1/E2 decay rates in Co XI, Ni XII and Cu XIII measured at a heavy-ion storage ring. Can. J. Phys. 2004, 37, 945–952.
240. Träbert, E. Experimental checks on computations for Cl-, S- and P-like ions of the iron group elements. J. Phys. B: At. Mol. Opt. Phys. 1996, 29, L217–L224.
241. Dumont, P.D.; Garnir, H.P.; Baudinet-Robinet, Y.; Kapenyak, M. Lifetime measurements in Ti IV-VII using transitions observed in beam-foil spectroscopy between 400 and 800 Å. J. Opt. Soc. Am. 1981, 71, 502–503.
242. Mohan, M.; Singh, A.K.; Jha, A.K.S.; Jha, P. Level energies, oscillator strengths, and lifetimes for transitions in Ti VI. At. Data Nucl. Data Tables 2007, 93, 105–126.
243. Garnir, H.P.; Dumont, P.D.; Tordoir, X. Atomic lifetime measurement of K III. unpublished work.

244. Lauer, S.; Liebel, H.; Vollweiler, F.; Schmoranzer, H.; Lagutin, B.M.; Demekhin, Ph.V.; Petrov, I.D.; Sukhorukov, V.L. Lifetimes of the ns1 np6 2S1=2 states of singly ionized argon, krypton and xenon. J. Phys. B: At. Mol. Opt. Phys. 1999, 32, 2015–2030.
245. Wilson, N.J.; Hibbert, A.; Bell, K.L. Oscillator Strengths for K III, Ca IV and Sc V. Phys. Scr. 2000, 61, 603–610.
246. Berrington, K.A.; Pelan, J.C.; Waldock, J.A. Oscillator strength for 3s23p5 - 3s3p6 in Cl-like ions. J. Phys. B: At. Mol. Opt. Phys. 2001, 34, L419–L424.
247. Saha, B.; Fritzsche, S. M1 and E2 transitions in Ar II. J. Phys. B: At. Mol. Opt. Phys. 2005, 38, 1161–1171.
248. Wagner, W.J.; House, L.L. Hartree-Fock computations of coronal forbidden lines in the Argon I isoelectronic sequence. Astrophys. J. 1969, 155, 677–686.
249. Wagner, W.J.; House, L.L. Emprically corrected computations of coronal visible lines from the 3p53d configuration. Astrophys. J. 1971, 166, 683–698.
250. Träbert, E.; Jupén, C.; Fritzsche, S. EUV line identifications and lifetime measurements in highly-charged ions of the iron group. Phys. Scr. T 1999, 80, 463–465.
251. Träbert, E.; Beiersdorfer, P.; Brown, G.V.; Chen, H.; Thorn, D.B.; Biémont, E. Experimental M1 transition rates in highly charged Kr ions. Phys. Rev. A 2001, 64, 042511.
252. Utter, S.B.; Beiersdorfer, P.; Träbert, E.; Clothiaux, E.J. Wavelengths of the 4s1/2-4p3/2 resonance lines in Cu-like heavy ions. Phys. Rev. A 2003, 67, 032502.
253. Vilkas, M.J.; Ishikawa, Y. Relativistic multireference many-body perturbation computations on multi-valence-electron systems: Benchmarks on Zn-like ions. Phys. Rev. A 2005, 72, 032512.
254. Ishikawa, Y.; Quiney, H.M.; Malli, G.L. Dirac-Fock-Breit self-consistent-field method: Gaussian basis-set computations on many-electron atoms. Phys. Rev. A 1991, 43, 3270–3278.
255. Malli, G.L.; Da Silva, A.B.F.; Ishikawa, Y. Universal Gaussian basis set for accurate ab initio relativistic Dirac-Fock computations. Phys. Rev. A 1993, 47, 143–146.
256. Blundell, S.A.; Johnson, W.R.; Safronova, M.S.; Safronova, U.I. Relativistic many-body computations of the energies of n = 4 states along the zinc isoelectronic sequence. Phys. Rev. A 2008, 77, 032507.
257. Blundell, S.A. computations of QED corrections in highly charged Zn-like ions. Can. J. Phys. 2009, 87, 55–65.
258. Safronova, U.I.; Safronova, M.S. Relativistic many-body computations of the oscillator strengths, transition rates and polarizabilities in Zn-like ions. J. Phys. B: At. Mol. Opt. Phys. 2010, 43, 074025.
259. Chen, M.H.; Cheng, K.T. A large-scale relativistic configuration-interaction approach: application to the 4s2 - 4s4p transition energies and E1 rates for Zn-like ions. J. Phys. B: At. Mol. Opt. Phys. 2010, 43, 074019.
260. Träbert, E.; Clementson, J.; Beiersdorfer, P.; Santana, J.A.; Ishikawa, Y. Measurements and computations of Zn-like heavy ions: an update. Can. J. Phys. 2011, 89, 639–645.
261. Träbert, E.; Clementson, J.; Beiersdorfer, P.; Santana, J.A.; Ishikawa, Y. Extreme-ultraviolet spectra of highly charged Pt ions with several valence-shell electrons: Observation and accurate computations. Phys. Rev. A 2010, 82, 062519.
262. Santana, J.A.; Ishikawa, Y.; Träbert, E. Relativistic MR-MP computations of the energy levels and transition probabilities in Ni- to Kr-like Pt ions. At. Data Nucl. Data Tables 2014, 100, 183–271.
263. Utter, S.B.; Beiersdorfer, P.; Träbert, E. Electron-beam ion-trap spectra of tungsten in the EUV. Can. J. Phys. 2002, 80, 1503–1515.
264. Träbert, E. EUV beam-foil observations of Cu-like ions through Ge-like ions of iodine. Phys. Scr. T 2011, 144, 014004.
265. Li, J.G.; Träbert, E.; Dong, C.Z. Energy levels, transition rates and lifetimes for low-lying levels in Cu-, Zn-, Ga- and Ge-like ions of iodine. Phys. Scr. 2011, 83, 015301.
266. Curtis, L.J.; Ellis, D.G. Alkali Spectra in the Promethium Isoelectronic Sequence. Phys. Rev. Lett. 1980, 45, 2099–2102.

267. Träbert, E.; Heckmann, P.H. Tentative Identification of the 5s-5p Transitions in Pm I-Like Au XIX. Z. Phys. D 1986, 1, 381–383.
268. Kaufman, V.; Träbert, E.; Heckmann, P.H.; Möller, G.; Lodwig, V.; Blanke, J.H. Search for intercombination lines in few-electron spectra of rare-earth sequence ions of Os through Au. Phys. Scr. 1990, 42, 705–713.
269. Vilkas, M.J.; Ishikawa, Y.; Träbert, E. Electric-dipole 5s-5p transitions in promethiumlike ions. Phys. Rev. A 2008, 77, 042510.
270. Kobayashi, Y.; Kato, D.; Sakaue, H.A.; Murakami, I.; Nakamura, N. Spectroscopic study of promethiumlike bismuth with an electron-beam ion trap: Search for alkali-metal-like resonance lines. Phys. Rev. A 2014, 89, 010501.
271. Kramida, A.E.; Shirai, T. Energy levels and spectral lines of tungsten, W III through W LXXIV. At. Data Nucl. Data Tables2009, 95, 305–474, Erratum: 2009, 95, 1051.
272. Beiersdorfer, P.; Osterheld, A.L.; Scofield, J.; Wargelin, B.; Marrs, R.E. Observation of magnetic octupole decay in atomic spectra. Phys. Rev. Lett. 1991, 67, 2272–2275.
273. Clementson, J.; Beiersdorfer, P.; Brown, G.V.; Gu, M.F. Spectroscopy of M-shell x-ray transitions in Zn-like through Co-like W. Phys. Scr. 2010, 81, 015301.
274. Clementson, J.; Beiersdorfer, P. Wavelength measurement of n = 3 to n = 3 transitions in highly charged tungsten ions. Phys. Rev. A 2010, 81, 052509.
275. Clementson, J.; Beiersdorfer, P.; Brown, G.V.; Gu, M.F.; Lundberg, H.; Podpaly, Y.; Träbert, E. Tungsten spectroscopy at the Livermore electron beam ion trap facility. Can. J. Phys. 2011, 89, 571–580.
276. Clementson, J.; Beiersdorfer, P.; Gu, M.F. X-ray spectroscopy of E2 and M3 transitions in Ni-like W. Phys. Rev. A 2010, 81, 052509.
277. Clementson, J.; Beiersdorfer, P.; Brage, T.; Gu, M.F. Atomic data and theoretical X-ray spectra of Ge-like through V-like W ions. At. Data Nucl. Data Tables 2014, 100, 577–649.
278. Träbert, E.; Beiersdorfer, P.; Pinnington, E.H.; Utter, S.B.; Vilkas, M.J.; Ishikawa, Y. Experiment and theory in interplay on high-Z few-electron ion spectra from foil-excited ion beams and electron beam ion traps. J. Phys.: Conf. Ser. 2007, 58, 93–96.
279. Kramida, A.; Ralchenko, Yu.; Reader, J.; NIST ASD Team. NIST Atomic Spectra Database (ver. 5.1), 2013. National Institute of Standards and Technology: Gaithersburg, MD, USA. Available online: http://physics.nist.gov/asd (accessed on 3 January 2014).
280. Dere, K.P.; Landi, E.; Young, P.R.; Del Zanna, G.; Landini, M.; Mason, H.E. CHIANTI—An atomic database for emission lines IX. Ionization rates, recombination rates, ionization equilibria for the elements hydrogen through zinc and updated atomic data. Astron. Astrophys. 2009, 498, 915–929.
281. Landi, E.; Del Zanna, G.; Young, P.R.; Dere, K.P.; Mason, H.E. CHIANTI—An atomic data base for emission lines. XII. Version 7 of thedatabase. Astrophys. J. 2012, 744, 99.
282. Mäckel, V.; Klawitter, R.; Brenner, G.; Crespo López-Urrutia, J.R.; Ullrich, J. Laser spectroscopy on forbidden transitions in trapped highly charged Ar13+ ions. Phys. Rev. Lett. 2011, 107, 143002.

CHAPTER 9

Estimating Relative Uncertainty of Radiative Transition Rates

Daniel E. Kelleher

Atomic Spectroscopy Group, Quantum Measurements Laboratory, National Institute of Standards and Technology (ret.); Gaithersburg, MD 20899-8422, USA

ABSTRACT

We consider a method to estimate relative uncertainties of radiative transition rates in an atomic spectrum. Few of these many transitions have had their rates determined by more than two reference-quality sources. One could estimate uncertainties for each transition, but analyses with only one degree of freedom are generally fraught with difficulties. We pursue a way to empirically combine the limited uncertainty information in each of the many transitions. We "pool" a dimensionless measure of relative dispersion, the "Coefficient of Variation of the mean," $\mathrm{CnV} \equiv s/(\bar{x}\sqrt{n})$. Here, for each transition rate, "s" is the standard deviation, and "$\bar{x}$" is the mean of "n" independent data sources. CnV is bounded by zero and one whenever the determined quantity is intrinsically positive.) We scatter-plot the CnV as a function of the "line strength" (here a more useful radiative transition rate than transition probability). We find a curve through comparable CnVs that envelops a specified percentage of the CnVs (e.g. 95%). We take this curve to represent the expanded relative uncertainty of the mean. The method is most advantageous when the number of determined transition rates is large while the number of independent determinations per transition is small. The transition rate data of Na III serves as an example.

Keywords: line strengths; normalized; relative uncertainty; transition probabilities

Our goal is to optimize uncertainty estimates for published radiative transition rates by "pooling" relative uncertainties of these rates within a spectrum. Most evaluated transition rates have been accurately determined by two or fewer sources (experimental and/or theoretical). In such cases it is generally difficult to make useful distribution-based estimates of uncertainty for individual transitions. Fortunately, there are many transitions in any given atomic spectrum. Here we propose a graphical method for empirically estimating relative uncertainties by "pooling" relative standard deviations of transition rates within a spectrum.

As we shall see, this "heterogeneous" pooling can result not only in improved uncertainty estimates of transition rates, but also in detecting outliers and interpolating transition rate uncertainties that have only one reference-quality determination. First we must find an appropriate measure of relative dispersion.

Consider a given transition with an ensemble $\{x_i\}$ of "n" independent determinations of radiative line strength. Each ensemble has a mean "$\bar{x}$" and unbiased standard deviation "s". The Coefficient of Variation, CV, is defined as:

$$C_V \equiv \frac{s}{\bar{x}} \quad (\bar{x} \neq 0) \tag{1}$$

For our purposes, it will be more useful to consider a slightly modified quantity.

Consider a large-n population $\{x_i\}$. Say we take different samples "j" from this population. The central limit theorem holds that the mean of such samples $\{\bar{x}_j\}$ will have a standard deviation of approximately $s/\sqrt{n}$, the standard deviation *of the mean*. Here, $\bar{x}_j = \sum_i x_{ij}/n_j$ is the jth sample mean and $s_j^2 = \sum_i (x_{ij} - \bar{x}_j)^2/(n_j - 1)$. Therefore, as a matter of nomenclature, we refer to:

$$C_V^n \equiv \frac{s}{\sqrt{n}\bar{x}} = \frac{C_V}{\sqrt{n}} \tag{2}$$

as the Coefficient of Variation *of the mean*. The properties of CnV are discussed in Kelleher [1]. This article presents the required proofs and considers in detail the advantages and limitations of heterogeneous pooling.

For a population $\{x_i\}$ of size n, CnV has the useful limit (with condition):

$$C_V^n \in [0,1] \text{ if } \{x_i > 0\}. \tag{3}$$

That is, the coefficient of variation of the mean is intrinsically bounded by zero and one if all determined quantities (such as line strengths (S)) are positive. This normalization property of CnV is useful. One knows automatically the quantitative significance of CnV with regard to one. Also, in Equation (3), the upper limit is independent of n. This makes it feasible to pool multiple sources with a different n value.

For each transition rate, when $n = 2$:

$$C_V^{n=2} = \frac{x_> - x_<}{x_> + x_<} \tag{4a}$$

Note that the upper bound on Equation (4a) is 1 whenever $x_>$ is positive.

In general, for each transition rate j having n_j independent determinations, and mean $\bar{x}_j$, the Coefficient of Variation of the mean is:

$$C_V^n(\mathrm{j}) = \frac{1}{\bar{x}_\mathrm{j}} \left[\frac{\sum_\mathrm{i} (x_\mathrm{ij} - \bar{x}_\mathrm{j})^2}{n_\mathrm{j}(n_\mathrm{j} - 1)} \right]^{1/2} \tag{4b}$$

As for a specific transition rate, we choose to pool the variation of the line strength (S), rather than transition probabilities or oscillator strengths. This is necessary because S is proportional to the radial matrix element only. Thus, transition energies and statistical weights do not enter (the former can be quite large). This allows us to pool transitions with different transition energies and statistical weights. (The term "transition rate" is used loosely here to refer collectively to the three radiative entities, S, A, and f. Each of the three can be converted to either of the others by using expressions in [2], for example.)

Below we describe the three steps of this empirical pooling method. We begin by scatter-plotting the CnV of different transitions of the same spectrum as a function of the logarithm of the mean line strength. Next, we fit a least-squares curve through the CnV data to determine the "slope" of the data. Finally, using this slope, we iteratively derive a second curve which envelops the specified fraction p of CnV data points. We take this envelope curve to represent the expanded relative uncertainty of the mean, Urelp+(logS¯¯). (The term "expanded" refers to the specified value of p (e.g. 95%); the "+" denotes the upper bound. To the extent possible, we use the terminology of the International Organization for Standardization [3]). This "GUM" does not consider the coefficient of variation. We use T_{p+} to indicate the upper confidence bound of the Coefficient of Variation, and Urelp+ to designate the expanded *relative* uncertainty of the mean. Such plots can also illuminate systematic trends and outliers.

In Figure 1, we plot the CnV for 192 radiative transition rates of the spectral lines of doubly ionized sodium (Na III). These rates were calculated by two different theoretical approaches: the "Multi-configuration Hartree–Fock" method, (Tachiev and Froese Fischer [4]), and the "Configuration Interaction" method (McPeake and Hibbert [5]). The CnV for each transition in this figure, given by Equation (4a), contains only one degree of freedom ($n = 2$, $\nu = 1$). Only transitions from energy levels *above* 415,000 cm^{-1} are included in this figure.

The two parameters of the lower fit curve are determined initially by least square fitting. Except in special cases, there is no *a priori* relationship between CnV and the sample mean. However, for CnV satisfying Equation (3), any fitting functions $\Phi_0$must have asymptotic bounds of one and zero. We have chosen:

$$\Phi_0(\bar{x}) = \frac{1}{2}\operatorname{erfc}\left[\beta^{-1}\log_{10}\left(\frac{\bar{x}}{\bar{x}_{1/2}}\right)\right]. \tag{5}$$

See Figure A1 in Appendix A. Here erfc is the complimentary error function, erfc = 1 – erf. An algorithm for computing erf is given in the appendix. x¯1/2 and β are fit parameters analogous to "intercept" and "slope", respectively. $\Phi_0(\bar{x})$ has asymptotic values of one and zero, consistent with Equation (3). This *ad hoc* fit curve has the equivalent functional form to the cumulative distribution function of the log-normal distribution.

$$C_V^n(j) \approx \frac{1}{2}\mathrm{erfc}\left[\beta^{-1}[\mathrm{LS}]\log_{10}\left(\frac{\bar{S}_j}{\bar{x}_{1/2}[\mathrm{LS}]}\right)\right] \quad \text{Least Squares (LS) fit to scatter-plot} \tag{6}$$

After the scatter-plot, the second step in our heterogeneous pooling procedure involves the determination of parameters $\beta[\mathrm{LS}]$, and $\bar{x}_{1/2}[\mathrm{LS}]$. We perform a Levenberg–Marquardt nonlinear LS fit for Equation (6) using the implementation in the "C" language by Press *et al.* [6]. This enables us to evaluate the two parameters for the LS curve (lower). (An equation like Equation (6) exists for each "Coefficient of Variation of the mean," CnV(j) and Line Strength S_j. In Equation (6), the "≈" symbol is meant to indicate a LS "best-fit" for the two LS fit parameters).

$$C_V^n(j) < \frac{1}{2}\mathrm{erfc}\left[\beta^{-1}[\mathrm{LS}]\log_{10}\left(\frac{\bar{S}_j}{\bar{x}_{1/2}[p]}\right)\right] \text{ for p \% of } C_V^n(j) \tag{7}$$

Finally, keeping this same fit value of β[LS], we estimate an expanded relative uncertainty of the mean, Urelp+(logS¯¯), by iterating x¯1/2[p] in Equation (7) to determine a second, higher curve such that a specified fraction p of the CnVs fall beneath it. (Using a different β for the LS and Urelp+(logS¯¯) curves would lead to the physically impossible consequence that the two curves would cross at some point).

This empirical method for "covering" any specified percentage of CnV does not require random data or the existence of any particular pdf. It does depend on the validity of heterogeneously pooling the CnV of different line strengths within the spectrum. We take the curve Urelp+ to be an empirical analog to the upper confidence bound of the mean for the normally-distributed Coefficient of Variation of the mean, Tp+/n√, with the distinction that Urelp+ incorporates, to some level of approximation, the effects of both random and systematic errors. (A website by Verrill [7] calculates Tp+ for Normal and log-Normal distributions). LS fit curves can generally be determined more precisely than, for example, 95% envelope curves, because the latter are heavily influenced by the small percentage of points lying outside them. If relative uncertainties for any "outliers" are assigned, this should be done independently of the heterogeneous pooling considered here.

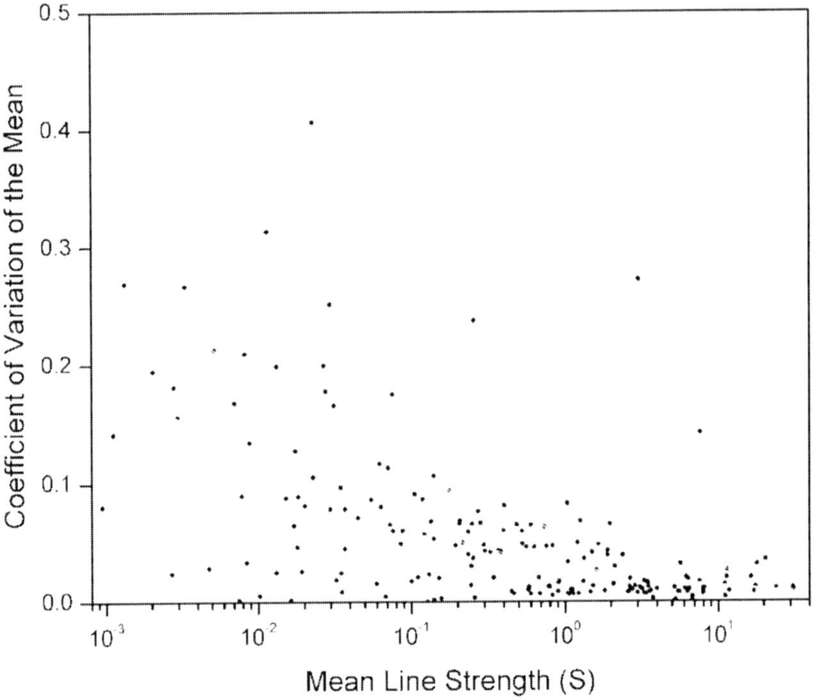

Figure 1. The mean line strength ($\bar{S}$) *vs.* the Coefficient of Variation of the mean, CnV, for radiative transition rates in Na III for which the energy of the upper level is *greater*than 415, 000 cm^{-1}. Seven outliers, three of which are off the scale, were given very small weights. Because only two data sources are used, the CnV for each transition is given by Equation (4a).

This heterogeneous pooling method has proven effective for critical evaluations of transition rates in recent NIST compilations of more than 90 different atomic and ionic spectra, for which we use p = 0.90. (See, for example, Kelleher and Podobedova [2].) We apply a correspondence table between the value on the envelope curve, $U_{p+}^{\mathrm{rel}}(\log\bar{S})$, and the letter-grade given in the compilation. We also interpolate this envelope curve when only one datum source is available (n = 1) for a given transition. (The number of degrees of freedom vanishes for CnV, but not for the curve $U_{p+}^{\mathrm{rel}}(\log\bar{S})$.

Thus far we have described a method for estimating the envelope curve that represents the expanded relative uncertainty of the mean as a function of the mean line strength, $U_{p+}^{rel}(\log\bar{S})$. The key assumption in the heterogeneous pooling of the coefficient of variation of the mean (relative standard deviations of the mean), is that pooled CnVs are comparable. Below we consider ways to discern whether any members CnV are not appropriate contributors to $U_{p+}^{rel}(\log\bar{S})$. They may belong to a separate subgroup or they may be outliers. The latter usually appear as isolated points on a CnV curve.

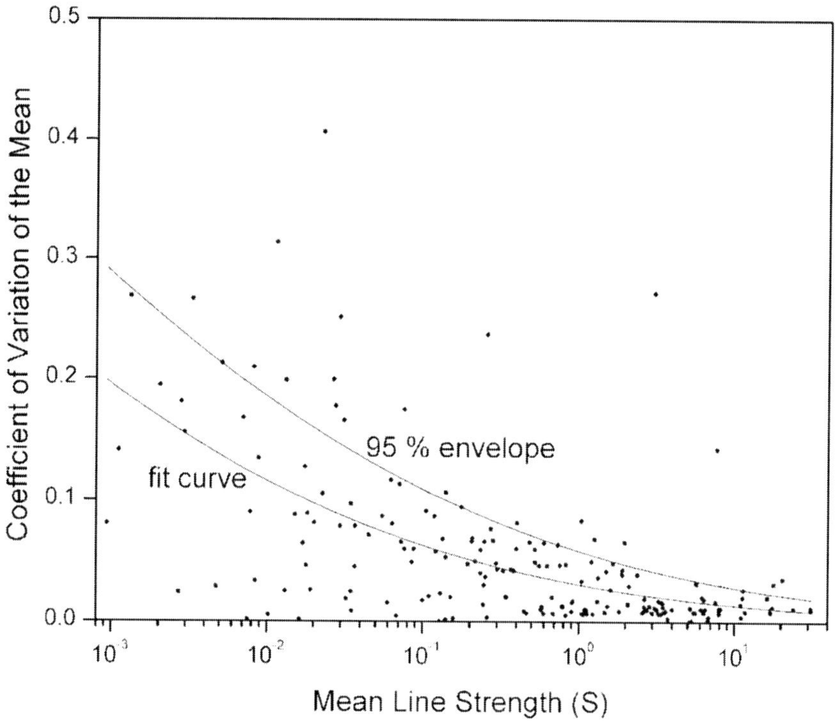

Figure 2. The CnV data points are the same as Figure 1, to which two curves have been added. Seven outliers, three of which are off the scale, were given small weights. Because only two data sources were used, the CnV for each transition is given by Equation (4a). For the lower curve, parameters β[LS] and x¯1/2[LS] in Equation (6) were evaluated by a LS fit, and β[LS] was found to be 4.2 ("slope"). Keeping the same β, x¯1/2[*p*] was then adjusted in Equation (7) until 95% of the points (excluding outliers) lie under the upper curve, for which x¯1/2[p] = 5.1 × 10^{-5} ("intercept"). Compare to Figure 3.

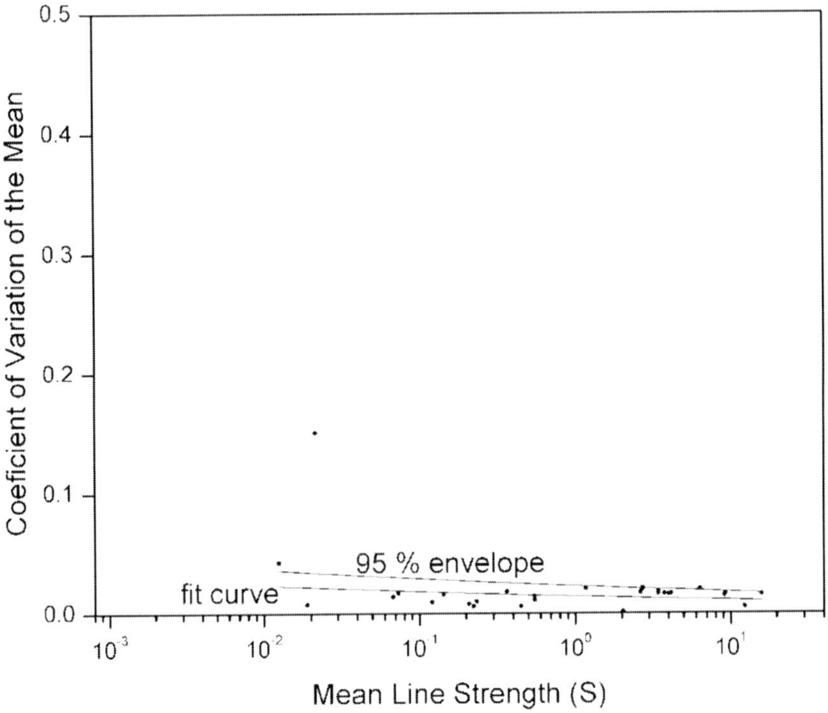

Figure 3. Coefficient of Variation of the mean (CnV) for radiative transitions in Na III in which the energy of the upper level is *less* than 415,000 cm^{-1}, in contrast to Figure 1and Figure 2. One value was weighted as an outlier. Each point is the CnV for a different radiative transition. These data were taken from the same sources as in Figure 2, and the scales are the same. Ninety-five percent of the points lie beneath the upper curve, for which $\beta = 13.8$ and $\bar{x}1/2 = 2.3 \times 10^{-19}$.

Comparison of Figure 2 with Figure 3 illustrates the importance of pooling comparable CnVs. The data in Figure 3 are limited to transitions from lower-lying levels, for which the energy of the upper level is *less* than 415,000 cm^{-1}. These levels belong to the same Na III spectrum, but are more widely spaced than the upper levels of the Figure 2 transitions. Thus level "mixing" is less for transitions from lower-lying levels. This results in smaller computational discrepancies between the two methods. If we had not separated out the data for Figure 3, we would have overestimated the relative uncertainty for this data.

Subsets of CnV can sometimes lie significantly below the envelope curve, Urelp+(logS¯). It may be that some of these are best treated separately from heterogeneous pooling, if justified by expert knowledge or self-consistency of CnV within the subset. We can test for subgroups of CnVs by computing for each transition rate the ratio of its CnV to the envelope curve (CnV/Urelp+), and sorting the ratio.

In the curve fitting, all CnV values are given a weight of one except those which lie more than a factor of three above or below the Urelp+(logS¯¯) envelope curve. Such "outliers" are given a weight inversely proportional to the square of their distance from this envelope curve, and thus have negligible influence on the fit. Distant outliers can destabilize a LS solution.

CnVs that lie above the pooled envelope (but do not qualify as outliers) can also be identified by inspection of the CnV plot, such as in Figure 2 and Figure 3. For the fraction 1-p of CnV values that lie above the envelope curve, we approximate their relative uncertainties by equating them to their individual CnVs.

In summary, we describe a pooling method to estimate expanded relative uncertainties of the mean for atomic transition rates. We consider cases where a small number of independent determinations have been made for a substantial number of transition rates from the same spectrum. Let CnV≡s/(x¯n√) where, for each transition, s is the standard deviation and x¯ is the mean of n independent determinations of the transition rate. Scatter-plotting CnVs from these sources *vs.* log, typically enables one to heterogeneously pool them. By finding a curve which covers a specified fraction "*p*" (e.g. 95%) of the CnV, one estimates the expanded relative uncertainty of the mean, Urelp+, as a function of log. It includes random and systematic errors. The Urelp+(logS¯¯) envelope curve can also facilitate the detection of outliers and interpolation of relative uncertainties for which only one transition strength has been determined. Estimating relative uncertainties for theoretical transition rates of Na III serves as an example of the method.

ACKNOWLEDGMENTS

The author gratefully acknowledges vital contributions by Mark Vangel and William Guthrie.

APPENDIX A:

Figure A1 displays Equation (5) as a function of $\bar{x}/\bar{x}_{1/2}$ for seven values of β ranging from 1.5 to 10. The only required properties of Equation (5) are that it has the appropriate asymptotes and has a weak monotonic dependence on $\bar{x}_{1/2}$. Except for poor data, smaller $\Phi_0(\bar{x})$ prevail (*i.e.*, CnV <≈ 0.3). Φ_0 has proven robust in fitting a wide range of different data types, including those that span essentially the entire CnV range of 0 to 1. Other functional forms may work better for different data sets. The whole procedure could also be performed graphically, without the explicit use of a parametric fitting function.

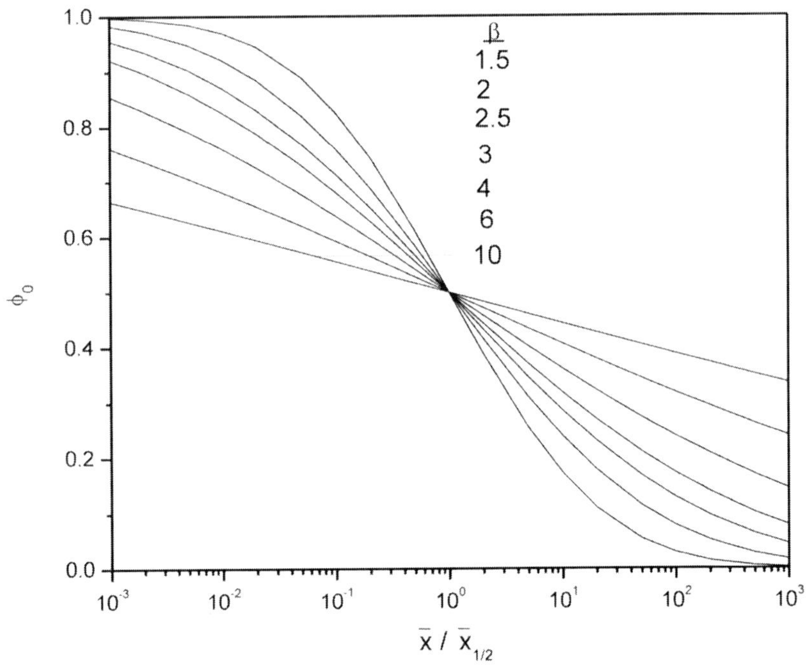

Figure A1. The fitting function $\Phi_0(\bar{x})$, Equation (5), for the Coefficient of Variation of the mean CnV as a function of $\bar{x}/\bar{x}_{1/2}$ for seven values of β. The mean is represented by $\bar{x}$ and $\bar{x}_{1/2}$ is the value of the mean for which CnV = 0.5.

Anomalous cases can arise, for example, when the CnV does not approach zero at asymptotically large values of $\bar{x}$. In some cases we add a term that is constant as $\bar{x}$ approaches zero and vanishes as $\Phi_0(\bar{x})$ approaches one, such as:

$$\Phi(\bar{x}) = \Phi_0(\bar{x}) + \gamma[1 - \Phi_0(\bar{x})]. \tag{8}$$

We recommend using Equation (8) only when the data clearly indicate an asymptote greater than zero. This can happen when one data set (usually inferior) lies consistently above the others. It is unlikely to be a good candidate for pooling. One should always plot the individual candidates for pooling, to check against systematic differences.

CONFLICTS OF INTEREST

The authors declare no conflict of interest.

REFERENCES

1. Kelleher, D.E.; (National Institute of Standards and Technology, Gaithersburg, USA). Empirical Estimates of Relative Uncertainty. Unpublished work. 2014.
2. Kelleher, D.E.; Podobedova, L.I. Atomic transition probabilities of sodium and Magnesium. A critical complication. *J. Phys. Chem. Ref. Data* **2008**, *37*, 267–706.
3. International Organization for Standardization. *Guide to the Expression of Uncertainty in Measurement (GUM)*; International Organization for Standardization: Geneva, Switzerland, 1995; Section 6, Annex G.
4. Tachiev, G.; Froese Fischer, C. Breit Pauli energy levels, lifetimes, and transition data: Na II and Na III. Available online: http://nlte.nist.gov /MCHF/ view.html (accessed on 30 December 2002).
5. McPeake, D.; Hibbert, A. Transitions in Na III. *J. Phys. B* **2000**, *33*, 2809–2832.
6. Press, W.H.; Teukolsky, S.A.; Vetterling, W.T.; Flannery, B.P. *Numerical Recipes in C*; Cambridge University Press: Cambridge, UK, 1992.
7. Verrill, S. Exact confidence bounds for a normal distribution coefficient of variation. Available online: http://www1.fpl.fs.fed.us/covnorm. dcd.html (accessed on 30 December 2002).
8. Hastings, C. *Approximations for Digital Computers*; Princeton University Press: Princeton, NJ, USA, 1955.

CHAPTER 10

The Equation of State of Nuclear Matter and Neutron Stars Properties

Khaled S. A. Hassaneen[1,2]

[1]*PhysicsDepartment, Faculty of Science, Taif University, Taif, KSA*

[2]*Physics Department, Faculty of Science, Sohag University, Sohag, Egypt*

ABSTRACT

The equation of state (EOS) of symmetric nuclear and pure neutron matter has been investigated extensively by adopting the non-relativistic Brueckner-Hartree-Fock (BHF). For more comparison, the extended BHF approaches using the self-consistent Green's function approach or by including a three-body force will be done. The EOS will be studied for different approaches at zero temperature. We can calculate the total mass and radius of neutron stars using various equations of state. A comparison with relativistic BHF calculations will be done. Relativistic effects are known to be important at high densities, giving an increased repulsion. This leads to a stiffer EOS compared to the EOS derived with a non-relativistic approach.

Keywords: Symmetric Nuclear Matter, Equation of State, Three-Body Force, Pressure, Incompressibility, Neutron Star Properties

1. INTRODUCTION

The properties of neutron stars depend on the equation of state (EOS) at densities up to an order of magnitude higher than those observed in ordinary nuclei. The EOS determines properties such as the mass range, the mass-radius relationship, the crust thickness and the cooling rate [1] -[3] . Data on the EOS can be obtained from many sources, such as studies of the monopole resonance in finite nuclei, high energy nuclear collisions, supernovae, and neutron stars. Supernovae simulations seem to require an EOS which is too soft to support some observed masses of neutron stars, whereas analyses of high energy nuclear collisions indicate a rather stiff EOS, predicting neutron star masses which are too large. Thus, no definite statements can be made about the EOS at high densities, except that it should probably be moderately stiff in order to support maximum neutron star masses in a range of approximately 1.4 $M_\odot$ to 1.9 $M_\odot$, where $M_\odot$ is the solar mass [4] .

Several theoretical approaches to calculations of the EOS have been considered. The hypothesis that strange quark matter may be the absolute ground state of the strong interaction [5] , has been used by Rosenhauer et al. [6] in the investigation of the possibility of interpreting pulsars as rotating strange stars. Other approaches introduce exotic states of nuclear matter, such as kaon [7] or pion condensation [8] .

The scope of this work is to derive the EOS from the underlying many-body theory, derived from realistic nucleon-nucleon (NN) interactions such as CD-Bonn potential [9] . The many-body approaches that be used are self-consistent Green's function (SCGF) approach [10] -[14] and the Brueckner-Hartree-Fock (BHF) approach [15] supplemented by three-body force [16] or contact-term interaction to give more repulsive EOS [17] . From this EOS we will study neutron star observables such as the mass-radius relationship. Baldo et al. [18] have shown that any realistic EOS must satisfy several requirements to be used in studding neutron star properties: 1) It must display the correct saturation point for symmetric nuclear matter (SNM); 2) It must give a symmetry energy consistent with nuclear phenomenology and well behaved at high densities; 3) The nuclear incompressibility for SNM at saturation must be compatible with the values extracted from phenomenology (210 ± 30 MeV [19]); 4) The speed of sound remains smaller than the speed of light, (causality condition), for all densities relevant in neutron stars, i.e. $v/c = \sqrt{dP/d\varepsilon} \leq 1$, where P is the pressure and ε is the energy density. The latter condition is automatically satisfied only in fully relativistic theory.

In the self-consistent Green's function approach, the binding energy as well as all single-particle observables in the nuclear matter are calculated from the exact in-medium single-particle propagator. The latter is obtained from the Dyson equation, where medium effects are taken into account by the irreducible self-energy that is obtained from an expansion in terms of the effective interaction obtained from the sum of all ladder diagrams. One important feature of the SCGF approach is that particles and holes are treated on an equal footing, whereas in BHF only intermediate particle $(k > k_F)$ states are included in the ladder diagrams. This feature assures that the thermodynamic consistency is satisfied in the SCGF approach, e.g., the Fermi energy or chemical potential of the nucleons equals the binding energy at saturation (i.e., the Hugenholz-van Hove theorem) [20] . In the low- density limit, the BHF approach and the SCGF approach coincide. As the density increases, the phase space for hole-hole propagation is no longer negligible, and this leads to an important repulsive effect on the total energy. Since particle-particle (pp) and hole-hole (hh) ladders are treated in a completely symmetrical way in the SCGF approach, the Green's function scheme is also suited for calculations at higher densities [12] -[14] . Furthermore, the SCGF generates realistic spectral functions, which can be used to evaluate the effective interaction and corresponding nucleon self-energy [10] [11] .

The paper is organized as follows. In the next section we will describe the formalisms within the non-relati- vistic approaches we have employed. Results for the EOS and pressure of symmetric and pure neutron matter using suggested models, together with neutron star matter observables will be presented in Section 3. A short summary and some conclusions will be given in Section 4.

2. THE THEORETICAL MODEL

2.1. BHF Approximation

Starting from realistic nucleon-nucleon (NN) interaction, we have to use more advanced many-body approximations like the BHF which have the capability to account for the effects of correlations, which are due to the strong tensor and short-range components of such realistic NN interaction. The single-particle energy of a particle in the BHF approximation corresponds to the Hartree-Fock expression using the G-matrix for the effective interaction. This means that the self-energy of a nucleon in nuclear matter with momentum k, isospin τ and energy ω in asymmetric nuclear matter is given by [11]

$$\sum_{\tau}^{\mathrm{BHF}}(\boldsymbol{k},\omega)=\int \mathrm{d}^3q\,\boldsymbol{kq}\left|G(\Omega)\right|\boldsymbol{kq}\,n_{\tau}^{0}(\boldsymbol{q}) \tag{1}$$

with the occupation probability of a free Fermi gas of protons $(\tau=p)$ and neutrons $(\tau=n)$ with a Fermi momentum $k_{F\tau}$

$$n_{\tau}^{0}(\boldsymbol{q})=\begin{cases}1 & \text{for} \quad |q|\le k_{F\tau}, \\ 0 & \text{for} \quad |q|>k_{F\tau},\end{cases} \tag{2}$$

This means for asymmetric nuclear matter with a total density ρ and asymmetry parameter α that

$$\rho=\rho_p+\rho_n, \qquad \alpha=\frac{\rho_n-\rho_p}{\rho} \tag{3}$$

With Fermi momenta for protons (k_{Fp}) and neutrons (k_{Fn}) which are related to the corresponding densities by

$$\rho_{\tau}=\frac{1}{3\pi^2}k_{F\tau}^{3} \tag{4}$$

The matrix elements in Equation (1) denote antisymmetrized matrix elements of the Brueckner G-matrix that are determined by solving the Bethe-Goldstone equation for a given realistic NN interaction V

$$\boldsymbol{kq}\left|G(\Omega)\right|\boldsymbol{kq}=\boldsymbol{kq}\left|V\right|\boldsymbol{kq}+\int \mathrm{d}^3p_1\mathrm{d}^3p_2\,\boldsymbol{kq}\left|V\right|\boldsymbol{p_1p_2}\,\frac{Q(p_1,p_2)}{\Omega-\left(\varepsilon_{p_1}+\varepsilon_{p_2}\right)+i\eta}\,\boldsymbol{p_1p_2}\left|G(\Omega)\right|\boldsymbol{kq} \tag{5}$$

The single-particle energies $\varepsilon_{p\tau}$ of the intermediate states should be the corresponding BHF single-particle energies which are defined in terms of the real part of the BHF self-energy of Equation (1) by

$$\varepsilon_{k\tau} = \frac{k^2}{2m} + \mathrm{Re}\left[\sum\nolimits_{\tau}^{\mathrm{BHF}}\left(\boldsymbol{k}, \omega = \varepsilon_{k\tau}\right)\right] \tag{6}$$

with a starting energy parameter $\Omega = \omega + \varepsilon_q$ in the Bethe-Goldstone Equation (5).

The Pauli operator $Q(p_1, p_2)$ restricts the intermediate states to particle states with momenta p_1, p_2, which are above the corresponding Fermi momentum. However, the single-particle spectrum is often parameterized in the form of an effective mass

$$\varepsilon_k \approx \frac{k^2}{2m^*} + U \tag{7}$$

so that a so-called angle-averaged propagator can be defined, which reduces the Bethe-Goldstone equation to an integral equation in one dimension. The exact Pauli operator has been treated in [21] .

2.2. Self-Consistent Green's Function

One of the drawbacks of the BHF approximation is the fact that it does not provide results for the equation of state, which are consistent from the point of view of thermodynamics. As an example we mention that BHF results do not fulfill, e.g., the Hugenholtz-Van Hove theorem [22] . This is due to the fact that the BHF approximation does not consider the propagation of particle and hole states on equal footing. An extension of the BHF approximation, which obeys this symmetry, is the self-consistent Green's function (SCGF) method using the so-called T-matrix approximation. During the last years techniques have been developed, which allow to evaluate the solution of the SCGF equations for microscopic NN interactions [23] -[25] . Those calculations demonstrate that for the case of realistic NN interactions, the contribution of particle-particle ladders dominates the contribution of corresponding hole-hole propagation terms. This justifies the use of the BHF approximation and a procedure, which goes beyond BHF and accounts for hole-hole terms in a perturbative way [26] . This leads to a modification of the self-energy in the BHF approximation by adding a hole-hole term of the form [26]

$$\Delta\sum_{\tau}^{2h1p}(k,\omega)=\sum\int_{k_{F\tau}}^{\infty}\mathrm{d}^3p\int_0^{k_{F\tau}}\mathrm{d}^3h_1\int_0^{k_{F\tau}}\mathrm{d}^3h_2\times\frac{kp\left|G(\Omega)\right|h_1h_2^2}{\omega+\varepsilon_p-\varepsilon_{h_1}-\varepsilon_{h_2}+i\eta} \tag{8}$$

The quasi-particle energy for the extended self-energy can be defined as

$$\varepsilon_{k\tau}^{qp}=\frac{k^2}{2m}+\mathrm{Re}\left[\Sigma^{\mathrm{BHF}}\left(k,\omega=\varepsilon_{k\tau}^{qp}\right)+\Delta\Sigma^{2h1p}\left(k,\omega=\varepsilon_{k\tau}^{qp}\right)\right] \tag{9}$$

The spectral functions for hole and particle strength, $S^h(k,\omega)$ and $S^p(k,\omega)$, are obtained from the real and imaginary part of the self-energy $\Sigma=\Sigma^{\mathrm{BHF}}+\Delta\Sigma^{2h1p}$

$$S_{\tau}^{h(p)}(k,\omega)=\pm\frac{1}{\pi}\frac{\mathrm{Im}\Sigma_{\tau}(k,\omega)}{\left[\omega-k^2/2m-\mathrm{Re}\Sigma_{\tau}(k,\omega)\right]^2+\left[\mathrm{Im}\Sigma_{\tau}(k,\omega)\right]^2} \tag{10}$$

where the plus and minus sign on the left-hand side of this equation refers to the case of hole $(h,\omega<\varepsilon_{F\tau})$ and particle states $(p,\omega>\varepsilon_{F\tau})$, respectively. The hole strength represents the probability that a nucleon with isospin i, momentum k, and energy ω can be removed from the ground state of the nuclear system with the removal energy ω, whereas the particle strength denotes the probability that such a nucleon can be added to the ground state of the system with A nucleons resulting in a state of the A + 1 particle system which has an energy of ω relative to the ground state of the A particle system.

Integrating the spectral distribution of the hole states yields the occupation probability

$$n^{\tau}(k)=\int_{-\infty}^{\varepsilon_{F\tau}}\mathrm{d}\omega S_{\tau}^{h}(k,\omega) \tag{11}$$

In the case of two body interactions, the hole spectral function gives access, through the Koltun sum-rule [27] , to the binding energy per particle,

$$\frac{E}{A}(\rho)=\frac{\deg}{\rho(2\pi)^3}\int\mathrm{d}^3k\int_{-\infty}^{\varepsilon_F}\frac{1}{2}\left(\frac{k^2}{2m}+\omega\right)S_h(k,\omega)\mathrm{d}\omega \tag{12}$$

with deg denoting the degeneracy of the single-particle level, which is 4 for nuclear matter. The kinetic energy per nucleon,

$$t/A(\rho)=\frac{\deg}{\rho(2\pi)^3}\int \mathrm{d}^3k\,\frac{k^2}{2m}\int_{-\infty}^{\varepsilon_F}\mathrm{d}\omega S_h(k,\omega)=\frac{\deg}{\rho(2\pi)^3}\int \mathrm{d}^3k\,\frac{k^2}{2m}n(k) \tag{13}$$

and the potential energy per particle has the form $v/A(\rho)=E/A(\rho)-t/A(\rho)$. Introducing the removal energy $\varepsilon_r(k)$,

$$\varepsilon_r(k)=\frac{\int_{-\infty}^{\varepsilon_F}\mathrm{d}\omega\,\omega S_h(k,\omega)}{\int_{-\infty}^{\varepsilon_F}\mathrm{d}\omega S_h(k,\omega)} \tag{14}$$

One can express the total binding energy per nucleon as

$$E/A(\rho)=\frac{\deg}{\rho(2\pi)^3}\int \mathrm{d}^3k\,\frac{1}{2}\left(\frac{k^2}{2m}+\varepsilon_r(k)\right)n(k) \tag{15}$$

In Brueckner-Hartree-Fock, $\varepsilon_r(k)=k^2/2m+u(k)$, i.e. it coincides with the quasi-particle energy, and $n(k)$ denotes the step function, $n(k)=\theta(k_F-k)$, therefore [28]

$$E/A^{\mathrm{BHF}}(\rho)=\frac{\deg}{\rho(2\pi)^3}\int \mathrm{d}^3k\left(\frac{k^2}{2m}+\frac{1}{2}u(k)\right) \tag{16}$$

2.3. Tolman-Oppenheimer-Volkov

Neutron stars are very interesting physical systems and their properties, such as masses and radii as function of the central density, can be derived from the equation of state (EOS) of the β-stable matter contained in them. The EOS is microscopically calculated from the sections a and b. After that, briefly we outline the derivation of neutron-star properties from its EOS. One starts from the Tolman-Oppenheimer-Volkov (TOV) equations for the total pressure P and the enclosed mass m [29] [30] ,

$$\frac{dP(r)}{dr}=\frac{Gm(r)\rho(r)}{r^2}\left(1+\frac{P(r)}{c^2\rho(r)}\right)\left(1+\frac{4\pi r^3P(r)}{c^2m(r)}\right)\left(1-\frac{2Gm(r)}{rc^2}\right)^{-1} \quad (17)$$

$$\frac{dm(r)}{dr}=4\pi r^2\rho(r) \quad (18)$$

where $P(r)$ is the pressure at radius r, $m(r)$ is the gravitational mass inside r, and G is the gravitational constant.

From the EOS pure neutron matter, we calculate the nuclear contribution P_{nucl} to the total pressure of stellar matter using the relation

$$P_{\text{nucl}}=\rho^2\frac{d(E/A)}{d\rho} \quad (19)$$

Then the total pressure and the total mass density ρ are given by

$$P=P_{\text{nucl}}+P_{\text{lep}} \quad (20)$$

$$\rho=\frac{1}{c^2}\left(\rho(E/A)+\rho m_N+\varepsilon_{\text{lep}}\right) \quad (21)$$

Being P_{lep} and ε_{lep} the lepton contribution to the total pressure and energy density, m_N is the nucleon mass and c is the speed of light in vacuum.

Starting with a central mass density $\rho(r=0)\equiv\rho_c$, we integrate out until the pressure on the surface equals the one corresponding to the density of iron. This gives the stellar radius R and the gravitational mass is then

$$M_G\equiv m(r)=4\pi\int_0^R drr^2\rho(r) \quad (22)$$

For the outer part of the neutron star we have used the equations of state by Baym et al. [31] , and for the middle density region (0.001 $fm^{-3} < \rho < 0.08$ fm^{-3}) we have used the results of Negele et al. [32] . In the high-density part of neutron star, ($\rho > 0.08$ fm^{-3}), the present EOS will be used.

3. RESULTS AND DISCUSSION

3.1. EOS, Pressure and Incompressibility of Nuclear Matter

All results of calculations, which refer to realistic NN interactions, have been obtained using the CD-Bonn interaction [9] . This includes all BHF and SCGF calculations. The effects which are related to the use of various NN interactions, which all fit the NN phase shifts, have been discussed before for BHF approach only [15] .

In Figure 1, the binding energy per particle of symmetric nuclear is plotted versus the density. The BHF results are indicated by a solid line, whereas the results of the full SCGF calculation, computed at zero temperature, are presented by a dashed line. In the case of symmetric nuclear matter, Figure 1, the empirical saturation point is given by the big square. Comparing the BHF and the SCGF results, the hole-hole contributions to the nucleon self-energy, which are included in the SCGF approach, yield a repulsive contribution to the energy per nucleon. One should keep in mind that this repulsion increases with increasing nuclear density. This shifts the saturation point to a lower density and binding energy per nucleon so that also the saturation point obtained with SCGF is within the Coester band [33] . Due to this effect, the saturation density in the SCGF approximation, $\rho_0^{SCGF} \approx 0.21\ \text{fm}^{-3}$ and $E_0^{SCGF} \approx -11.4\ \text{MeV}$, is considerably smaller than the saturation density in the BHF approach that has the value, $\rho_0^{BHF} \approx 0.39\ \text{fm}^{-3}$ and $E_0^{BHF} \approx -24\ \text{MeV}$. After all the saturation density is reduced by almost 50%; however, both values are not near the empirical one ($\rho_0 = 0.16\ \text{fm}^{-3}$; $E_0 = -16$ MeV). In the low-density limit, the BHF approach and the SCGF approach coincide. As the density increases, the phase space for hole-hole propagation is no longer negligible, and this leads to an important repulsive effect on the total energy. Since particle-particle and hole-hole ladders are treated in a completely symmetrical way in the SCGF approach, the Green's function scheme is also appropriated for calculations at higher densities [12] [13] .

As we see, non-relativistic calculations, based on purely two-body interaction, fail to reproduce the correct saturation point for symmetric nuclear matter. This well-known deficiency is commonly corrected by introducing three-body force (3BF). Relevant progress has been made in the theory of nucleon 3BF, but a complete theory is not yet available. A realistic model for nuclear 3BF has been introduced by the Urbana group [34] .

There is another method used to achieve saturation properties in nuclear matter. One has to supplement the effective interaction or the self-energy of BHF and SCGF calculations by a simple contact interaction (CT), which we have chosen following the notation of the Skyrme interaction to be of the form

$$\Delta H/\rho = CT = \frac{1}{2}t_0\rho + \frac{1}{12}t_3\rho^{1+\alpha} \tag{14}$$

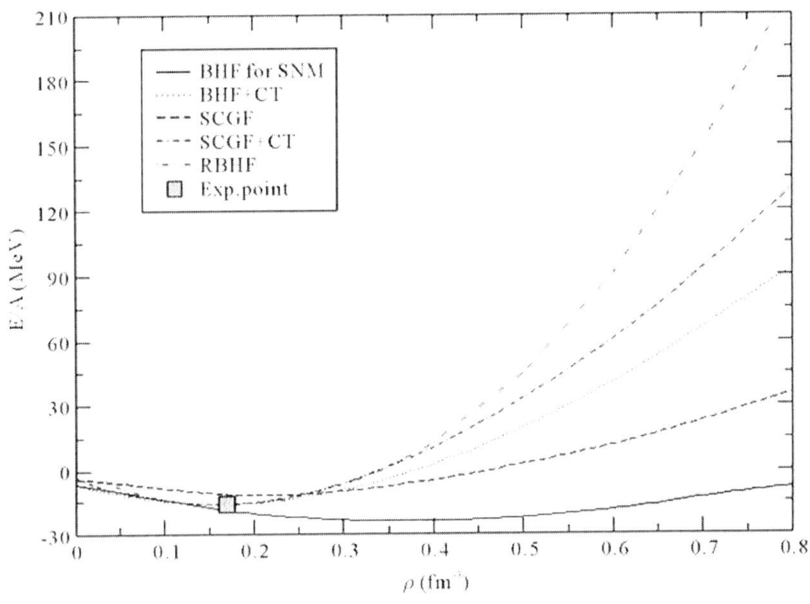

Figure 1. The energy per particle E/A is plotted vs. the density ρ for symmetric nuclear matter within different models described in the text using CD-Bonn potential. The results are compared with RBHF [37] by Müther et al., whereas the empirical saturation point is given by the big square.

where ρ is the matter density, t_0, t_3 and α are parameters. The parameter t_0 and t_3 represent the zero range and 3-body strength while the exponent α determines the high density behavior. The Hamiltonian density of the contact term, which is used in the last equation, has no isospin dependence. More parameters used in [35] [36] to give a more quantitative estimate of the needed correction to the BHF results.

Now the nuclear many-body approach plus the above contact term is fit to reproduce the saturation point ρ_0 = 0.16 fm^{-3} and E_0 = −16.0 MeV at fixed α (typically $\alpha = 0.5$). The results for these fitting parameters are listed in Table 1, whereas the corresponding EOS versus density curves are displayed in Figure 1 using dotted line for BHF + CT and double dash-dotted line for SCGF + CT. Also, the EOS's using all discussed approaches are listen in Table 2 For all cases, the fit yields an attractive two-body contact interaction and a repulsive t_3 term. Comparing the suggested approaches with the relativistic Brueckner-Hartree-Fock (RBHF) approach [37] , we find that RBHF approach leads to the stiffer EOS around the saturation density as well as at higher densities than the others. The SCGF and the RBHF calculations yield rather similar results after the contact terms are included up to density equals to 2.5ρ_0, but the difference increases at high densities, which are a little bit softer than the BHF + CT results.

We will extend the analysis to pure neutron matter EOS, which is more suitable for neutron star studies, at densities up to about five times the saturation one. Moreover, we consider the calculations for two nucleon-nucleon potential, i.e., CD-Bonn interaction likes symmetric nuclear matter EOS. The results for the EOS of the pure neutron matter, obtained by including only two-body force, is reported (solid line for BHF and dashed line for SCGF approaches) in Figure 2. The results of both BHF and SCGF approaches plus a simple contact interaction, with the same parameters defined in Table 1, are illustrated by dotted and double dash-dotted curves, respectively. As for symmetric nuclear matter, all present EOS's for neutron matter are compared with the RBHF calculations introduced by Müther et al. [37] . Comparing the BHF and the SCGF results without the contact term, it gets clear that the inclusion of hh excitations makes EOS of neutron matter more repulsive especially at high density. When the contact term is added to the BHF and the SCGF approaches, the results for EOS are more comparable stiffness. Also, these observations are shown in Table 3. As one can see that, the SCGF and the RBHF calculations yield rather similar results after the contact terms are introduced up to density greater than twice of saturation density. Furthermore the relative differences between two approaches are smaller.

The pressure of nuclear matter is defined in terms of the energy per nucleon as in Equation (16). In Figure 3

Table 1. Parameters t_0 and t_3 as obtained for the fit to the saturation point $\rho_0 = 0.16$ fm^{-3} and E/A = −16.0 MeV at $\alpha = 0.5$ for BHF and SCGF approaches.

Parameters	BHF	SCGF
t_0 (MeV fm^3)	−221	−327
t_3 (MeV fm$^{3+3\alpha}$)	3750	3917

Table 2. The nuclear binding energy per nucleon E/A in MeV is shown vs. the nucleon density ρ in fm^{-3} for symmetric nuclear matter. The values are the results of a non-relativistic BHF, SCGF, BHF + CT and SCGF + CT calculations with CD- Bonn potential.

Model ρ (fm^{-3})	BHF	SCGF	BHF + CT E/A (MeV)	SCGF + CT
0.08	−12.373	−7.209	−13.364	−12.729
0.16	−18.28	−10.754	−16.013	−16.061
0.24	−21.966	−11.285	−13.676	−12.647
0.32	−23.624	−8.914	−7.073	−3.313
0.40	−23.889	−4.499	2.853	10.672
0.48	−22.749	1.249	15.91	28.315
0.56	−20.169	7.863	31.979	48.948
0.64	−16.667	15.722	50.427	72.789
0.72	−12.811	25.164	70.589	100.05
0.80	−8.822	35.524	92.167	129.96

Table 3. Same as Table 2, but for pure neutron matter.

Model ρ (fm^{-3})	BHF	SCGF	BHF+CT E/A (MeV)	SCGF+CT
0.08	8.094	10.43	7.103	4.91
0.16	11.871	17.572	14.137	12.265
0.24	16.247	25.886	24.538	24.525
0.32	21.704	35.563	38.255	41.165
0.40	28.559	46.614	55.301	61.784
0.48	36.583	59.33	75.242	86.396
0.56	45.576	73.946	97.724	115.03
0.64	55.989	90.175	123.083	147.241
0.72	68.164	107.572	151.563	182.458
0.80	81.354	125.569	182.344	220.006

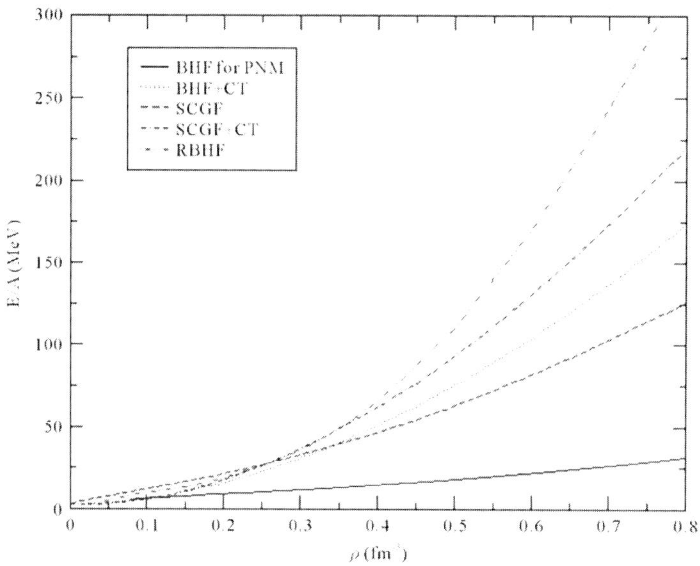

Figure 2. The energy per particle E/A is plotted vs. the density ρ for pure neutron matter within different models described in the text using CD-Bonn potential. The results are compared with RBHF calculations by Müther et al. [37] .

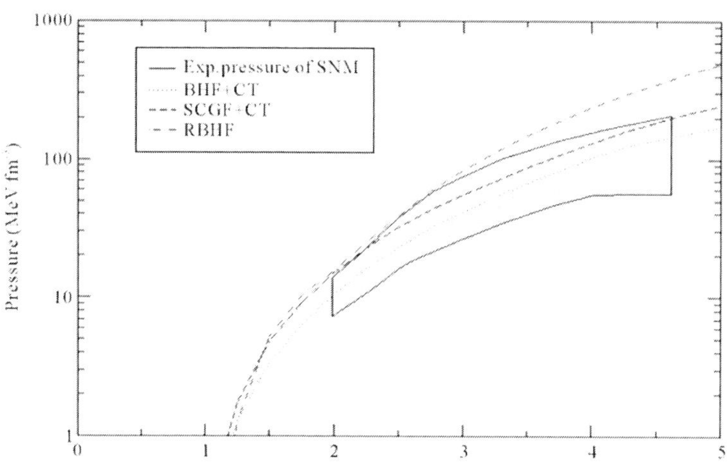

Figure 3. The pressure of SNM is illustrated as a function of ρ/ρ_0. The dotted line represents the BHF + CT; the dashed line represents the SCGF + CT whereas the dash-dotted line represents the RBHF model. The area enclosed by the continuous line corresponds to the region of pressures consistent with the experimental flow data [38] for SNM.

and Figure 4, we show the pressure from the BHF + CT and SCGF + CT approaches based on EOS for symmetric matter and neutron matter. These values of the pressure are compared with RBHF results by Müther et al. [37] . The area within the double dash-dotted lines indicates EOS consistent with the elliptic flow measurements reported in Ref. [38] of symmetric nuclear matter. For the case of neutron matter, two different parameterizations for the symmetry energy, $a_s(\rho)$, are assumed in the analysis. The two pressure contours appear in Figure 4 correspond to the weakest (lower contour with double dot-dashed line) and the strongest (higher contour with double dash-dotted line) density dependence for $a_s(\rho)$ suggested by Prakash et al. [39] . It is fair to say that SCGF + CT predictions produce a reasonable amount of repulsion like RBHF calculations, whose generate too much pressure, while the EOS compared to that BHF + CT predictions are less repulsive.

There is another important characteristic of the EOS and it enters in the discussion of a variety of phenomena such as supernovae explosions or heavy ion collisions, that it is the incompressibility K. It measures the stiffness of the EOS, usually defined as a slope of the pressure at saturation point:

$$K = 9\left.\frac{\partial P(\rho)}{\partial \rho}\right|_{\rho=\rho_0} = 9\rho^2 \left.\frac{\partial^2 E/A(\rho)}{\partial \rho^2}\right|_{\rho=\rho_0} \quad (28)$$

The experimental value of the incompressibility of symmetric nuclear matter at its saturation density ρ_0 has been determined to be 210 ± 30 MeV [19] . At the saturation density, the values of the incompressibility and the binding energy per nucleon obtained for symmetric nuclear matter are summarized in Table 4. In the case of RBHF, although the EOS of nuclear matter is characterized by a relatively low value of K, the behavior with increasing density is fairly stiff. This is reflected in the relatively large values of the pressure in the region (2 - 5) ρ_0 as one sees from Figure 3.

3.2. Neutron Star Properties

Using the methods just described we obtain for each model the EOS for cold pure neutron matter. The TOV general relativistic equations for a spherically symmetric (non-rotating) neutron star are solved, then the gravitational mass of the star M_G is obtained as a function of both the stellar radius R and a central density ρ_c. The dependence of the neutron star masses on the stellar radius R for three models is shown in Figure 5. Results with the RBHF by Müther et al. [37] are included for comparison. The maximum masses for the three models as function of ρ_c are illustrated in Figure 6. From Figure 5 and Figure 6 we find a maximum mass of neutron star is $M_{max} \approx 1.98\ M_{\odot}$, in the case of BHF + CT calculation, at a central density of $\rho_c = 2.82 \times 10^{15}$ gm/cm^3 with

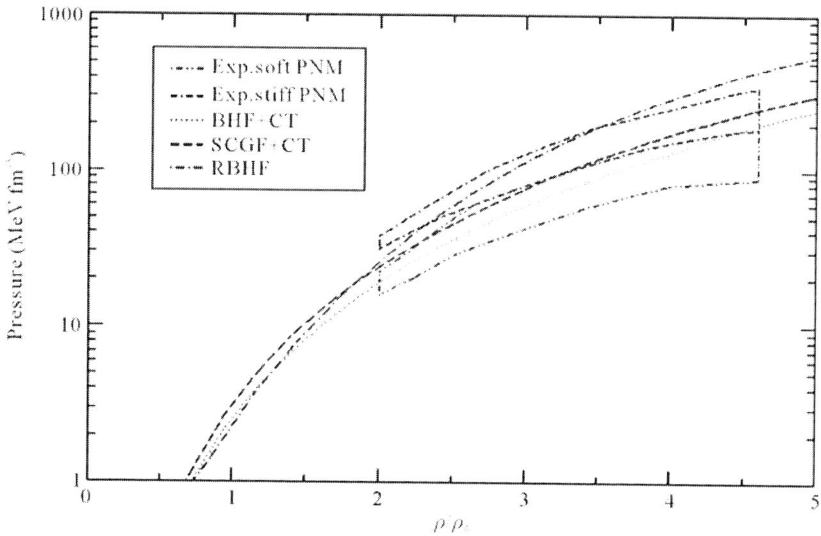

Figure 4. The pressure of PNM is shown as a function of ρ/ρ_0. The dotted line represents the BHF + CT, the dashed line represents the SCGF + CT whereas the dash- dotted line represents the RBHF model. The areas enclosed by the double dash-dotted and the double dot-dashed lines show the pressures regions for PNM consistent with the experimental flow data after inclusion of the pressures from asymmetry terms with soft PNM and stiff PNM density dependencies, respectively [38] .

Table 4. The main bulk properties of the nuclear matter that be extracted from the equation of state at saturation points. These values are the saturation density ρ_0, saturation energy E_0, and the incompressibility K. These values are constructed using different models.

Model	ρ_0 (fm^{-3})	E_0 (MeV)	K (MeV)
BHF + CT	0.16	−16.01	209
SCGF + CT	0.16	−16.06	266
RBHF [37]	0.157	−16.0	200

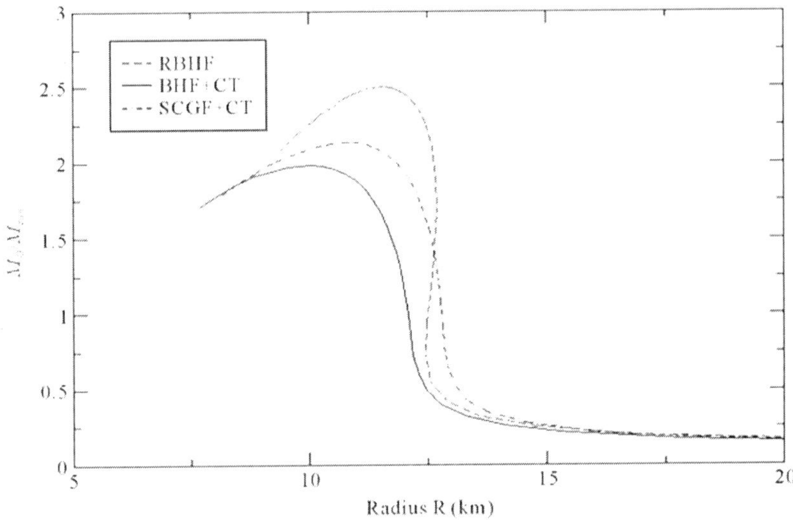

Figure 5. Predicted neutron-star gravitational masses M_G (in units of $M_{\odot}$= M_{sun}) plotted as a function of the stellar radius (in km). Two different Equation of states are considered, those for pure neutron matter and compared with previous RBHF results by Müther et al. [37] . The calculations are performed with CD-Bonn potential. Solid line is for BHF + CT, the double dash-dotted line is for SCGF + CT whereas the dashed line is for RBHF model.

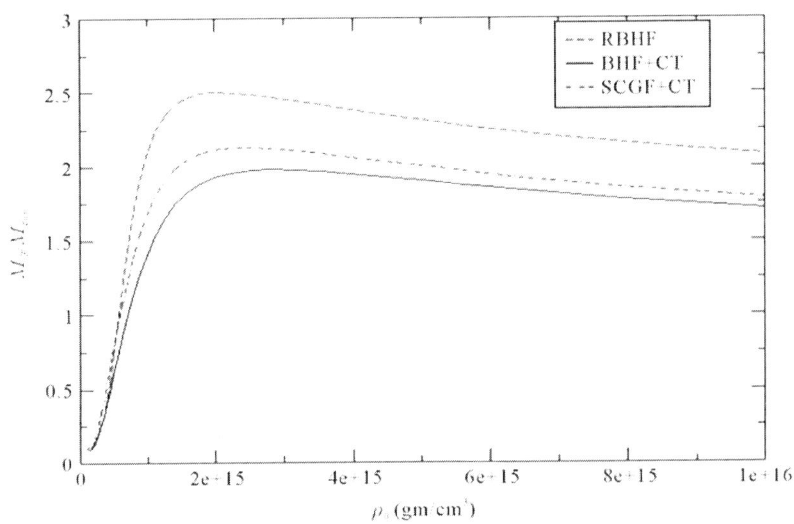

Figure 6. Neutron-star gravitational masses M_G (in units of $M_{\odot}$ = M_{sun}) plotted as a function of the central density ρ_c (in units of gm/cm³) for different models described in the text. The notation is as in Figure 5.

a radius R ≈ 10.05 km. This value is consistent with the recent observation of a (1.97 ± 0.04) $M_\odot$ neutron star (Demorest et al. [40]). Also, it was reported that a new massive NS, PSR J0348 + 0432 with (2.01 ± 0.04) $M_\odot$, was observed very recently by Antoniadis et al. [41] . Whereas a maximum mass of $M_{max} \approx 1.98\ M_\odot$ at a central density of $\rho_c = 2.52 \times 10^{15}$ gm/cm^3 with a radius R ≈ 10.5 km for SCGF+CT model. The present non-relativistic calculations are compared with RBHF results [37] where a maximum mass of $M_{max} \approx 2.5\ M_\odot$ at a central density of $\rho_c = 1.995 \times 10^{15}$gm/cm^3 with a radius R ≈ 11.58 km.

One can see that the results from our non-relativistic equation of state may look more reasonable than those from the relativistic one. The present results are more compatible with recent microscopic calculations of neutron matter based on nuclear interactions derived from chiral effective field theory [42] . They have constructed three representative EOSs (soft, intermediate, and stiff), consistent with the constraints from nuclear physics and observation. Generally, simple construction of neutron stars from the EOS of pure neutron matter is not enough for a consistency check between the theoretical neutron star maximum mass and the observed neutron star masses. If it is possible to duly deduce the information on the EOS of nuclear matter by ignoring other constituents than nucleons, it would be reasonable to focus on low mass neutron stars (see, e.g., Sotani et al. [43]).

4. CONCLUSIONS

Comparing the non-relativistic approaches from realistic interaction among themselves, we find that the effect of the hole-hole interaction has been found to be important; it significantly accounts for more repulsion at large densities. Also, it enhances the nuclear matter E/A and at the same time reduces the saturation density ρ_0, as compared with the BHF results. If we do not consider the contact term these characteristics lead to better values for the saturation point. In general, microscopic BHF or SCGF calculations cannot simultaneously reproduce nuclear matter E/A and ρ_0. To improve the situation, one may either extend to three-body force to the effective interaction or include relativistic corrections RBHF, then the EOS of pure neutron matter can be used to determine the structure of neutron stars.

Using the present EOS's of pure neutron matter at absolute zero, the TOV equation of general relativity, from which the gross properties of neutron stars follow, e.g., mass and radius can be solved numerically. It turns out that both maximum stable masses and radii depend crucially on the stiffness of the suggested EOS. Neutron stars including realistic EOS give the following general results: star model calculated with a stiff EOS have a lower central density, a larger radius than do stars of the same mass computed from a soft EOS. There is another important consequence; NS parameters such as the total mass and radius, are sensitive to microscopic model calculations. Also, the results from non-relativistic equation of state may look more reasonable than those from the relativistic one.

ACKNOWLEDGEMENTS

Khaled S. A. Hassaneen, The present research support from Taif University, Kingdom of Saudi Arabia under Project No. 1-434-2667 is acknowledged.

REFERENCES

1. Weber, F., Glendenning, N.K. and Weigel, M.K. (1991) Astrophysical Journal, Part 1, 373, 579-591.
2. Pethick, C.J., Ravenhall, D.G. and Lorenz, C.P. (1995) Nuclear Physics A, 584, 675-703.
3. Lorenz, C.P., Ravenhall, D.G. and Pethick, C.J. (1993) Physical Review Letters, 70, 379.
4. Thorsett, S.E., Arzoumanian, Z., McKinnon, M.M. and Taylor, J.H. (1993) Astrophysical Journal, Part 2, 405, L29- L32.
5. Witten, E. (1984) Physical Review D, 30, 272.
6. Rosenhauer, A., Staubo, E.F., Csernai, L.P., Øvergård, T. and Østgaard, E. (1992) Nuclear Physics A, 540, 630-645.
7. Brown, G.E., Lee, C.H., Rho, M. and Thorsson, V. (1994) Nuclear Physics A, 567, 937-956.
8. Migdal, A.B., Saperstein, E.E., Troitsky, M.A. and Voskresensky, D.N. (1990) Physics Reports, 192, 179-437.
9. Machleidt, R. (2001) Physical Review C, 63, Article ID: 024001.
10. Frick, T., Hassaneen, Kh.S.A., Rohe, D. and Müther, H. (2004) Physical Review C, 70, Article ID: 024309.
11. Hassaneen, Kh.S.A. and Müther, H. (2004) Physical Review C, 70, Article ID: 054308.
12. Gad, Kh. and Hassaneen, Kh.S.A. (2007) Nuclear Physics A, 793, 67-78.
13. Hassaneen, Kh.S.A. and Gad, Kh. (2008) Journal of the Physical Society of Japan, 77, Article ID: 084201.
14. Mansour, H., Gad, Kh. and Hassaneen, Kh.S.A. (2010) Progress of Theoretical Physics, 123, 687-700.
15. Hassaneen, Kh.S.A., Abo-Elsebaa, H.M., Sultan, E.A. and Mansour, H.M.M. (2011) Annals of Physics, 326, 566-577.
16. Baldo, M. and Shaban, A. (2008) Physics Letters B, 661, 373-377.
17. Gögelein, P., van Dalen, E.N.E., Gad, Kh., Hassaneen, Kh.S.A. and Müther, H. (2009) Physical Review C, 79, Article ID: 024308.
18. Baldo, M., Bombaci, I. and Burgio, G.F. (1997) Astronomy and Astrophysics, 328, 274-282.

19. Haensel, P., Potekhin, A.Y. and Yakovlev, D.G. (2007) Neutron Stars 1, Equation of State and Structure. Springer, New York.
20. Hugenholz, N.M. and Van Hove, L. (1958) Physica, 24, 363-376.
21. Schiller, E., Müther, H. and Czerski, P. (1999) Physical Review C, 59, 2934.
22. Hassaneen, Kh.S.A. and Mansour, H.M.M. (2013) Journal of Nuclear and Particle Physics, 3, 77-96.
23. Dewulf, Y., Dickhoff, W.H., Van Neck, D., Stoddard, E.R. and Waroquier, M. (2003) Physical Review Letters, 90, Article ID: 152501.
24. Bożek, P. (2002) Physical Review C, 65, Article ID: 054306.
25. Frick, T. and Müther, H. (2003) Physical Review C, 68, Article ID: 034310.
26. Grange, P., Cugnon, J. and Lejeune, A. (1987) Nuclear Physics A, 473, 365-393.
27. Koltun, D.S. (1974) Physical Review C, 9, 484.
28. Hassaneen, Kh.S.A. (2013) Physics Research International, 2013, Article ID: 415605, 6 p.
29. Oppenheimer, J. and Volkoff, G. (1939) Physical Review, 55, 374.
30. Tolman, R.C. (1934) Proceedings of the National Academy of Sciences of the United States of America, 20, 169-176.
31. Baym, G., Pethick, C. and Sutherland, P. (1971) Astrophysical Journal, 170, 299.
32. Negele, J.W. and Vautherin, D. (1973) Nuclear Physics A, 207, 298-320.
33. Day, B.D. (1981) Physical Review Letters, 47, 226.
34. Carlson, J., Pandharipande, V.R. and Wiringa, R.B. (1983) Nuclear Physics A, 401, 59-85.
35. Hassaneen, Kh.S.A. and Mansour, H.M.M. (2013) Journal of Modern Physics, 4, 37-41.
36. Mansour, H.M.M. and Hassaneen, Kh.S.A. (2014) Physics of Atomic Nuclei, 77, 290-298.
37. Müther, H., Prakash, M. and Ainsworth, T.L. (1987) Physics Letters B, 199, 469-474.
38. Danielewicz, P., Lacey, R. and Lynch, W.G. (2002) Science, 298, 1592-1596.
39. Prakash, M., Ainsworth, T.L. and Lattimer, J.M. (1988) Physical Review Letters, 61, 2518.
40. Demorest, P.B., Pennucci, T., Ransom, S.M., Roberts, M.S.E. and Hessels, J.W.T. (2010) Nature, 467, 1081-1083.
41. Antoniadis, J., Freire, P.C.C., Wex, N., Tauri, T.M., Lynch, R.S., van Kerkwijk, M.H., et al. (2013) Science, 340, 1233232-1:9.
42. Hebeler, K., Lattimer, J.M., Pethick, C.J. and Schwenk, A. (2013) The Astrophysical Journal, 773, 11, 14 p.
43. Sotani, H., Iida, K., Oyamatsu, K. and Ohnishi1, A. (2014) arXiv:1401.0161v2 [astro-ph.HE]

CHAPTER 11

From Space-Time Quantization to Dark Matter

Auguste Meessen

UCL, Louvain-la-Neuve, Belgium

ABSTRACT

We generalize relativistic quantum mechanics and the Standard Model of elementary particle physics by considering a finite limit for the smallest measurable length. The resulting theory of Space-Time Quantization is logically consistent and accounts for all possible particle states by means of four new quantum numbers. They specify possible variations of wave functions at the smallest possible scale in space and time, while states of motion are defined by their large-scale variations. This theory also provides insight into the nature and properties of dark matter particles. It can facilitate their detection and identification because of a very strict conservation law.

Keywords: Dark Matter, Elementary Particles, Relativistic Quantum Mechanics, Space-Time Quantization

1. INTRODUCTION

Although more than 80% of all matter in our universe is dark matter (DM), it has not yet been possible to identify this type of particles. We know only that they cannot interact with electromagnetic fields, but are subjected to gravitational forces. They have thus to be electrically neutral and to carry some mass. To understand the nature and properties of DM particles is "one of the most fundamental open questions in cosmology and particle physics" [1] . It is related to the equally basic problem of identifying the nature of dark energy and the cause of the accelerated expansion of our universe [2] . Moreover, we know that the Standard Model (SM) of elementary particle physics describes very remarkable facts, but does not explain them. This "spectroscopy" is similar to that of atomic physics before the development of quantum mechanics.

All these problems seem to call for a generalization of present day theories, as this happened already when it appeared that classical mechanics contained simplifying assumptions. They were corrected by the theory of relativity and quantum mechanics, but we kept the concept of a "space-time continuum". This allows us to use differential equations, but physical laws have to be verifiable. They imply thus the belief that it should be possible to measure always smaller and smaller lengths, without ever reaching a finite limit. Where does this conviction come from and are the justifications really compelling?

At the beginning of rational thinking, it was obvious that a length has to be measured by comparing it to a juxtaposition of some invariant unit of length. To achieve utmost precision, it would have to be an extremely small, undividable length. This led to the idea that any length is an integer multiple of some elementary length a. Adjacent sides of a square and the length of its diagonal would then respectively be equal to na and ma, where n and m are integer numbers. Because of the Pythagorean Theorem, we get thus $m^2 = 2n^2$, whatever the value of a may be. Since the square of an odd number is odd, it appears that m is an even number. Setting $m = 2q$, where q is an integer number, it appears that $n^2 = 2q^2$. The number n would thus also be even, but when n and m can be divided by 2, it is not true that a is undividable. It was concluded that the divisibility of lengths has to be unlimited.

This proof was perfectly rational and justified the mathematical concept of "infinitely" small intervals. It had even great cultural impact, since it supported the idea that logical reasoning is able to grasp what is out of reach of observations. It was therefore generally accepted that space is continuous, but this claim was flawed. Indeed, when the smallest measurable length $a \neq 0$, precisely measured lengths of two adjacent sides of a square are both equal to na. However, the length of the diagonal can then be calculated. It does not have to be equal to ma, where m is an integer number.

To distinguish measured values from calculated ones did not seem necessary as long as physical laws were considered as statements about reality, telling us how it is. Today, we know that they are concerned with knowledge that we can get about reality by means of measurements. This is very different for two basic reasons: 1) nature can impose restrictions on our measurements and 2) knowledge can be expressed in very subtle ways. This applies especially to space and time. Newton postulated the existence of an "absolute space" that is motionless and an "absolute time", flowing always and everywhere in the same way. Space and time were thus assumed to be some kind of substances. These postulates were sufficient to account for a vast domain of observations, but had to be corrected when this domain was enlarged.

There are also electromagnetic phenomena. Einstein realized that they require that the velocity c of light in vacuum has to be independent of the chosen reference frame. This was incompatible with the Galilean velocity addition law and with Newton's concept of space and time. Einstein solved this conflict, by considering that space and time are only defined by possible results of measurements. Nature could thus impose that measurements of lengths and durations have to yield the same value c, when the velocity of light in vacuum is measured with respect to inertial reference frames. Because of the principle of relativity, Newton's laws of motion had then to be modified for very great velocities.

Planck discovered another universal constant. Its value h defines the smallest energy quantum hν for light waves of frequency ν, but it is also the quantum of action. It determines thus the smallest possible value of a specific combination of space-time coordinates and kinematical observables. This led to "quantization rules" for possible states of motion of very light material particles, but these rules could only be explained by introducing the concept of wave functions. They can store an enormous amount of information, because of their possible variations in space and time. Relativistic Quantum Mechanics (RQM) accounts even for annihilation and creation processes, but something is missing. RQM provides no information about possible types of elementary particles that can be transformed into one another.

Why are wave functions (for one particle) and fields (for any number of identical particles) unable to store this information? Could this be due to the postulate that space and time are continuous? Is it a logical necessity or not? To answer this question, we considered that Nature could impose another restriction and constructed a theory of Space-Time Quantization (STQ) that generalized RQM to account for c, h and a. This is the value of the smallest measurable length, which is not yet known, but it would be sufficient to show that $a \neq 0$ leads to logical inconsistencies, to prove that space and time are continuous. Actually, it was necessary to replace differential equations by finite-difference equations. This led to modifications for very rapidly varying ψ functions, but not to logical inconsistencies. The existence of a quantum of length $a \neq 0$ cannot be excluded.

Could this be relevant for elementary particle physics? We were surprised to discover that variations of ψ functions in space and time can be described in a more detailed way when $a \neq 0$. At the smallest possible scale, it is even necessary to introduce four new quantum numbers. It appeared then also that they allow us to distinguish elementary particles from one another [3] [4] . Even their possible transformations can be specified. The agreement with a huge amount of very remarkable empirical data, accumulated during more than half a century, means that STQ is real.

Nevertheless, the concept of a space-time continuum is so deeply rooted in our culture, that it is natural to resist any change in this regard. It has been objected, for instance, that the existence of a finite quantum of length is forbidden by the Lorentz transformation for space-time coordinates. It accounts for the constancy of c, but presupposes that space and time are continuous. The constancy of c, h and a for different reference frames is insured by a generalized Lorentz transformation.

Another objection can result from the usual, intuitive concept of motions. It is based on the fact that there are continually existing material objects. Their center of mass can thus only move from a point A to another point B by passing through a continuous array of intermediate points. This would also be true for any point-like material particle in non-relativistic quantum mechanics, but how do we define knowledge in regard to motions? Quantum mechanics accepts that space-time coordinates can be measured with absolute precision, but it is sufficient to know the probability distribution for possible results, to calculate the average position of a point-like particle at any particular instant. Even when measurable positions are quantized in a given reference frame, the calculated average position does not have to coincide with one of these values. STQ is thus compatible with continuous motions.

It has also been stated that the Planck length is the smallest measurable distance. Planck wanted only to attract attention on the fact that h, c and the gravitational constant G yield a system of units of universal validity. The Planck length is a natural unit for quantum gravity, like the Bohr radius in atomic physics, but the concept of a quantum of length is more general. To claim a priori that it has to be equal to the Planck length would be arbitrary, like the assumption that $a = 0$.

String theories try to account for the degrees of freedom of elementary particle physics by assuming the existence of additional dimensions of space. They would not be observable, if they were wrapped up like the surfaces of extremely thin filaments, which can vibrate. This ingenious paradigm remains compatible with the usual concept of a space-time continuum, but STQ explains elementary particle physics in a simpler way. The normal 4-D space-time is sufficient when $a \neq 0$.

The organization of this article is the following. In Section 2, the construction of the theory of STQ is briefly reviewed and justified in regard to possible states of motion. In Section 3, we generalize this theory to account also for particle states. The aim is to present the proposed theory in a compact, unified and more complete way. In Section 4, we show that STQ explains and generalizes the Standard Model of elementary particle physics. This accounts also for the nature and properties of DM particles. In Section 5, we indicate how their detection and identification can be facilitated by a very strict and general conservation law. Section 6 presents conclusions and points towards further applications.

2. SPACE-TIME QUANTIZATION

2.1. The Generalized Energy-Momentum Relation

The quantum-mechanical description of motions is much more detailed than the classical one, but we are still speaking of the momentum p and energy E of a particle, although these observables acquired a radically new meaning. When their values are exactly known for any free particle that is moving along the x-axis of a given inertial frame, we express this knowledge by means of the wave function

$$\psi(x,t) = A e^{i(kx-\omega t)} \tag{1}$$

where $p = \hbar k$ and $E = \hbar\omega$. The amplitude A is constant and $\hbar = h/2\pi$. The classically defined values of E and p were related to one another by means of Einstein's relation, where m_o is the rest mass:

$$(E/c)^2 = p^2 + (m_o c)^2 \tag{2}$$

Relativistic Quantum Mechanics (RQM) postulated that this relation remains true, although it does not include h. Louis de Broglie discovered, indeed, that the wave-par- ticle duality can be generalized without modifying (2). Actually, he defined the new observables $p = h/\lambda$ and $E = h\nu$. Their values are only determined with absolute precision when the ψ function varies always and everywhere with the same wavelength λ and the same frequency ν. This is an idealization, but it is sufficient that $\psi(x,t)$ results from a linear superposition of harmonic functions like (1) to get a wave-packet. The values of p and E are then not exactly known, but their average values are still related to one another by (2). We can even define local values $\hat{p}$ and $\hat{E}$ at any particular point in space and time. They depend only on the variations of $\psi(x,t)$ at the smallest possible scale around this point. They should there be the same as if p and E were exactly known. The usual concept of a space-time continuum implies therefore that

$$\hat{p}\psi = -i\hbar\partial_x\psi \quad \text{and} \quad \hat{E}\psi = i\hbar\partial_t\psi$$
$$\hat{p}^2\psi = -\hbar^2\partial_x^2\psi \quad \text{and} \quad \hat{E}^2\psi = -\hbar^2\partial_t^2\psi \tag{3}$$

$\partial_x\psi$ and $\partial_t\psi$ designate partial derivatives of ψ with respect to x and t. For any precisely defined state of motion (1), we get thus $\hat{p} = p$ and $\hat{E} = E$. It is customary to speak of operators, but their physical meaning is more important than formal rules. Actually, we define new observables and RQM postulates that $\hat{p}$ and $\hat{E}$ have also to satisfy (2). This yields the Gordon-Klein equation, which is a second-order differential equation. However, when $a \neq 0$, we have to adopt the definition

$$\hat{p}^2\psi(x) = -\hbar^2 \frac{\psi(x+a)+\psi(x-a)-2\psi(x)}{a^2} \tag{4}$$

For an exactly known value of p, it follows from (1) that $\hat{p}^2$ is equivalent to setting

$$\hat{p} = \hbar\frac{\sin(ka/2)}{(a/2)} \to p \ \text{ when } \ a \to 0 \tag{5}$$

In general, the local value $\hat{p}$ depends not only on $p = h/\lambda$, but also on the quantum of length a. The instantaneous value of $\left(\hat{E}/c\right)^2$ is defined in the same way, by considering the second order finite derivative of ψ with respect to ct. The Gordon- Klein equation has thus to be replaced by a finite difference equation. It is equivalent to replacing Einstein's relation (2) by the generalized energy-momentum relation

$$\sin^2\left(\pi aE/hc\right) = \sin^2\left(\pi ap/h\right) + \sin^2\left(\pi aE_o/hc\right) \quad (6)$$

This is a new physical law, since it differs from (2) when $a \neq 0$. All possible states of motion of any free particle are specified in any given inertial frame in terms of p and E. These values are subjected to the condition that $E = E(p)$. It is reduced to Einstein's relation (2) when space and time are continuous ($a = 0$) or when $a \neq 0$, but extremely small. For usual values of E and p, it is then possible to approximate the sine-functions by their arguments. The essential difference between (2) and (6) is therefore that Einstein's relation was assumed to be valid for arbitrarily high energies, while the generalized energy-momentum relation implies that the arguments of the sine-functions are limited by $\pm\pi/2$. Because of the de Broglie relations $p = h/\lambda$ and $E = h\nu$, this means that we can only consider wavelengths $\lambda \geq 2a$ and frequencies $\nu \leq c/2a$. It would be meaningless to consider oscillations of $\psi(x,t)$ at smaller scales than measurable ones. They can be imagined, but are physically irrelevant.

It appears thus that $p \leq h/2a$ and $E \leq hc/2a$ when $a \neq 0$. The rest energy is $E_o = E(0) = m_o c^2$. The curves $E = E(p)$ resulting from (6) coincide thus with Einstein's hyperbolas (2) for usual values of E and p. However, they shrink when $E_o \to hc/2a$. This can be graphically shown [4] , but we can immediately deduce from (6) that when the rest-energy $E_o = hc/2a$, there remains only one possible state: p = 0 and E = E_0. In quantum mechanics, absolute rest means that the Ψ function has everywhere the same value ($\lambda = \infty$). It follows that $E_u = hc/2a$ is the total energy content of our universe. Indeed, if a material body did have such a high rest energy, there would be no energy left that could appear in the form of kinetic energy. The actual value of $E_u = hc/2a$ is unknown, but energies cannot be infinite when $a \neq 0$. What would happen for motions at extremely high energies?

Quantum mechanics describes the average motion of any particle by considering a wave-packet. It results from a superposition of harmonic functions like (1), where $\omega = \omega(k)$. The group velocity is then $v = \mathrm{d}w/\mathrm{d}k = \mathrm{d}E/\mathrm{d}p$. Since Einstein's relation (2) remains valid in RQM, the velocity $v \leq c$ even for arbitrarily high energies E. However, according to (6), the slope of the curves $E = E(p)$ is greater than c for extremely high energies [4] . It is important in regard to principles that superluminal velocities are not excluded when $a \neq 0$, but this allows us also to test the logical consistency of STQ.

2.2. The Generalized Lorentz Transformation

The principle of relativity requires that all physical laws are independent of the chosen reference frame. The same state of motion is thus defined for a given free particle by $E(p)$ in one inertial frame and by $E'(p')$ in another one. The correspondence results from the invariance of (6). It can be expressed by considering local values for the momentum and energy variables or sine-functions of E and p. The resulting "generalized Lorentz transformation" accounts for constant values of c, h and a, but the derivatives $v = \mathrm{d}E/\mathrm{d}p$ and $v' = \mathrm{d}E'/\mathrm{d}p'$ yield a generalized velocity addition law. The relative velocity of the reference frames has to be defined in terms of the average velocity of a material object that is at rest in one frame. In the other frame, it could move at superluminal velocities.

The principle of causality requires that the sequence of events is the same for any observer, since causes have to precede their effects for any observer. So-called "tachyons" are hypothetical particles, assumed to obey (2) with an imaginary rest mass [5] . They can thus only move at superluminal velocities, but the invariance of (2) would impose the usual Lorentz transformation. For superluminal velocities, it contradicts the principle of causality. We verified that the generalized Lorentz transformation does not violate this principle [6] . It justifies even Mach's principle, requiring that inertial frames should be related to the whole universe.

3. FOUNDATION OF ELEMENTARY PARTICLE PHYSICS

3.1. New Quantum Numbers and Particle States

Since the possibility of STQ is not excluded, we wanted to know if this could be related to elementary particle physics. Actually, it appeared that ψ functions are able to store more information than we thought. To see why this is possible, we consider ideally precise measurements of the coordinate x along an arbitrarily chosen reference axis. When "x" is a possible result, "−x" is also one, since the orientation of the x-axis could be reversed. However, the distance 2x has to be an integer multiple of a. It can be an even or odd number. Thus,

$$x = 0, \pm a, \pm 2a, \cdots \quad \text{and} \quad x = \pm a/2, \pm 3a/2, \tag{7}$$

There are two lattices of lattice-constant a. The first one will be called the "normal lattice", since we are accustomed to the idea that x should be measured by starting at the chosen origin. There is also an "intercalated lattice" and the wave function $\psi(x)$ of any point-like particle has to be defined for all possible values of x. The probability

distribution $|\psi(x)|^2$ cannot allow for any ambiguity. It has to be single valued and to

be compatible with a smooth variation for the continuum approximation ($a \to 0$). Nevertheless, there is a degree of freedom, since $\psi(x)$ can be positive or negative on the intercalated lattice with respect to its values on the normal lattice (Figure 1(a)). There are even more than two possibilities, since $\psi(x)$ is a complex function. It can be modulated by a sign-function $S_x = \exp(i\theta_x)$, to differentiate $\psi(x)$ anywhere on the intercalated lattice with respect to $\psi(x)$ on the normal lattice (Figure 1(b)). This allows for $S_x = \pm 1$ when $\theta_x = u_x \pi$ and $u_x = 0, \pm 1, \pm 2 \cdots$

These considerations apply also to $\psi(x, y, z, ct)$ for any reference frame. It is only necessary that the three spatial axes are not coplanar to allow for independent measurements. The time-axis is freely chosen. We get thus four quantum numbers (u_x, u_y, u_z, u_{ct}), which can be positive or negative integer numbers. Every set of u- quantum numbers defines a possible "particle state". They can be specified by imaging 4 panels with pointer needles that can only be rotated by half-turns towards the left or the right. We can also define particle states by vectors in the 4-D space of u-quantum numbers. The homogeneity and isotropy of space implies that the u-quantum numbers are always and everywhere identical in the whole universe for a given type of particles. They are even identical for any, arbitrarily chosen reference frame, because of the principle of relativity. Inertial frames are only necessary to define states of motion of free particles by means of the energy-momentum relations (2) or (6).

Actually, we are expressing knowledge concerning "particles states" by means of possible variations of ψ functions at the smallest possible scale in space and time. Quantum mechanics defined "states of motion" by means of large-scale variations of ψ functions, specified in terms of possible wavelengths and frequencies. Since large- scale and small-scale variations can be combined by multiplexing, much more knowledge can be stored in ψ functions than allowed by a space-time continuum.

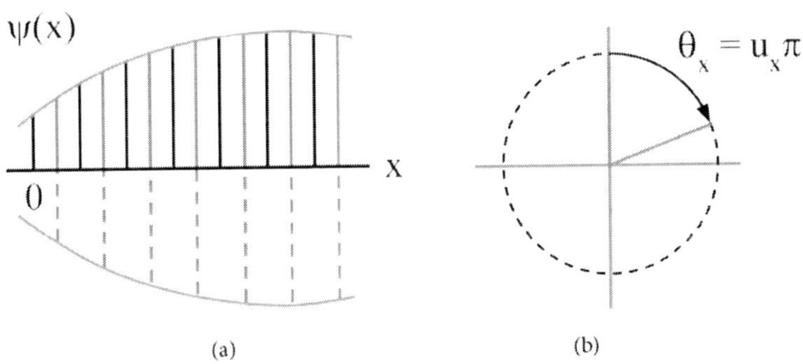

Figure 1. (a) STQ yields a degree of freedom for $\psi(x)$, since its sign can be the same on the intercalated lattice than on the normal lattice, but also opposite. The normal lattice includes the origin of the reference axis. (b) The quantum number u_x is defined by the angle θ_x. Its value determines the sign-function, which modulates $\psi(x)$ everywhere on the intercalated lattice.

3.2. The Generalized Dirac Equation

We discovered the new degrees of freedom by applying Dirac's procedure. He replaced the Gordon-Klein equation by an equivalent set of first-order differential equations. Since we believed that space-time coordinates have to be measured by starting at the origin of the chosen frame, we defined first-order finite derivatives in terms of $\psi(x)$ and $\psi(x \pm a)$. However, because of (7), the local value $\hat{p}$ of the momentum p along the x-axis can be defined in a unique and more precise way when

$$\hat{p}\psi(x) = -i\hbar \frac{\psi(x+a/2) - \psi(x-a/2)}{a}$$

When the momentum p is well defined, it follows from (1) that

$$\hat{p} = \hbar \frac{\sin(ka/2)}{(a/2)} S_x \to pS_x \quad \text{when} \quad a \to 0 \tag{8}$$

This result is similar to (5), but accounts also for the sign-function $S_x = \exp(iu_x\pi)$. The concept of a space-time continuum ($a = 0$) led to (3), but this concealed the existence of additional degreees of freedom. Even (4) did only account for states of motion, while (8) depends also on possible particle states. For usual states of motion, where the energy $E \ll hc/2a$, we get Einstein's relation (2), but for local values like (8). It has to be replaced by the generalized Dirac equation

$$(E/c)S_{ct} = \alpha_x p_x S_x + \alpha_y p_y S_y + \alpha_z p_z S_z + i\beta(m_o c) \tag{9}$$

Dirac's equation was equivalent to setting the four sign functions equal to 1. That was sufficient to discover already 4 possible states, since the matrices $\alpha_x, \alpha_y, \alpha_z$ and β yield four coupled relations. Considering electrons, Dirac knew that their spin has two possible orientations along any given z-axis. He attributed the two other possibilities to the fact that Einstein's relation (2) allows for positive and negative energies. The curves $E = E(p)$ would be hyperbolas, indeed. They allow for $E \geq m_o c^2$ and $E \leq -m_o c^2$, but if negative energy states were possible for electrons, these particles would be unstable, which is unacceptable.

To eliminate transitions towards negative energy states, Dirac took into account the fact that electrons are fermions. They have to occupy different energy states. Thus, he assumed the existence of an infinite number of electrons, occupying all possible negative energy states. This "Dirac Sea" would produce no observable effects, unless an electron is excited from a negative energy state to a positive energy state. The resulting "hole" would behave like a particle of rest-mass m_o, but its charge is +e instead of −e. Dirac predicted therefore that a photon could create an electron-positron pair. This was experimentally confirmed and provided the first example of particle-antiparticle pairs. Dirac's assumption that $m_o c$ can be positive or negative was equivalent to setting $S_{ct} = S_x = S_y = S_z = \pm 1$ in (9). However, there are much more possibilities, although free particles can only have positive energies.

3.3. Symmetries and Conservation Laws

Any symmetry results from modifications that yield related images. According to STQ, particle states are defined by means of u-quantum numbers, which can be positive or negative. Any vector in the 4-D space of u-quantum numbers has thus an image, repre- sented by the opposite vector. It defines another particle state, but the rest masses are identical. They depend on the cloud of virtual particles that can be created and absorbed by a given particle and thus on all its u-quantum numbers. This yields the particle-antiparticle symmetry

$$\left(u_x, u_y, u_z, u_{ct}\right) \quad \leftrightarrow \quad \left(-u_x, -u_y, -u_z, -u_{ct}\right) \tag{10}$$

Figure 1(b) shows that by inversing the orientation of the x-axis, we change the sign of u_x. Since the parity operator P reverses the orientation of all spatial axes, it changes the sign of u_x, u_y and u_z. The operator T for time reversal changes only the sign of u_{ct}, but the combined operator $C = PT$ converts any particle into its antiparticle. Modulation of ψ functions at the smallest possible scale in space and time reminds us of the distinction between acoustical and optical phonons in ionic crystals, but ψ functions require no material medium. They are tools for expressing knowledge. Although elementary particles are entities that carry energy and momentum, electric charge and other attributes, they can also be viewed as possible "excitations of space and time".

States of motion and particle states are respectively defined by large-scale and small scale variations of ψ functions in space and time. Although the spin may seem to be an intrinsic property of particles, it is defined by large-scale variations of their ψ function. Indeed, when the generalized angular momentum component $J_z = m\hbar$ along a given z-axis is precisely known, the ψ function varies around this axis like $f(\varphi) = \exp(im\varphi)$, where φ is the azimuthal angle with respect to this axis. The pro- bability distribution has to be single-valued, but this allows for $f(\varphi + 2\pi) = \pm f(\varphi)$. Thus, m can be an integer or half-integer number and the ensemble of all possible elementary particles is divided in two classes: bosons and fermions. This distinction is independent of possible modulations of ψ at the smallest possible scale in space and time. For every particle state of a fermion, there is a particle state for a boson that is characterized by the same set of u-quantum numbers and vice-versa. Representing fermion states by round brackets and boson states by square brackets, this means that

$$(u_x, u_y, u_z, u_{ct}) \quad \leftrightarrow \quad [u_x, u_y, u_z, u_{ct}] \tag{11}$$

This correspondence defines supersymmetry (SUSY), but does not require that supersymmetric images have the same mass. STQ yields also information about possible transformations of elementary particles. Usually, one particle is annihilated by creating two different particles or vice-versa. Since the ψ functions of particles that coexist in the initial or final state are multiplied by one another, their phase factors are then added, but the result has to be identical for the initial and final state. This yields the well-known conservation laws for the energy and momentum variables. They are compatible with the energy-momentum relation (2) or (6) for two successive transformations, since the intermediate state is then subjected to Heisenberg's uncertainty relations. They result from the fact that the phase factor of (1) contains products like px and Et. However, the sign-functions depend only on u-quantum numbers. We get thus a very strict conservation law for the sum of u-quantum numbers before and after any transitions between particle states. The sum of u-quantum numbers has to be identical for the initial and final states, separately for every space-time axis. This is equivalent to a vector addition law in the four-dimensional space of u-quantum numbers, as well for $[u_x, u_y, u_z, u_{ct}]$ as for (u_x, u_y, u_z, u_{ct}) states. This allows for

$$[0,0,0,0] = (u_x, u_y, u_z, u_{ct}) + (-u_x, -u_y, -u_z, -u_{ct})$$

Photons are in the $[0,0,0,0]$ state and can be annihilated by creating a particle- antiparticle pair, but a particle and its antiparticle can also annihilate one another to create a photon. If the Big Bang did create particles and antiparticles in equal proportions, they would have annihilated one another. The observed cosmic matter-antimat- ter asymmetry raises thus a fundamental problem. It can be solved [7] , but we will now simply accept that the Big Bang created only those fermions that we call "particles". Electrons and quarks are particles, for instance.

The conservation law allows also for $(u_x, u_y, u_z, u_{ct}) = [0,0,0,0] + (u_x, u_y, u_z, u_{ct})$. This accounts for the emission and absorption of photons, but it is well-known that these processes are only possible for electrically charged particles. Since electromagnetic waves correspond to vector fields, they are coupled with possible variations of ψ functions along the three spatial axes. The electric charge of a particle has thus to depend on (u_x, u_y, u_z), but permutations of these quantum numbers should be irrelevant, since the spatial reference axes can be chosen in a completely arbitrary way. This yields a symmetry or kinship between particles that have the same electric charge Q, defined in units e by the average value of the spatial u-quantum numbers:

$$Q = (u_x + u_y + u_z)/3 \tag{12}$$

4. THE GENERALIZED STANDARD MODEL

4.1. Elementary Fermions and DM Particles

The first generation of the Standard Model (SM) corresponds to spin $1/2$ fermions, characterized by the triplet (u_x, u_y, u_z), where these quantum numbers can be equal to 0, ± 1, while $u_{ct} = 0$. For orthogonal u_x, u_y and u_z axes, this yields the cubic lattice of Figure 2. Every lattice-point corresponds to a different particle state. The positron

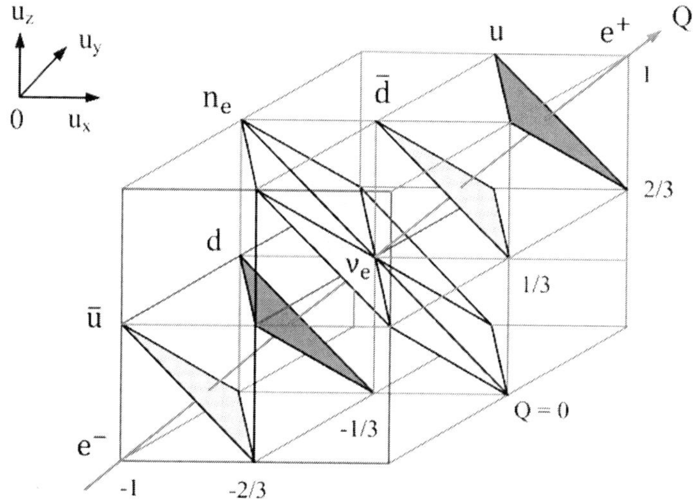

Figure 2. STQ accounts for the Standard Model, since particle states are specified by four quantum numbers (u_x, u_y, u_z, u_{ct}). The first generation correspond to states where u_x, u_y and u_z are equal to 0, ± 1, while $u_{ct} = 0$. Leptons are represented by points on the Q axis. Quarks and antiquarks correspond to the vertices of red and green triangles. STQ yields also 8 possible states for elementary Dark Matter particles, represented by the vertices and the center of the hexagon.

e^+ corresponds to the $(1,1,1)$ state and the electron e^- to $(-1,-1,-1)$. The elec- tron-neutrino ν_e and its antiparticle $\bar{\nu}_e$ are both in $(0,0,0)$ states. All known leptons are thus represented by points that belong to the Q axis. Quarks correspond to the vertices of the red triangles. There are three possible states for up-quarks (u), since they are of type $(0,1,1)$ with 3 possible permutations, while $Q = 2/3$. Down-quarks (d) allow for 3 states of type $(-1,0,0)$ and $Q = -1/3$. Antiparticle states are defined by (10) and represented by opposite points with respect to the center $(0,0,0)$. Antiquarks have thus also 3 possible states, represented by the vertices of the green triangles.

Since DM particles are electrically neutral and were not produced by accelerators, they could not be included in the SM. However, the vertices of the yellow hexagon correspond to states of type $(1,-1,0)$ with 6 possible permutations. The charge Q = 0. We add two $(0,0,0)$ states, represented by superposed points at the center of the hexagon. They differ from the $(0,0,0)$ states for ν_e and $\overline{\nu}_e$, since they belong to the hexagon, although they are also situated on the Q axis. These states correspond to elementary DM particles, but this octet is constituted of 4 particles and 4 antiparticles.

By analyzing empirical results it was possible to distinguish 3 generations of elementary particles. They result from $u_{ct} = 0, \ \pm 1$. These generations display the same family structure, including always two types of quarks that have 3 possible states. They were said to be red (R), green (G) and blue (B) color states. The SM could not explain why there are 3 generations and 3 color states [8] , but STQ reveals that the 3 color states result from the fact that space is three-dimensional and that different generations are associated with the time variable.

When Figure 2 is viewed along the Q axis, the triangles and the hexagon yield Figure 3. The color code for quarks is chosen to be similar to the repartition of R, G and B colors in the color plane of colorimetry. The superposed triangles for u-quarks and d- quarks yield superposed color states when we adopt the following convention:

$(0,1,1)$, $(1,0,1)$, $(1,1,0)$ for $u(R)$, $u(G)$, $u(B)$

$(-1,0,0)$, $(0,-1,\ 0)$, $(0,0,-1)$ for $d(R)$, $d(G)$, $d(B)$

Antiquarks are said to have anticolors. The antired $\overline{R}$ state corresponds thus to $(0,-1,-1)$ for the $\overline{u}$ antiquark and to $(1,0,0)$ for the $\overline{d}$ antiquark. The 8 possible states of elementary DM particles are represented in Figure 2 by points that belong to the plane Q = 0. Since it is perpendicular to the Q-axis, like the planes for quarks, they can be considered as defining “neutral quarks”. For simplicity, we called them narks [3] [4] . They are the supersymmetric partners of gluons. The existence of gluons was deduced from empirical results, indicating that a quark can change its color by creating a gluon. According to usual conventions, $R \to G + R\overline{G}$.

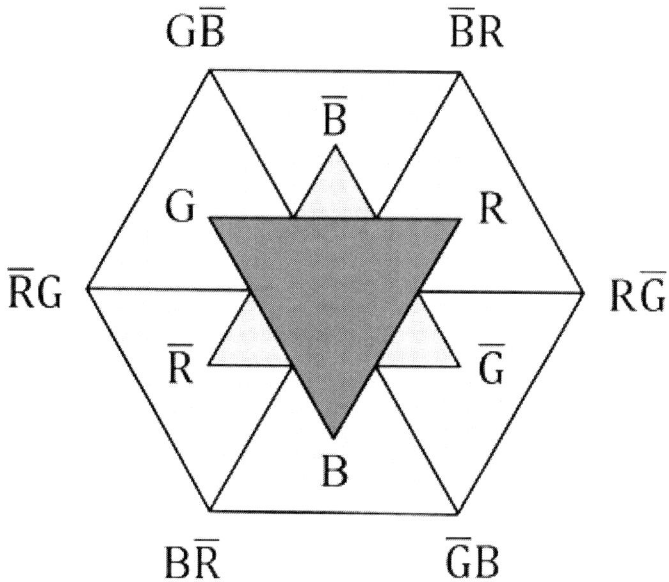

Figure 3. Quarks have 3 possible color states: R, G and B. Antiquarks have anticolors. Colored gluons and colored narks or antinarks are characterized by a color and an anticolor.

Because of the conservation law for u-quantum numbers, this means that $(0,1,1)\rightarrow(1,0,1)+[-1,1,0]$ for a u-quark and $(-1,0,0)\rightarrow(0,-1,0)+[-1,1,0]$ for a d-quark. We have thus to attri- bute the color $R\bar{G}$ to the $[-1,1,0]$ state of gluons. We adopt the same color code for narks, to get superposed color states in Figure 3.

We distinguish narks from antinarks, since these fermions were already created dur- ing the Big Bang. They are thus subjected to the cosmic matter-antimatter asymmetry. This can be explained, since quarks created gluons, which could be transformed into narks. Thus $[-1,1,0]\rightarrow(0,1,-1)+(-1,0,1)$, for instance. This allows us to adopt the following color code:

$(1,-1,0)$, $(0,1,-1)$, $(-1,0,1)$ for $\bar{R}G, \bar{G}B, \bar{B}R$ states (narks)

$(-1,1,0),(0,-1,1),(1,0,-1)$ for $R\bar{G}, G\bar{B}, B\bar{R}$ states (antinarks)

Possible color states for narks and antinarks are represented by opposite points in Figure 3, but they are situated in the same plane Q = 0. There are also two $(0,0,0)$ states for narks and antinarks. They are "color-neutral" and represented by two superposed point at the center of the hexagon.

Symmetry considerations are very useful to analyze incomplete empirical data, since they can reveal "empty places" that might be filled later on. This happened for Mendeleyev's table of chemical elements, which could only be explained by quantum mechanics. Gluons are bosons that were essential ingredients of the SM. It has thus been suggested that supersymmetric partners of gluons might also exist. These fermions were called "gluinos" $(\tilde{g})$, but considered as purely hypothetical entities. STQ justifies their existence in a direct way and we relate them to DM. The octet of possible states for gluons corresponds to an octet of possible states for elementary DM particles, but for bosons, no distinction is made between particles and antiparticles. It should be noted that we consider gluons and narks as single elementary particles, although they are defined by a color and an anticolor.

4.2. Other Generations and Excited States

The SM considers 3 generations of spin $1/2$ fermions. They are due to $u_{ct} = 0, \ \pm 1$. We considered n_e narks for the first generation of the SM in Figure 2, since we have also to expect the existence of n_μ and n_τ narks, according to the correspondence

$$\left(u, d, e^-, \nu_e, n_e\right) \quad \leftrightarrow \quad \left(c, s, \mu^-, \nu_\mu, n_\mu\right) \quad \leftrightarrow \quad \left(t, b, \tau^-, \nu_\tau, n_\tau\right)$$

There are thus quarks of different "flavor" (u, d, c, s, t and b). All these states are represented by superposed vertices of red triangles with the color code of Figure 3. The orientation of the Q axis was chosen by setting $Q = -1$ for the electron. We are thus also free to choose the orientation of the u_{ct}-axis, so that $u_{ct} = -1$ for the second generation [4] . STQ generalizes the SM not only because of the hexagon in Figure 2. Since $u_{ct} = \pm 2$ is also possible, there are more than 3 generations, although the rest-energy of these quarks and narks may be too high for their production by means of existing accelerators. Even the Big Bang did eventually produce only quarks and narks of the first generation.

We can also consider particle states that correspond to the upper left and lower right vertices of the front side of the large cube in Figure 2.

Combining them with the lower rear left vertex of this cube, we get a large triangle. Its vertices define states of type $(1,-1,-1)$ with 3 possible permutations. The electric charge $Q=-1/3$. They can be considered as the first excited states of d quarks, since they imply more modulations of ψ functions on intercalated lattices with respect to the normal one. The first excited states of u quarks are of type $(2,0,0)$. They allow also for 3 permutations and $Q=2/3$. The triangles for the excited states d^* and u^* of d and u quarks and their antiparticles are represented by means of interruped lines in Figure 4. The color code is the same as for the corresponding quarks.

Even excited states of positrons and electrons are possible. For the electron, they are of type $(0,-1,-2)$ with 6 possible permutations and $Q=-1$. For the positron, they are of type $(0,1,2)$. They are thus colored leptons and represented by points that are situated at the vertices of superposed hexagons in Figure 3. However, in Figure 2 these hexagons would be situated in the planes $Q=\pm 1$ and be centered on the Q axis. Since electrons and positrons are light particles, we have to expect that their excited states are created more easily than those of quarks and antiquarks.

4.3. Bosons and Possible Interactions

We mentioned that narks are supersymmetric partners of gluons (g), but (11) is more general. The first generation of spin $1/2$ fermions corresponds thus to an ensemble of spin 1 bosons:

$$(\nu_e,\bar{\nu}_e)\leftrightarrow(\gamma,Z) \qquad e^+\leftrightarrow W^\pm \qquad n_e\leftrightarrow g \tag{13}$$

Electromagnetic interactions result from exchanging photons (γ), while weak interactions require exchanges of Z or $W^\pm$ bosons. Photons and Z particles are in $[0,0,0,0]$ states, but the rest-mass of photons is zero, while Z and $W^\pm$ bosons are heavy and lead to short-range forces. Weak interactions allow for transformations like $u(R)\to d(R)+W^+$, immediately followed by $W^+\to e^+ +\nu_e$ to account for energy and momentum conservation. Every one of these transformations satisfies the conservation law for u- quantum numbers, since $(0,1,1)=(-1,0,0)+[1,1,1]$ and $[1,1,1]=(1,1,1)+(0,0,0)$.

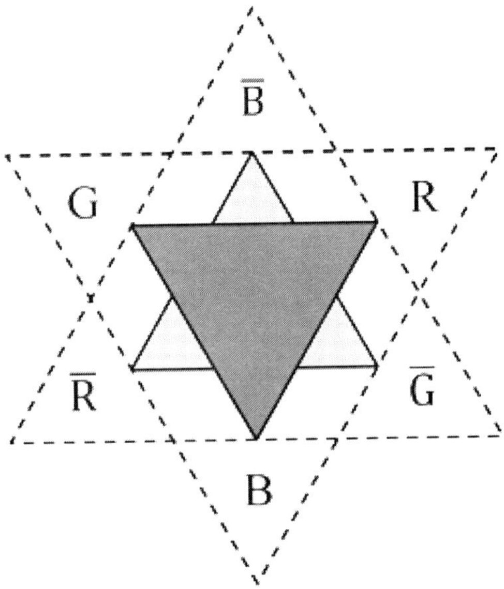

Figure 4. Excited states of quarks and antiquarks correspond to vertices of surrounding triangles.

Weak interactions can thus be mentally represented by transitions that are paral- lel to the Q axis in Figure 2. Since strong interactions result from exchanges of gluons, they imply transitions that are perpendicular to the Q axis. For instance, $u(R)\rightarrow u(G)+g(R\overline{G})$. This applies also to transitions inside the plane Q = 0. For instance, $(-1,1,0)=(0,1,-1)+(-1,0,1)$ or $R\overline{G}\rightarrow \overline{G}B+\overline{B}R$. In short, one vertex of the hexagon in Figure 3 can be replaced by two neighboring ones, but narks are then converted into antinarks and vice-versa. Quantum chromo dynamics (QCD) applies thus also to DM particles.

It is even possible to get transitions between different generations, since $u_{ct}=\pm 1$ should be expected for gluons. Transformations of particle states could also corres- pond to oblique transitions with respect to the Q axis in Figure 2. They would convert quarks into antiquarks of different color and quarks or antiquarks into leptons. STQ accounts directly for bosons of type $[0,2,2]$ and $[0,1,1]$, while Grand Unified Theories (GUT) led to the concept of X and Y bosons in a more complicated way [9] .

Supersymmetric partners of leptons were called sleptons ($\tilde{\ell}$). They are usually considered as being hypothetical bosons, but (13) indicates that they are already known for $u_{ct} = 0$. In addition to spin 1 bosons, there are also spin 0 and spin 2 bosons, corresponding to scalar and tensor fields. There are thus Higgs bosons and gravitons in other states than $[0,0,0,0]$.

4.4. Hyperons and Composite DM Particles

Hyperons are composed of quarks or antiquarks, which are strongly bound to one another by exchanging gluons. This yields different types of composite particles, repre- sented in Figure 5, where occupied color states are represented by black dots for quarks and by white dots for antiquarks. We can mentally represent an exchange of gluons between quarks or antiquarks by shifting two dots in opposite directions. Neighboring dots can thus be exchanged, but opposite shifts of two dots along parallel lines are sufficient. It is only necessary that every shift occurs within the same Q plane (for normal gluons with $u_{ct} = 0$). This applies also to quarks of different flavors (u, d, c, s, t, b), since two associated shifts are then possible in different Q planes.

A proton (uud) requires that the three quarks are in different color states, since they are fermions. We could thus get $u(R)$, $u(G)$ and $d(B)$ at a particular instant. The resulting particle state is $(0,1,1)+(1,0,1)+(0,0,-1)=(1,1,1)$ and $Q=1$. All composite particles have to be color-neutral, since no spatial reference axis can be privileged for intrinsic properties of these particles. For mesons, this is achieved statistically by means

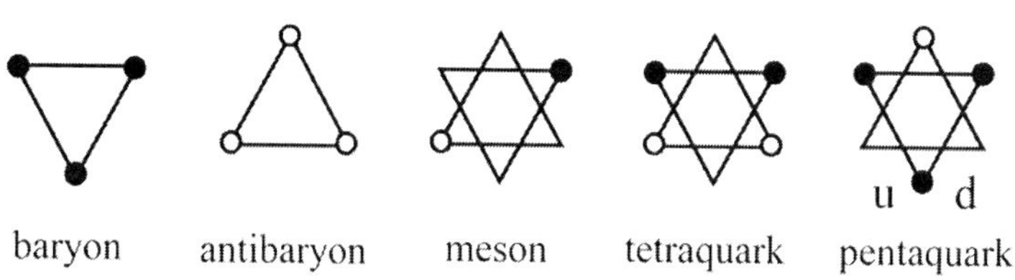

Figure 5. Hyperons are constituted of quarks (black dots) or antiquarks (white dots). They are strongly bound to one another by exchanging gluons (antiparallel shifts of two dots).

of all possible gluon exchanges. We include tetraquarks and pentaquarks in Figure 5. The discovery of ($u\bar{u}b\bar{b}$) and ($d\bar{u}c\bar{c}$) tetraquarks was puzzling, since they might be viewed as molecular associations of two mesons. Actually, all quarks and antiquarks are equally bound to one another, since dots can be shifted in different Q planes.

The ($uddu\bar{s}$) pentaquark seemed to be so exotic that its very existence was doubted [10] , but it was confirmed [11] . This is justified by the fact that 2 dots can be superposed for quarks of the same color when they belong to different Q-planes. A new type of tetraquark has been discovered [12] . It may be considered as a combination of a strange B meson ($s\bar{b}$) and a π meson ($u\bar{d}$), but even 4 quarks could be strongly bound to one another when they are of different flavor. More exotic compound particles could be created by occupying excited states, like those of Figure 4.

Composite DM particles are constituted of narks and/or antinarks, which are strongly bound to one another by exchanging gluons. Since the resulting particles are similar to nucleons, but electrically neutral, we call them "neutralons". Figure 6 shows various possibilities in terms of occupied states. Narks are represented by black diamond-shaped dots and antinarks by white ones, since their color states are different from those of quarks and antiquarks. Gluon exchanges can still be represented by opposite shifts of two different dots, but a nark is then transformed into an antinark and vice-versa. We represent neutralons by the symbol N_n where n specifies the number of narks or antinarks that are "glued" together. Since narks and antinarks are spin $1/2$ particles, we get bosons when n is even and fermions when n is odd.

The creation of cosmic DM particles was subjected to the general matter-antimatter asymmetry and compound particles have to be color neutral. N_1 is thus only possible for a nark in the $(0,0,0)$ state. The N_3 neutralon is composed of 3 narks, but gluon exchanges transform narks into antinarks and vice-versa. N_5 and N_7 neutralons could have been formed by combining N_1 with adequate bosons, but because of gluon exchanges, the central state can also be occupied by an antinark. The N_2 boson is only stable as long as opposite dots are shifted along the contour of the hexagon. When they were shifted towards the center, they are represented by slightly separated diamonds, but the nark and its antinark will annihilate one another. Cosmic DM particles are thus only stable when they are fermions.

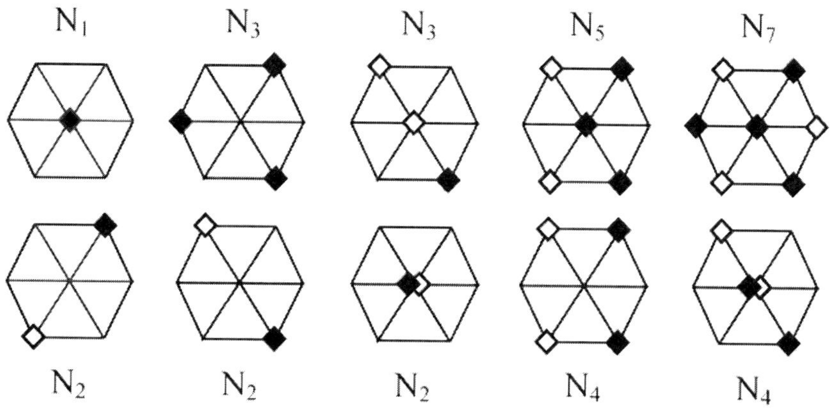

Figure 6. Neutralons are compound DM particles, constituted of narks (black diamond-shaped dots) or antinarks (white dots). They have to occupy different states, but are glued together.

4.5. Hybrid Particles

During the Big Bang, quarks and narks interacted with one another by exchanging gluons, but they were still relatively free until they became definitely bound to one another. Eventually, the cosmos was populated with nucleons (composed of quarks) and neutralons (containing only narks). Quarks and narks got separated, since binding results from superposition of ψ functions and quantum-mechanical resonance effects. The deepest negative energy states and strongest bonds will thus be achieved for elementary particles of the same type. However, during the Big Bang, there could also appear hybrid particles. Figure 7 shows a H_o particle that combines a single nark with a quark and an antiquark. Color states like $\bar{B}R$, B and $\bar{R}$ did insure color- neutrality, conserved by specific exchanges of gluons. The resulting particle was not sufficiently stable to survive. The cosmic DM gas contains thus no or nearly no hybrid particles. Nevertheless, it is conceivable that H_1 or H_3hybrid particles could sometimes result from combining a nucleon with N_1 or N_3 DM particles.

4.6. Interactions between DM Particles

Nucleons interact with one another by exchanging π mesons. This accounts for nuclear forces, which allow for scattering and the constitution of nuclei. Neutralons interact also with one another, but they exchange N_2 bosons.

This process is represented for a particular case by the Feynman graph of Figure 8. A nark that belongs to a neutralon A can interact with a nark that belongs to another neutralon B. They are there for instance in $\bar{B}R$ and $\bar{R}G$ states.

For simplicity, we represent bosons by straight red lines and fermions by black lines. Every vortex joins two black lines and one red line to insure spin conservation. Arrows will only be used for charged particles, but all transformations obey the conservation law for u-quantum numbers. The nark of A leaves this neutralon as an antinark by creating a gluon. This $R\bar{G}$ gluon is transformed into two narks.

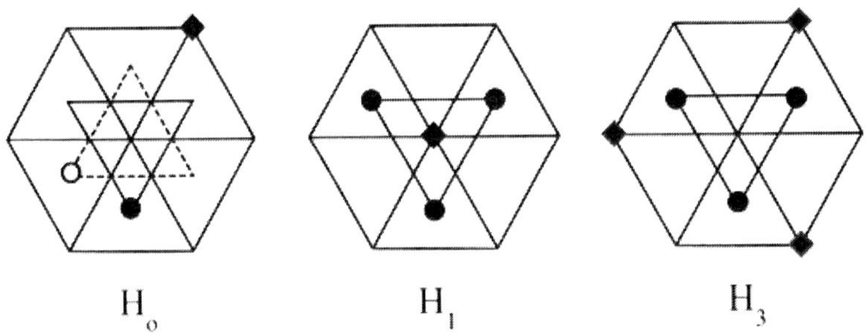

Figure 7. Quarks and narks can be "glued" together to constitute unstable hybrid particles.

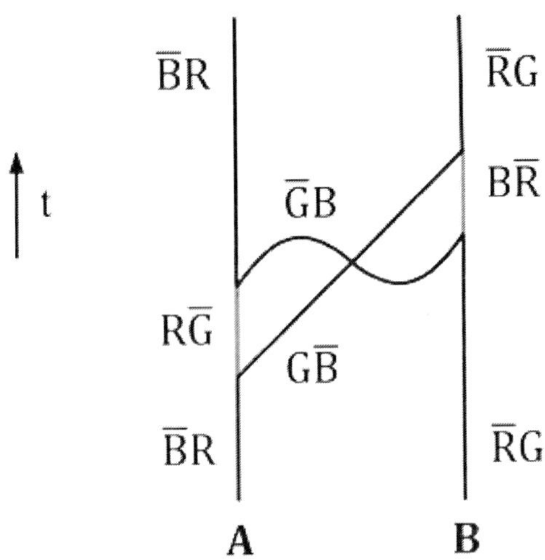

Figure 8. Interactions between DM particles result from exchanges of N_2bosons.

This restores the initial color state and creates an N_2 boson. Its nark $\bar{G}B$ is captured by the nark of B to create a $B\bar{R}$ gluon. Indeed, $(1,-1,0)+(0,1,-1)=[1,0,-1]$. This gluon merges with the initially emitted antinark and changes its color: $[1,0,-1]+(0,-1,1)=(1,-1,0)$. The color-neutrality is preserved for both neutralons and the apparently magical coordination of these processes results from the coexisting fields of all real and virtual particles. They influence one another.

We recall that nucleons are moving inside nuclei in the average field, created by its interactions with neighboring nucleons. This allows for shell-model structuring, excitations, fusion and fission processes. That has also to be expected for neutralons in compound DM particles. There will thus be "magic numbers" for more stable DM particles. However, the cold cosmic DM gas does usually not allow for sufficiently powerful collisions to excite DM particles. Elastic scattering will predominate, because of energy quantization. It requires only that DM particles come close enough to one another to be deviated by exchanging N_2 bosons. The cosmic DM gas is thus similar to usual gases. It has some pressure. Because of their mass, DM particles are also subjected to gravitational forces. This accounts for many astrophysical observations [13] .

The concept of self-interacting cold DM has independently been advanced for the same reason [14] . It has been confirmed by simulations [15] [16] , but the basic mechanism is that of Figure 8. A large cross-section for elastic scattering would imply a relatively low mass of N_2 bosons and a sufficient long lifetime. However, collisions between compound DM particles can also be hard enough to cause individual or collective excitations of neutralons. DM physics is similar to nuclear physics, but we have to stress the fact that the cosmic DM gas allows for fusion and fission processes, since they are very important for cosmology [7] .

5. DETECTION AND IDENTIFICATION OF DM PARTICLES

5.1. Possible Signals of Cosmic DM Particles

Their detection requires special instruments, but they exist already and did even provide some enigmatic results. NASA's Fermi Large Area Telescope (LAT) detected 130 GeV gamma-rays coming from galaxies. Their spatial distribution suggests that they are produced in galactic DM halos [17] [18] . According to STQ and the resulting concept of DM particles, this type of signals could be due to collisions between DM particles that confer enough energy to a particular nark to allow for the creation of a pair of gluons. Such a process is illustrated in Figure 9. The nark changes its color from $\bar{B}R$ to $G\bar{B}$, by creating a $R\bar{G}$ gluon, but the nark does immediately recover its initial color by creating a $\bar{R}G$ gluon. For the represented sequence, this $\bar{R}G$ gluon is annihilated first, by creating an electron and an exited position: $[1,-1,0]=(-1,-1,-1)+(2,0,1)$. This colored lepton and the $R\bar{G}$ gluon create together a normal positron: $(2,0,1)+[-1,1,0]=(1,1,1)$. The electron emits a photon, while the electron-positron pair is annihilated by creating another photon. This yields a $\gamma\gamma$ pair and could occur quite frequently where the density of DM particles is high.

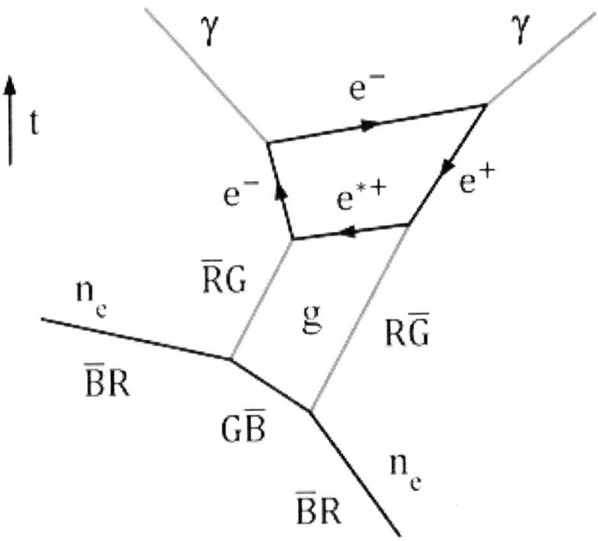

Figure 9. Cosmic DM particles could be detected by means of gamma rays, when a nark creates a pair of gluons. It is converted into an electron-positron pair by means of an excited lepton.

The Alpha Magnetic Spectrometer (AMS-02), attached to the International Space Station, revealed a surprising excess of positrons with respect to expectations for cosmic rays [19]. The e^+ fraction of the electron-positron flux increases from about 10 GeV up to 100 GeV and remains then nearly constant up to 350 GeV and even up to 500 GeV. It has been suggested that these positrons could result from dark matter decays. Pulsars can produce positrons of various energies, but the observed positron excess suggests a more general origin. The detection of an antiproton excess [20] would also be difficult to explain. To account for the threshold effect and the high energies of positrons in terms of DM particles, we can consider that collisions between two DM particles could lead to their fusion. This would liberate energy, since binding implies deeper negative energy states. This energy could be sufficient to fragment neutralons inside the resulting compound DM particle. When $N_5 \rightarrow N_3 + N_2$ for instance, the N_2 boson could subsists for a short while by internal exchanges of gluons. Eventually, the nark and its antinark will annihilate one another.

They could then create a Z boson, when the excitation energy E^* is superior to the required energy E_Z. This $[0,0,0]$ boson can decay into an e^+e^- pair (left part of Figure 10), but the metastable N_2 boson was excited inside the compound DM particle, while the electron and positron can leave the parent particle with a combined energy $E = E^* - B$ (right part of Figure 10). The threshold would then result from the minimal excitation energy E_Z. Since the Z boson can decay by creating hadrons [21] , this would also account for antiprotons. This scenario is hypothetical, but indicates that the conservation law for u-quantum numbers can be helpful to understand possible signals.

5.2. Direct Detection or Production

Since so-called "direct detection" of DM particles would be very important, several searches are presently undertaken by means of sophisticated equipment in underground facilities. This endeavor is partly inspired by experimental neutrino research and presumes that DM particles can interact with nucleons. The resulting nuclear recoil might then be detectable by some energy transfer, causing minute effects of various types. However, it is usually assumed that DM particles are WIMPS or some other hypothetical particles. Would the concept of DM particles resulting from STQ also allow for interactions with nucleons? A single free nark can interact with a single free quark, by exchanging gluons, but this was only possible during the Big Bang. In the present universe, narks and quarks are separately bound together. They are constantly exchanging gluons, but inside different entitities.

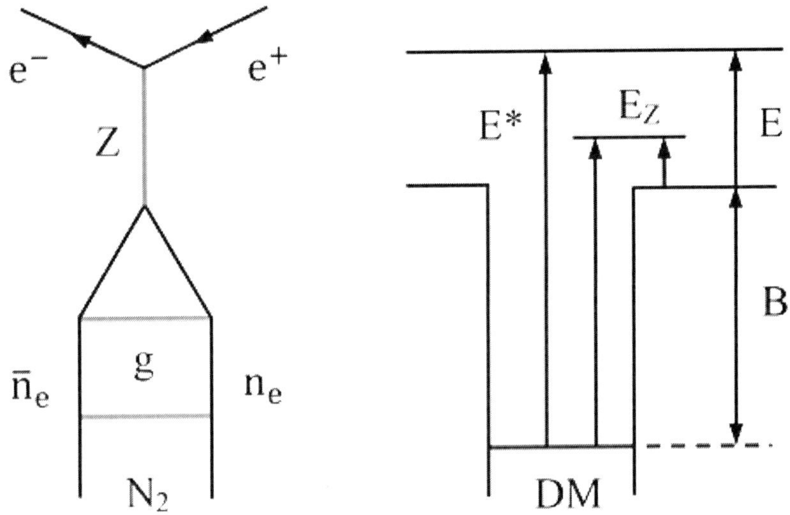

Figure 10. The positron excess, detected by AMS-02, could result from fission of DM particles, creating an N_2 boson of energy E^* inside the compound particle. The N_2 boson is converted into a Z boson that can decay into and electron-positron pair. This yields a threshold for E.

Since compound particles have to remain color-neutral, a nark that belongs to a DM particle and a quark that belongs to a nucleon could exchange one color-neutral gluon, a pair of gluons, an N_2 or Z boson. These intereactions are not impossible, but will be softer and much less probable than strong interactions between quarks inside nucleons and narks inside neutralons. Astrophysical evidence indicates also that interactions between cosmic DM and baryonic matter are negligible and so far, even the most sensitive DM detector did not yield any evidence of interactions between DM particles and nucleons [22] . However, we will show that DM particles can interact with electrons [13] .

Could STQ justify the hope to discover physics beyond the SM by means of proton-proton collisions at the Large Hadron Collider of CERN? One possibility would be to detect $\gamma\gamma$ pairs, produced by a process like that of Figure 9. It is sufficient to replace there the nark by a quark. It is electrically charged, but transformations like $R \to G + R\bar{G}$ and $G \to R + \bar{R}G$ would also create a pair of gluons. When sufficient energy is available, they could yield an electron-positron pair by means of an excited electron or positron. The resulting $\gamma\gamma$ pair would then be detectable. This is possible as well for d-quark as for u-quarks, but resonances may occur at different energies.

Another process that could also produce $\gamma\gamma$ pairs is represented in Figure 11. It includes the relevant u-quantum numbers for checking their conservation. We consider here a u-quark that creates a nark and its antinark. The resulting N_2 boson can produce an electron-positron pair that is annihilated. Such a resonance would be very remarkable, since it implies the creation of DM particles. Moreover, it would involve super symmetric partners of a quark and an excited lepton.

The discovery of a mysterious diphoton resonance at about 750 GeV was announced in 2015 and then reported to be only a statistical fluctuation. The actual data is more complex [23] [24] and will require further studies. If it were possible to detect a virtual N_2 boson, the resonance could determine its mass. Anyway, we are entering an era where detection and identification of DM particles is becoming a realistic goal. The Chinese satellite DAMPE (Dark Matter Particle Explorer) was launched in December 2015 and is specifically equipped to yield complementary data [25] . The possibility of analyzing signals in terms of a conservation law, resulting from STQ may also be useful.

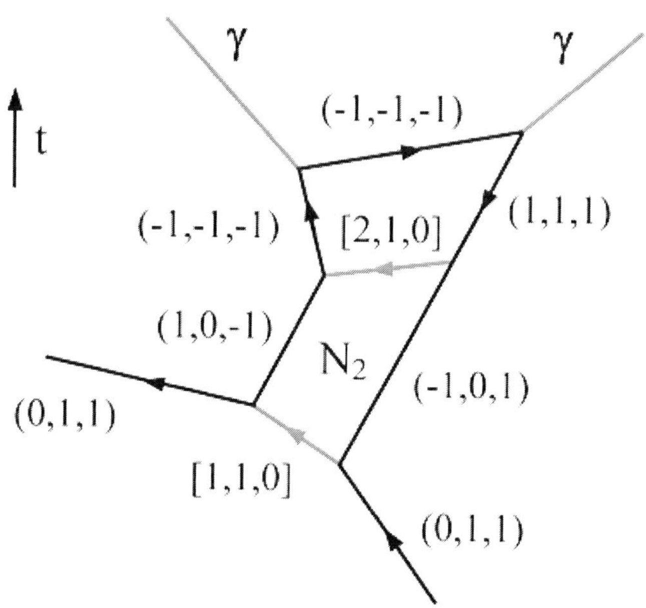

Figure 11. Conceivable creation of DM particles by proton-proton collisions.

6. CONCLUSIONS AND APPLICATIONS

We started this investigation with a simple, but basic question: Are space and time continuous or not? We considered thus the value a of the smallest measurable length as being unknown and constructed a theory of Space-Time Quantization (STQ). It generalized Relativistic Quantum Mechanics to account for c, h and a. The highest possible energy of free particles in inertial frames is then $hc/2a$. It is not infinite when $a \neq 0$. This allows for superluminal velocities at extremely high energies, but there are no logical inconsistencies. We could thus not exclude that space and time are quantized.

It appeared then that the existence of a finite limit for the smallest measurable length accounts also for elementary particle physics. Empirical data have been described and organized by means of the standard model. This required the introduction of phenomenologically defined concepts, but the resulting spectroscopy could not be explained. We ignored even why there are elementary particles and why they are characterized by quantized observables, obeying specific rules. This can be clarified when space and time are quantized. It is necessary, however, to realize that ideally precise measurements of spacetime coordinates do yield different lattices for possible results.

The "normal lattice" of lattice-constant a includes the origin of the reference frame, but there are also other lattices. They are displaced with respect to the normal lattice by $a/2$ along the chosen reference axes. The directions and orientations of these axes can be chosen in a completely arbitrary way, but space is three-dimensional. Quantum- mechanical ψ functions have to be defined on all these lattices and can differ on the intercalated lattices with respect to the normal lattice. The essential point is that quantum-mechanical ψ functions are epistemological tools, which allow us to store information by means of possible variations of $\psi(x, y, z, ct)$ for any given reference frame. The continuum assumption concealed some possibilities.

Large-scale variations define "states of motion" in terms of possible wavelengths and frequencies. By multiplexing, they are compatible with different types of variations at the smallest possible scale in space and time. They define "particle states" in terms of four quantum numbers, specifying relative phases for $\psi(x, y, z, ct)$ functions on intercalated lattices with respect to the normal lattice. It appears also that transformations of elementary particles into one another and resulting interaction are subjected to very strict conservation laws for these quantum numbers.

Although we are accustomed to consider that particles carry some energy, electric charge and other observable properties, they can also be viewed as being excitations of space and time. The four new quantum numbers (u_x, u_y, u_z, u_{ct}) are always and everywhere identical in the whole universe for a given type of particles. They are also independent of the chosen reference frame. These consequences of STQ agree with a great amount of already known empirical results in elementary particle physics. We understand for instance why the Standard Model of elementary particle physics required that quarks have 3 possible color states and that there are (at least) three generations of elementary particles. We can thus conclude from actual observations that STQ is real. It is not necessary to assume the existence of additional dimensions to account for elementary particle physics, since the usual 4-D space-time is sufficient when $a \neq 0$. It is not even necessary to know the actual value of the smallest measurable length.

STQ generalizes the Standard Model and accounts in particular for the nature and basic properties of DM particles. STQ could thus also be helpful for the detection and identification of DM particles. It opens a wide field of research in cosmology and astrophysics. We will provide some examples [7] [13] . Here, we have shown that it is possible to enlarge the foundations of physics by means of a natural extension of the theory of relativity and quantum mechanics.

CITE THIS PAPER

Meessen, A. (2017) From Space-Time Quantization to Dark Matter. Journal of Modern Physics, 8, 35- 56.

REFERENCES

1. Bahcall, N.A. (2015) Proceedings of the National Academy of Sciences of the United States of America, 112, 12243-12245.
2. Turner, M.S. (2001) Dark Energy and the New Cosmology.
3. Meessen, A. (1999) Foundations of Physics, 29, 281-316.
4. Meessen, A. (2011) Space-Time Quantization, Elementary Particles and Dark Matter.
5. Tachyon.
6. Meessen, A. (1978) Foundations of Physics, 8, 399-415. http://www. meessen.net/AMeessen/STQ1978.pdf
7. Meessen, A. (2017) Accelerated Expansion of Space, Inflation and Dark Energy. (To Be Published)
8. Francis, M.R. (2015) The Mystery of Particle Generations. http://www.symmetrymagazine.org/article/august-2015/the-mystery-of-particle-generations

9. Georgi, H. and Glashow, L.S. (1974) Physical Review Letters, 32, 438-441. http://pcbat1.mi.infn.it/~battist/astrop/su5.pdf
10. Carter, K. (2006) The Rise and Fall of the Pentaquark. http://www.symmetrymagazine.org/article/september-2006/the-rise-and-fall-of-the-pentaquark
11. LHCB Collaboration. (2015) Physical Review Letters, 115, Article ID: 072001.
12. The D0 Collaboration (2016) Physical Review Letters, 117, Article ID: 022003.
13. Meessen, A. (2017) Astrophysical Applications of Dark Matter Theory. (To Be Published)
14. Spergel, D.N. and Steinhardt, P.J. (2000) Observational Evidence for Self-Interacting Cold Dark Matter.
15. Kruezi, L. (2015) The Case for Complex Dark Matter. Interview of Bullock, J., Quanta Magazine. https://www.quantamagazine.org/20150820-the-case-for-complex-dark-matter/
16. Bullock, J. (2014) Self-Interacting Dark Matter. Center for Galaxy Evolution. https://www.cfa.harvard.edu/events/2014/sackler/index/talks/Harvard2014_Bullock.pdf
17. Weniger, C. (2012) Journal of Cosmology and Astroparticle Physics, 2012, 1-23.
18. Su, M. and Finkbeiner, D.P. (2012) Strong Evidence for Gamma-Ray Line Emission from the Inner Galaxy.
19. AMS Collaboration (2013) Physical Review Letters, 110, Article ID: 141102.
20. AMS-02 (2015) "AMS Days at CERN" and Latest Results.
21. Beringer, J., et al. (2012) Physical Review D, 86, Article ID: 010001. http://pdg.lbl.gov/2012/listings/rpp2012-list-z-boson.pdf
22. LUX (Large Underground Xenon) Collaboration (2016) Results from a Search for Dark Matter in the Complete LUX Exposure.
23. The CSM Collaboration (2016) Physical Review Letters, 117, Article ID: 051802.
24. ATLAS Collaboration (2016) Search for Resonances in Diphoton Events at $\sqrt{s} = 13$ TeV with the ATLAS Detector. JHEP 09 001.
25. Spaceflight Dynamics (2016) Dark Matter Particle Explorer (DAMPE). http://spaceflight101.com/spacecraft/dark-matter-particle-explorer/

CHAPTER 12

The Quantum Oscillatory Modulated Potential I—The Hydrogen Atom

Waldemar Wolney Filho

Instituto de Física, Universidade Federal de Goiás, Goiânia, Brazil

ABSTRACT

In this work we are presenting a modified Coulomb potential function to describe the interaction between two microscopic electric charges. In particular, concerning the interaction between the proton and the electron in the hydrogen atom. The modified potential function is the product of the classical Coulomb potential and an oscillatory function dependent on a quantized phase factor. The oscillatory function picks up only selected points along the Coulomb potential, creating potential wells and barriers around the nucleus of the atom. The new potential reveals us new features of the hydrogen atom. Searching for a manner to determine the phase factor, we are using the concept of the de Broglie particle wavelike behavior and the quantum analogue of the virial theorem for describing the bound motion of a particle in a central force field. This procedure is a kind of feedback action, where we are making use of well established concepts of the quantum mechanics aiming to determine the phase factor of the new interaction potential.

Keywords: Potential; atom; electron; oscillator fuction

1. INTRODUCTION

This work is the first of a series of four papers, for which we are revisiting, reexamining and questioning some points about the foundations of the atomic theory developed, mostly between the years 1910-1930, by many distinguished physicists: Niels Bohr, Erwin Schroedinger, Werner Heisenberg, Alberl Einstein, Max Born, Louis de Broglie and many others. What motivate us to produce this work are, in our opinion, some unclear and questionable points, that we believe still exist, particularly, concerning the applicability of the classical Coulomb potential for describing the interaction between two microscopic electric particles. For instance, in the hydrogen atom, the Coulomb potential, used by Bohr and, years later, in the quantum mechanics (Schroedinger) to describe many of the known properties of the current atomic model, is an increasing smooth function of the distance between the proton and the electron. Being so, how can the electron possibly know that certain positions around the nucleus of the atom are more likely to be occupied than others? [1] What is the real cause for this behavior for the electron? It is difficult to imagine the electron making regular probabilistic choice to occupy certain particular positions in the vicinity of the nucleus of the atom, based solely in the characteristics presented by the Coulomb potential. The observed discrete values of the total energy of the electron, suggest us the existence of potential wells and barriers in the form of closed shells surrounding the nucleus of the atom. The question of how and why the electron is able to jump from one state of equilibrium to another, emitting or receiving energy radiation (light), could be better explained if we can accept the idea that the electron is tunnelling through these potential barriers.

The stability of the atom is an interesting question that deserves further explanation.[1] The de Broglie wavelength associated to the waving nature of a microscopic charged particle is an attribute of the mass, as postulated by Louis de Broglie in 1923, or is it an attribute of the charge of the particle?[2] Another interesting question is concerned to the actual movement of the electron around the nucleus of the atom. Is the electron orbiting the nucleus of the atom as predicted by Bohr theory? Is it behaving like a cloud of negative charge around the nucleus of the atom? Or yet is it oscillating radially in quantum wells like a harmonic oscillator?

The Coulomb potential belongs to what is known as classical physics. Since long ago it is a fact well known that it correctly describes the energy interaction between two punctual macroscopic electric charges. Also, it is well known that the behavior of a particle in the micro-world of the quantum physics is not the same as that in the macro-world of classical physics. What we want to say is that, the classical Coulomb potential reveal us only the macro-behavior of the interaction between electric particles

This make us to suspect the existence of a different interaction potential, capable of describing more accurately the physics of the atoms, and also be able to answer the questions formulated above. We believe that the choice of any potential function that proposes to be candidate to describe the interaction between two microscopic charges, in some way, needs to carry by itself the seeds of the quantum physics of the atom, that is, the quantization of the angular momentum and energy, both concepts, intimately related to the Heisenberg uncertainty principle. Besides, in some way, the new potential must be responsible to the wavelike nature of the particles.

2. THE HYDROGEN ATOM

In this paper, trying to answer the questions formulated above, let us postulate that in the hydrogen atom, the interaction energy between the proton (nucleus of the atom) and the electron, considered as spinless particles, can be expressed by the product of the Coulomb potential and one oscillatory function, written as

$$U_{osc.}(r) = -\frac{1}{4\pi \in_0}\frac{Ze^2}{r}\cos^2(\xi r). \tag{1}$$

The parameter ξ, with dimension of inverse of distance, represents a radial-space frequency. We believe that the phase factor ξr of the oscillatory function is the key point to answer the questions formulated preously. As we will see soon, this phase factor is directly related to the Heisenberg uncertainty principle and the quantization of the angular momentum.

For fixed values of the parameter ξ ($\xi \neq 0$), the function $\cos^2(\xi r)$ oscillates between zero and one, while the function $U_{osc.}(r)$ oscillates between zero and the classical Coulomb potential energy. The equation (1), tell us that when $\xi \to 0$, $\cos^2(\xi r) \to 1$, the new potential function becomes equal to the classical Coulomb potential.

$$U_{cou.}(r) = -\frac{1}{4\pi \in_0}\frac{Ze^2}{r}. \tag{2}$$

However, we need to observe that if we make $\xi = 0$, we have $U_{osc.}(r) = U_{cou.}(r)$ for all values of the distance r between the proton and the electron in the hydrogen atom. This implies that the linear momentum of the electron $p = \hbar\xi$ is null. That is, the electron is at rest in the LFR (Laboratory Frame of Reference). This condition strongly contraries one of the basilar principles of the quantum mechanics, the Heisenberg uncertainty principle $\Delta r \Delta p_r \geq \hbar.$ That is, if we make the uncertainty of the linear momentum $\Delta p_r = \hbar\Delta\xi = 0$, we have $\Delta r \rightarrow \infty$, and the radial position of the electron becomes completely undetermined. On the other hand, if we make $\Delta r = 0$, we have $\Delta p_r = \infty$, making the kinetic energy of the electron infinitely large [2-4]. This clearly shows the inadequacy of the Coulomb potential to describe the interaction between the proton and the electron in the hydrogen atom, except at particular points of the space in the vicinity of the nucleus of the atom. The phase factor of the oscillatory function selects only specific points of the Coulomb potential, making it to satisfy the Heisenberg uncertainty principle only at points, where the condition $\xi r = n\pi$ is satisfied. This constitutes a quantized relation between the normal and the reciprocal space.

Looking for the determination of the parameter ξ, we will make use of the quantum-mechanical analogue of the virial theorem and the de Broglie relation between the linear momentum of the particle and the wave number k. With this procedure, we are making use of well known quantum concepts to determine the phase factor of the new potential. The quantum analogue of the virial theorem can be set in the form [4]

$$\langle T \rangle = \frac{1}{2}\langle \mathbf{r}.\nabla U(r) \rangle \tag{3}$$

and the required de Broglie relation is

$$p = \hbar k \tag{4}$$

To facilitate the application of equation (1), it turns out to be convenient that the parameter ξ may be expressed in terms of known quantities such as, the charge and mass of the electron and some physical constants. With the use of equations (1) and (3), the average kinetic energy of the electron in the hydrogen atom is found to be expressible as

$$\left\langle \frac{p^2}{2m} \right\rangle = \frac{1}{2}\left(\frac{Ze^2}{4\pi \in_0}\right)\left\langle \left[\frac{1}{r}\cos^2(\xi r) + \xi \sin(2\xi r)\right]\right\rangle \tag{5}$$

With the use of equation (4) it follows that

$$\langle k \rangle = \left[\left(\frac{1}{4\pi \in_0}\right)\frac{Zme^2}{\hbar^2}\right]^{\frac{1}{2}} \left\langle \left[\frac{1}{r}\cos^2(\xi r) + \xi \sin(2\xi r)\right]^{\frac{1}{2}} \right\rangle. \tag{6}$$

We will select the desired particular points along the Coulomb potential, observing that every time the product ξr satisfies the quantization relation,

$$\xi r = n\pi, \; n = 1, 2, 3, \cdots, \tag{7}$$

the Coulomb potential becomes equal to the oscillatory potential and the equation (6), may be written as

$$\frac{\langle k \rangle}{n} = \left[\frac{1}{4\pi \in_0}\frac{Zme^2}{\hbar^2 n^2}\right]^{\frac{1}{2}} \left\langle \left(\frac{1}{r}\right)\right\rangle^{\frac{1}{2}}, \tag{8}$$

or

$$\frac{\langle k \rangle}{n} r = \left[\frac{1}{4\pi \in_0}\frac{Zme^2}{\hbar^2 n^2}\right]^{\frac{1}{2}} \left\langle \left(\frac{1}{r}\right)\right\rangle^{\frac{1}{2}} r. \tag{9}$$

The equations (7) and (9) become consistent if we set $\xi = \langle k \rangle / n$ and

$$\left\langle \left(\frac{1}{r}\right)\right\rangle = \left[\left(\frac{1}{4\pi \in_0}\right)\frac{Zme^2}{\hbar^2 n^2}\right], \tag{10}$$

what permit us to rewrite the equation (9) as

$$\langle \xi \rangle r = \left\langle \left[\frac{1}{4\pi \in_0}\frac{\pi Zme^2}{\hbar^2 n^2}\right]\right\rangle r. \tag{11}$$

Comparing the equations (7) and (11) we can identify the space-frequency ξ as a quantized quantity given by

$$\langle \xi_n \rangle = \frac{1}{4\pi \in_0} \frac{\pi Z m e^2}{n^2 \hbar^2}. \tag{12}$$

In the hydrogen atom we have $Z=1$ and, in its ground state, $n=1$. The quantity $\left(4\pi \in_0 \hbar^2 / me^2\right)$ may be identified as the first Bohr radius $\langle r_1 \rangle = a_0 = 0.529$ Å of the electron. The equation (12) may be rewritten as

$$\langle \xi_n \rangle = \frac{\pi}{n^2 a_0}. \tag{13}$$

This implies that for two atomic bound particles, such as the proton and the electron in the hydrogen atom, the average value of the linear momentum of the electron is also a quantized quantity

$$\langle p_n \rangle = \left(\frac{1}{4\pi \in_0}\right) \frac{me^2}{n\hbar}. \tag{14}$$

Using equation (13), the new potential function may be rewritten as

$$U_n(r) = -\frac{1}{4\pi \in_0} \frac{e^2}{r} \cos^2(\xi_n r). \tag{15}$$

We are naming this potential by Quantum Oscillatory Modulated Potential-QOMP; because, it is quantized, it oscillates with k-space frequency ξ_n, and it is partially modulated by the classical Coulomb potential. Using the condition given by equation (7) and the equations (1) and (5), the average total energy of the electron with respect to the nucleus is

$$\begin{aligned}\langle T+U \rangle &= \left\langle \frac{1}{2}\left(\frac{1}{4\pi \in_0}\right) e^2 \frac{1}{r} - \frac{1}{4\pi \in_0} Ze^2 \frac{1}{r} \right\rangle \\ &= -\frac{1}{2}\left(\frac{1}{4\pi \in_0}\right) Ze^2 \left\langle \frac{1}{r} \right\rangle. \end{aligned} \tag{16}$$

Taking the average value $\langle 1/r \rangle$, as given by equation (10), the energy of the electron becomes quantized and may be written as

$$E_n = -\frac{1}{2}\left(\frac{1}{4\pi \in_0}\right)^2 \frac{me^4}{n^2 \hbar^2}, \tag{17}$$

as would be expected, it is the same relation as obtained with Bohr theory and also with the solution of the time-independent Schroedinger wave equation for the hydrogen atom when the Coulomb potential is used. But, differently from these two theories, that in different circumstances, make explicit use of the quantization of the angular momentum to obtain the total energy of the electron, in this work, the same result is achieved without making explicit use of this concept. Implicitly, the quantization of the angular momentum is contained in the phase factor of the oscillatory function, equation (7).

The figure 1 shows a graphic of the potential $U_1(r)$. As shown in the figure, the electron in the hydrogen atom occupies the first well with position and total energy satisfying the equations (10) and (17), respectively. The figure 2 shows the graphics of $U_n(r)$ for $n=1,2$ and 3. The values of the energies shown in the graphic determined by the positions of the electron are, respectively, −13.6 eV, −3.4 eV and −1.5 eV. The same results are predicted by equation (17) [4,5].

The new potential present us with several interesting features: 1) Its waving character shows a sequence of periodic wells and barriers with variable depth and height; 2) The first well is at the position most likely to be occupied by the electron in the ground state of the atom; 3) The quantization of the energy may be attributed to the existence of concentric quantum shells, formed by wells and barriers, created by the interaction of the nucleus of the atom and the electron; 4) According to this potential, the electron bound to the nucleus can only jumps from one quantum well to another by tunneling through finite potential barriers. The tunneling effect (a purely quantum mechanical phenomenon) now appears naturally as necessary condition for emission or absorption of energy (light) by the atom; 5) Another new interesting aspect of the oscillatory potential, observed in the figure 2, and predicted by equations (13) and (14) is the fact that the de Broglie wavelength associated to the electron is equal to the first Bohr radius of the electron in the hydrogen atom. Thus, the average distance between the proton and the electron, for any excited state of the atom, may be written as

$$\langle r_n \rangle = n^2 \langle r_1 \rangle. \tag{18}$$

Or, equivalently, in terms of the de Broglie wavelengths we may write

$$\langle \lambda_n \rangle = n^2 \langle \lambda_1 \rangle. \tag{19}$$

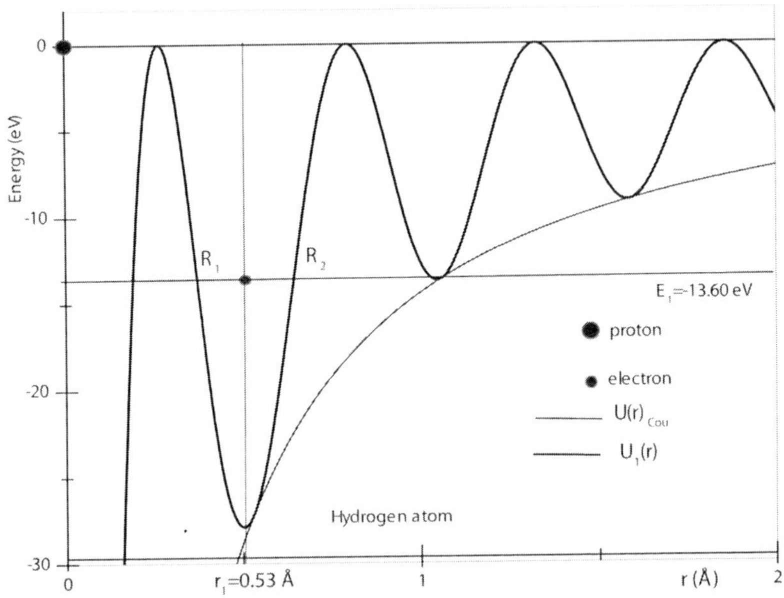

Figure 1. The Coulomb potential $U(r)_{Coul}$ and the quantum oscillatory modulated potential $U_1(r)$, showing the position and energy of the electron in the ground state of the hydrogen atom. The positions R_1 and R_2 are classical turning points for the electron.

This result is completely different from that predicted by the quantization rule $2\pi r = n\lambda$ for each allowed orbit for the electron, as postulated by Louis de Broglie, to explain the quantization of the angular momentum of the electron in the Old Quantum Theory [5,6].

Figure 2 shows that the averaged positions occupied by the electron in the first three quantum wells of $U_1(r)$, $U_2(r)$ and $U_3(r)$ are, respectively, $\langle r_1\rangle = 0.53$ Å, $\langle r_2\rangle = 4\langle r_1\rangle$ and $\langle r_3\rangle = 9\langle r_1\rangle$. In this work we are arguing that the Coulomb potential is not totally appropriated to describe correctly the potential interaction between two microscopic electric charges. However, as is well known, the solution of the time independent Schroedinger wave equation predicts correctly the quantization of the angular momentum and energy of the electron, when Coulomb potential is used in the Hamiltonian of the protonelectron system. This occur because, in the solution of the radial part of the Schroedinger differential wave equation, without noticing

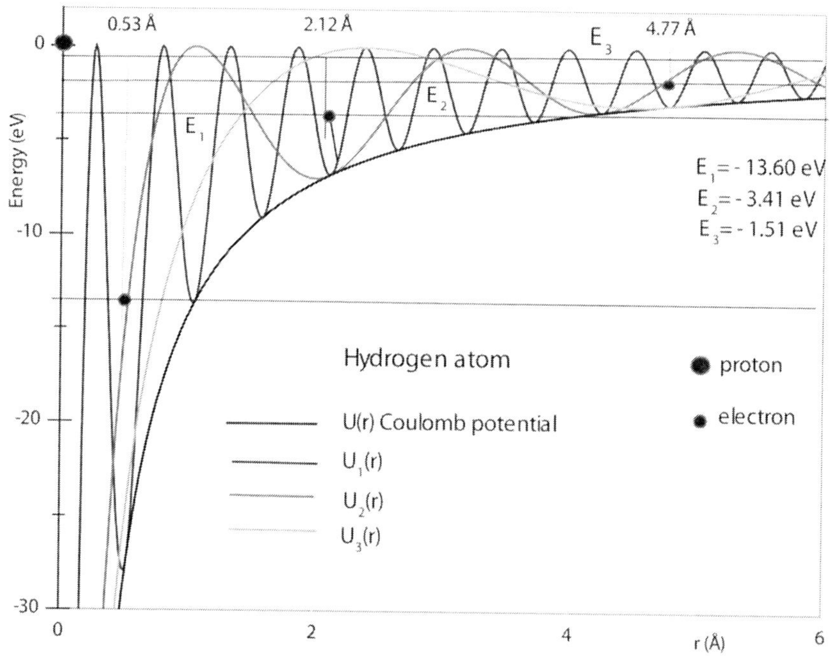

Figure 2. The Coulomb potentials $U(r)_{Coul}$, the quantum oscillatory modulated potentials $U_n(r)$ and the energy levels and positions of the electron in the hydrogen atom for n = 1, 2 and 3.

it, we are making use the quantization rule $\xi r = n\pi$ for the phase factor of the oscillatory function of the new potential. That is, implicitly, we are making $\cos^2(\xi_n r) = 1$ and, consequently, selecting only the points of minima of the new potential, where both interaction potentials are coincident, see figure 2. In these selected points the Coulomb potential satisfies the Heisenberg uncertainty principle. In terms of a mathematical treatment, the quantization of the phase factor ξr allow us to see that the Coulomb potential is a particular case of the oscillatory potential, by transforming the Schroedinger radial differential equation

$$\frac{1}{\rho^2}\frac{\mathrm{d}^2 S(\rho)}{\mathrm{d}\rho^2} + \frac{2}{\rho}\frac{\mathrm{d}S(\rho)}{\mathrm{d}\rho} + \left[\frac{\delta}{\rho}\cos^2\left(\xi_n \frac{1}{2\beta}\rho\right) - \frac{l(l+1)}{\rho^2} - \frac{1}{4}\right]S(\rho) = 0, \tag{20}$$

obtained with the use of the Quantum Oscillatory Modulated Potential in the Hamiltonian of the protonelectron system into the simpler and familiar radial differential equation [4,5]

$$\frac{1}{\rho^2}\frac{\mathrm{d}^2S(\rho)}{\mathrm{d}\rho^2}+\frac{2}{\rho}\frac{\mathrm{d}S(\rho)}{\mathrm{d}\rho}+\left[\frac{\delta}{\rho}-\frac{l(l+1)}{\rho^2}-\frac{1}{4}\right]S(\rho)=0, \tag{21}$$

obtained with the use of the classical Coulomb potential, with the condition $\xi_n\rho/2\beta=n\pi$ or $\cos^2(\xi_n\rho/2\beta)=1$. The solution of equation (21) is well known and it is found in many books of modern physics literature [4-7]. Its solution shows that the total energy of the electron in the hydrogen atom is the same as predicted by equation (17). Up to this point we can also state that the Coulomb potential is restricted to describe correctly the interaction between two microscopic electric charges only at particular points of the space around nucleus of the atom, points where the quantization relation $\xi_n r=n\pi$ and, consequently, the Heisenberg uncertainty principle are satisfied.

3. A NEW ELECTRIC INTERACTION MICROSCOPIC FORCE

The electric force between two macroscopic electric point charges Q_1 and Q_2, separated by a distance r, is given by the well known classical Coulomb formula

$$F_{coul.}=\pm\frac{1}{4\pi\epsilon_0}\frac{Q_1Q_2}{r^2} \tag{22}$$

which is perfectly valid as far as we consider the two charges built up of a very large number of protons or electrons, but still considered as point charges. However, the electric force between two microscopic charges such as proton-proton, proton-electron, or electron-electron, originated from the new potential function, has the form of a wave packet, written as

$$\begin{aligned}F_n(r)&=-\frac{\mathrm{d}}{\mathrm{d}r}U_n(r)\\&=\pm\frac{1}{4\pi\epsilon_0}Ze^2\left[\frac{1}{r^2}\cos^2(\xi_n r)+\frac{\xi_n}{r}\sin(2\xi_n r)\right],\end{aligned} \tag{23}$$

The forces $F_n(r)$ oscillate with the same k-space frequencies ξ_n. For the hydrogen atom, figure 3 shows a graphic of $F_1(r)$, as predicted by equation (23). Since this force is the gradient of the energy function $U_1(r)$, which presents a minimum at position $\langle r_1 \rangle$, it becomes null at this point. We must observe that with the use of the classical Coulomb potential this situation never occurs, the interaction force between the proton and the electron is always attractive. The acceleration due to the electric force being null at this position, eliminates the argument or defiance against the Classical Laws of the Electrodynamic that an electron revolving the nucleus of the atom, lose energy by radiation [8-10]. The force being null at the equilibrium position $\langle r_1 \rangle = a_0$, means that the electron cannot be orbiting the nucleus of the atom as predicted by Bohr theory. Thus, according to the QOMP, the angular momentum of the electron in its ground state is null, in agreement with the prediction of the Schroedinger radial differential wave equation (21). On the other hand, the electron cannot be a static physical object due to the Heinsenberg uncertainty principle. The angular moment of the electron being null ’ implies that the movement of the electron, inside a quantum well must be radial, bouncing back and forth, confined by two potential barriers.

This argument is reinforced by observing that in figure 3, the interaction force between the proton and the electron is repulsive to the left of the position r_1 and attractive to the right of this position. That is, the electron is always being pushed towards the equilibrium position, acting like a harmonic oscillator. However, being accelerated towards the equilibrium position, again, the electron should lose energy by radiation and the problem of stability of the atom continues to be unexplained. The stability of the atom will be object of a deeper discussion in the second paper of this series of work. **Figure 4** shows the behavior of the electric forces $F_n(r)$ for $n = 1, 2$ and 3. The graphics show that the waving character of interaction electric force decreases when n increases, in agreement with the Bohr correspondence principle. The reasons for this behavior are two: 1) The destructive interference of the waving pattern of the individual charges, protons and electrons occur because, when macroscopic charges[3] are built up, these particles do not occupy the same position in space and the waving pattern of the force must be destroyed by an interference process; 2) The actual measurements of the interaction

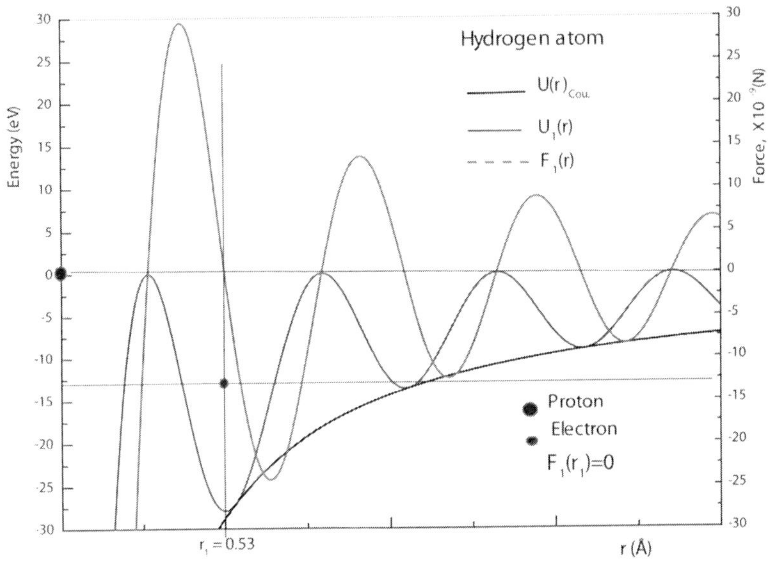

Figure 3. The interaction oscillatory force $F_1(r)$, the quantum modulated potential $U_1(r)$ and the Coulomb potential $U(r)_{coul}$ in the ground state of the hydrogen atom. The figure also shows that the force is null at the equilibrium position r_1.

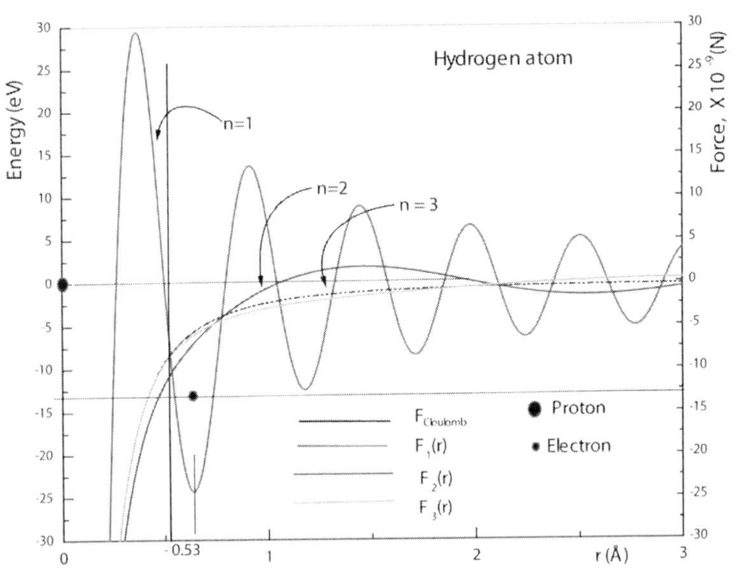

Figure 4. The Coulomb force $F(r)_{coul}$ and the interaction oscillatory forces $F_n(r)$ between the proton and the electron in the hydrogen atom, for n = 1, 2 and 3.

force between two macroscopic particles are usually performed, using large ionized atoms, for which the escaped electrons come from outer shells of the atoms, making the new electric force to approximate quickly to the classical Coulomb force as indicated in figure 4.

4. CONCLUSION

Waldemar Wolney Filho We can summarize this work as follows: For the hydrogen atom, the oscillatory potential function is able to predict correctly the values of the total energy and position of the electron in the atom, without the necessity of solving the time-independent Schroedinger wave equation. In addition, it shows several peculiarities of the hydrogen atom, such as: The existence of quantum wells and barriers, forming closed shells around nucleus of the atom. These wells constitute the most probable positions for the electron to occupy when moving around the nucleus of the atom; The emission or absorption of energy by the atom happens when the electron moves from one state of equilibrium to another by tunneling through these potential barriers created by the oscillatory potential; The new potential shows that the electric force between the proton and the electron oscillates, presenting negatives and positives peaks, the positive peaks represent regions where the interaction force between the proton and the electron is repulsive and the negative peaks where it is attractive, that is, the electron is always been pushed to the equilibrium position, which is a necessary condition for the stability of the atom. The new potential predicts that de Broglie wavelength associated to the electron is equal to the radius of Bohr $r_1 = a_0$, what is quite different of the condition postulated by Louis de Broglie for the quantization of the angular momentum of the electron in the Bohr Old Quantum Theory. According to the new potential, the electron is not orbiting the nucleus of the atom but oscillating radially, confined by two potential barriers. Whats more, this work presents strong evidences that the Coulomb potential function is not adequate to describe the correct interaction between two microscopic electric charges, because, except at particular points around the nucleus of the atom, it is in disagreement with one of the basic principles of quantum mechanics-the Heisenberg uncertainty principle. The new potential is much more informative than the classical Coulomb potential and more appropriated to accomplish the function for the interaction between two microscopic electric charges.

REFERENCES

1. R. P. Feymann, R. B. Leighton and M. Sands, "Lectures on Physics," Addison Wesley, Reading, 1966.
2. C. Cohen-Tannoudji, B. Diu and F. Laloë, "Quantum Mechanics," John Wiley & Sons, New York, 1997.
3. L. Gilder, "The Age of Entanglemen," Alfred A. Knopf, New York, 2008.
4. W. W. Filho, "Mecânica Quântica," Editora da UFG, Goiânia, 2002.
5. R. M. Eisberg, "Fundamentals of Modern Physics," John Wiley & Sons Inc., New York, 1961.
6. P. A. Tipler, "Modern Physics," Worth Publishers, Inc. Irving, 1986.
7. L. Pauling and E. B. Wilson, "Introduction to Quantum Mechanics," McGraw Hill, New York, 1935.
8. W. Hauser, "Introduction to the Principles of Electromagnetism," Addison Wesley Publishing Company Inc., Boston, 1970.
9. E. Merzbacher, "Quantum Mechanics," John Wiley & Sons Inc., New York, 1976.
10. V. A. Fock, "Fundamentals of Quantum Mechanics," Mir Publishers, Moscow, 1982.

NOTES

[1]The stability of the atom will be treated in the second paper of this series.

[2]This subject will be considered in the third paper and, for an uncharged particle, possessing an electric dipole moment, the subject will treated in the fourth paper of the series.

[3]A macroscopic charge of 1 μC = 1×10^{-6} C, contains about 10^{13} electronic charges.

CHAPTER 13

The Quantum Oscillatory Modulated Potential—Part II —The Stability of the Hydrogen Atom

Waldemar Wolney Filho

Instituto de Física, Universidade Federal de Goiás, Goi^ania, Brasil

ABSTRACT

The purpose of this work is to show the stability of the hydrogen atom with useing the Quantum Oscillatory Modulated Potential and the Heisenberg equations of motion, postulating that the electron in the hydrogen atom is behaving as a quantum harmonic oscillator. With the electron confined between two potential barriers, created by the new potential function, we are considering that at absolute temperature the power absorbed or emitted by the electron per unit of time can be used to determine the zero point energy of the oscillator. Assuming that electron is only exchanging energy with the nucleus of the atom we are making use of the operators of creation and annihilation of a photon to explain how the energy between the proton and the electron can be exchanged to keep the atom a stable system.

Keywords: Quantum Oscillator; Hydrogen Atom; Electron

1. INTRODUCTION

In 1913, the Dane physicist Niels Bohr, to assure the stability of the atoms as a nuclear planetary model, such as proposed two years earlier, by the New Zealand physicist Ernest Rutherford, postulated the quantization of the angular momentum of the electrons when revolving the nuclei of the atoms. About twelve years later the Bohr theory was supplanted by the quantum mechanics, a new theory, independently formulated by Erwin Schroedinger and Werner Heisenberg [1,2]. Both theories have wider applications to more complex atomic systems. In the Schroedinger theory, the concept of orbit for the electron was ignored and its position around the nucleus was formulated in terms of probability density. In the application of the new quantum theory for the hydrogen atom, the question of the stability of the atom was set aside. However, we need to taken into consideration that an electron bound to the nucleus of the atom is always under the action of a central force field [3]. Therefore, whatever be its movement in the vicinity of the nucleus of the atoms, in someway, it needs to be accelerated and, according to the Classical Law of the Electrodynamics, losing energy by radiation [4]. Since the atoms are stable systems, this does not occur and the question of the stability of the atoms needs to be explained.

In this work we are proposing to solve the problem of the stability of the atoms, making use of the Quantum Oscillatory Modulated Potential[1]—QOMP. This potential was formulated to explain in different mathematical bases and concepts the interaction between the proton and the electron in the hydrogen atom. The new potential is represented by the formula

$$U_n(r) = -\frac{1}{4\pi\varepsilon_0}\frac{e^2}{r}\cos^2(\xi_n r), \text{ with } \xi_n = \frac{\pi}{n^2\langle r_1\rangle}, \tag{1}$$

where the $\langle r_1\rangle = a_0$ is the Bohr radius for the hydrogen atom. The potential $U_n(r)$ is quantized, oscillatory and presents potential wells and barriers, partially modulated by the Coulomb potential. As gradient of the new potential, the interaction force between the proton and the electron takes the form

$$F_n(r) = -\frac{e^2}{4\pi\epsilon_0}\left[\frac{1}{r^2}\cos^2(\xi_n r) + \frac{\xi_n}{r}\sin(2\xi_n r)\right]. \tag{2}$$

These forces have the form of wave packets around the nucleus of the atom. The **Figure 1** shows the graphic of this interaction force between the proton and the electron in the hydrogen atom in its ground state. It displays positive and negative peaks and is null at the position more likely to be occupied by the electron. Let us observe that with the Coulomb potential, the interaction force between the proton and the electron is always attractive and never becomes null, except at infinite distance from the nucleus of the atom. With the use of the QOMP, it is shown that the interaction force at the equilibrium position r_1 is null and the electron cannot be orbiting the nucleus of the atom as predicted by Bohr theory. Then, the most plausible movement of the electron is the radial, what makes the angular momentum of the electron be null. Thus, in its ground state, the electron must behaves like a harmonic oscillator, bouncing back and forth radially, confined by two potential barriers, making the electron to be accelerated towards the equilibrium position and, consequently, losing energy by radiation [4,5]. In the next section we will use the concept of the quantum harmonic oscillator to investigate the behavior of the oscillating electron inside the first quantum well produced by the new potential.

2. THE HYDROGEN ATOM AS A QUANTUM HARMONIC OSCILLATOR

Let us assume that the departure of first potential well from the potential of a harmonic oscillator is small. The origin of the quantum harmonic oscillator potential is taken at the bottom of the first potential well, at the position $r_1 = 0.53$ Å and nearly 13.59 eV bellow the energy of the electron in the ground state. See **Figure 2**. Symbolically the movement of the electron inside the quantum well is represented by arrows as shown in the same figure. The Hamiltonian for the quantum harmonic oscillator may be written as [2]

$$\hat{H} = \hbar\omega \hat{a}^{\dagger}\hat{a} + \frac{1}{2}\hbar\omega, \tag{3}$$

where $\hat{a}^{\dagger}$ and $\hat{a}$ are, respectively, the operators of creation and annihilation of a photon (a quantum unit of energy $\hbar\omega$). The term $(1/2)\hbar\omega$ in the Hamiltonian operator represents the zero point energy, it is the minimum energy of the quantum harmonic oscillator. The operators for position $\hat{r}$ and momentum $\hat{p}$ of the electron may be expressed in terms of the operators $\hat{a}^{\dagger}$ and $\hat{a}$, written as

$$\hat{r} = \frac{1}{2\alpha}\left(\hat{a} + \hat{a}^{\dagger}\right) \tag{4}$$

$$\hat{p} = \frac{1}{2i\beta}\left(\hat{a} - \hat{a}^{\dagger}\right), \tag{5}$$

where $\alpha = (m\omega/2\hbar)^{1/2}$ and $\beta = (1/2\hbar m\omega)^{1/2}$. These operators when applied to the eingenstates $|\phi_n\rangle$ of the quantum harmonic oscillator, produce the following well known relations

$$\hat{a}^{\dagger}|\phi_n\rangle = \sqrt{n+1}\,|\phi_{n+1}\rangle, \tag{6}$$

$$\hat{a}|\phi_n\rangle = \sqrt{n}\,|\phi_{n-1}\rangle, \tag{7}$$

$$\hat{a}|\phi_0\rangle = 0. \tag{8}$$

The eingenstates $|\phi_n\rangle$ of the quantum harmonic oscillator are orthonormal. That is, they obey the condition

$$\langle\phi_m|\phi_n\rangle = \delta_{mn}.$$

Moreover, $\hat{a}^{\dagger}$ and $\hat{a}$ are not commuting operators, they satisfy the relation

$$\left[\hat{a}, \hat{a}^{\dagger}\right] = 1, \tag{9}$$

what means that the kets $|\phi_n\rangle$ are not eigenstates of these two operators. That is, they cannot act simultaneously on the eingstates of the harmonic oscillator. As simple harmonic oscillator the binding energy of the electron in the atom in its ground state $E_1 = -13.59$ eV and can be totally kinetic at equilibrium position r_1 or totally potential at the classical turning points R_1 and R_2. See **Figure 2**. According to Equations (3) and (8) the zero

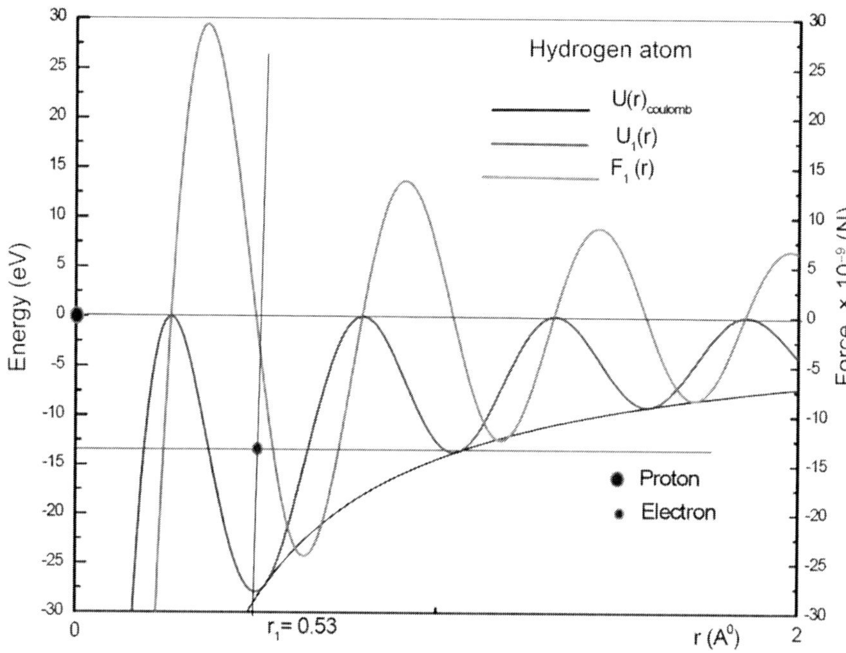

Figure 1. The coulomb potential $U(r)_{Coul.}$, the quantum oscillatory modulated potential $U_1(r)$ and electric oscillatory force $F_2(r)$ for the hydrogen atom in its ground state. The vertical line represents the position where the oscillatory force is null.

point energy of the harmonic oscillator is given by the expectation value [2,6]

$$E_0 = \langle \phi_0 | H | \phi_0 \rangle = \frac{1}{2}\hbar\omega. \tag{10}$$

To determine the angular frequency ω in Equation (10) and, consequently the zero point energy of the quantum harmonic oscillator, we will make the assumption, that in the ground state of the hydrogen atom, the amount of energy absorbed or emitted by the electron per unit of time, at absolute temperature, $T = 0$, can be expressed by a formula obtained by Silva and França [7,8] with the use of the stochastic electrodynamics. The power absorbed by the electron may be expressed by

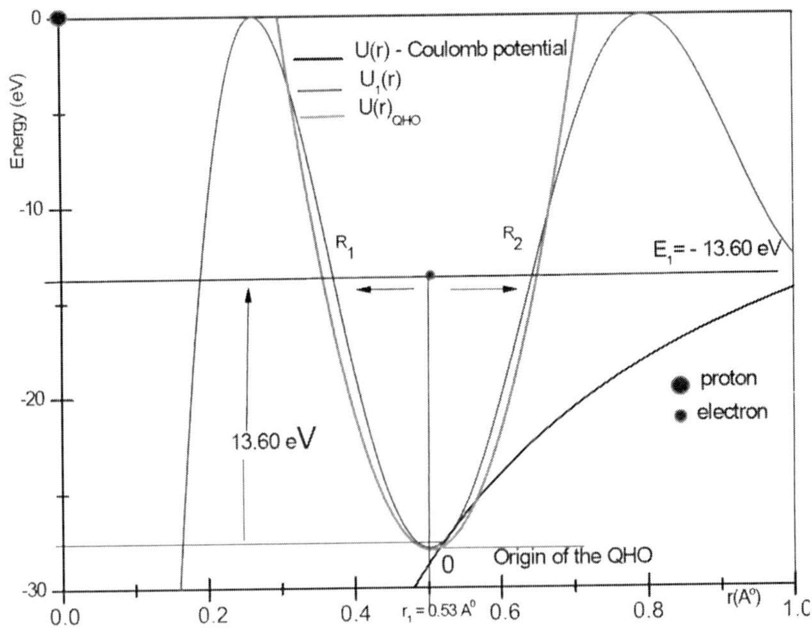

Figure 2. In the ground state of the hydrogen atom, the electron is oscillating, bouncing back and forth like a quantum harmonic oscillator. The positions R_1 and R_2 represent classical turning points.

$$P_{abs.}\left(T=0\right) \approx \frac{2\hbar e^5}{m^{\frac{3}{2}} c^3 r^{\frac{9}{2}}}, \tag{11}$$

where c is the velocity of light. However, differently of Silva and França, that for deduction of Equation (11), assumed that the electron is orbiting the nucleus of the atom, as in Bohr theory, we are arguing that the electron is oscillating radially in the first potential well, with average distance equal to the radius of Bohr, $\langle r \rangle = a_0$, as predicted by the QOMP.

For the hydrogen atom, in its ground state, we are assuming that at absolute temperature, electron is only be able to exchange energy with the nucleus of the atom, with which it is interacting. This image is the analogue to two masses M and m, with $M \gg m$, tied at the end of a spring, oscillating with an angular frequency ω and exchanging energy between them. With the use of Equation (11), we can write,

$$P_{abs.}(T=0) \approx \frac{\hbar\omega^2}{2\pi}, \tag{12}$$

where ω is the lowest angular frequency corresponding to the zero point energy of the quantum harmonic oscillator. From Equation (12) we have

$$\omega \approx \sqrt{\frac{2\pi P_{abs.}(T=0)}{\hbar}}. \tag{13}$$

Considering that, at absolute temperature $(T=0)$, the most probable position for the electron is $\langle r \rangle = 0.53$ Å, from Equations (11) and (13) we can estimate the value of the angular frequency of the quantum harmonic oscillator in its lowest state. The predicted value is $\omega \approx 47\ s^{-1}$. Thus, the amount of energy absorbed or emitted by the electron when moving between two adjacent quantum states $|\phi_n\rangle$. and $|\phi_{n\pm1}\rangle$ is $\hbar\omega = 3.08\times10^{-14}$ eV. A photon with energy of this magnitude would be very difficult to be experimentally observed. The zero point energy of the quantum oscillator is $(1/2)\hbar\omega = 1.54\times10^{-14}$ eV, sometimes called the vacuum state energy. With the electron oscillating between the classical turning points R_1 and R_2, corresponds to the quantum harmonic oscillator be in an excited state $|\phi_n\rangle$, with energy $E_n = n\hbar\omega = 13.6$ eV. Thus, the number of eigenstates $|\phi_m\rangle$ existing between the energy E_n and E_0, the lowest energy level is $n \approx 44\times10^{13}$. This very large number of eigenstates is a consequence of the very small magnitude of quantum unit of energy exchanged between the proton and the electron in the atom, in each quantum transition. In the next section we will consider the time evolution of the electron in the hydrogen atom.

3. THE TIME EVOLUTION OF THE ELECTRON IN THE HYDROGEN ATOM

The time evolution of the electron as a quantum harmonic oscillator can be investigate with the use the Heinsenberg equations of motion for the operators $\hat{a}$ and $\hat{a}^{\dagger}$ given by [2]

$$\frac{\mathrm{d}}{\mathrm{d}t}\hat{a} = \frac{i}{\hbar}\left[\hat{H}, \hat{a}\right] = -i\omega\hat{a} \tag{14}$$

$$\frac{\mathrm{d}}{\mathrm{d}t}\hat{a}^{\dagger} = \frac{i}{\hbar}\left[\hat{H}, \hat{a}^{\dagger}\right] = i\omega\hat{a}^{\dagger} \tag{15}$$

whose solution may written as

$$\hat{a}(t) = \hat{a}(0)e^{-i\omega t} \tag{16}$$

$$\hat{a}^{\dagger}(t) = \hat{a}^{\dagger}(0)e^{i\omega t}. \tag{17}$$

Thus, the operators of position $\hat{r}(t)$ and momentum $\hat{p}(t)$, given by Equations (4) and (5) take the form

$$\hat{r}(t) = \frac{1}{2\alpha}\left[\hat{a}(0)e^{-i\omega t} + \hat{a}^{\dagger}(0)e^{i\omega t}\right] \tag{18}$$

$$\hat{p}(t) = \frac{1}{2i\beta}\left[\hat{a}(0)e^{-i\omega t} - \hat{a}^{\dagger}(0)e^{i\omega t}\right]. \tag{19}$$

We need to observe that the expectation values of $\hat{r}(t)$ and $\hat{p}(t)$ will not oscillate with angular frequency ω, because for any well defined energy eigenstates, characterized by $|\phi_n\rangle$, the expectation values $\langle\phi_n|\hat{r}(t)|\phi_n\rangle$ and $\langle\phi_n|\hat{p}(t)|\phi_n\rangle$ vanish. In order to see more clearly the action of the operators $\hat{a}(t)$ and $\hat{a}^{\dagger}(t)$, let us consider a linear combination of two adjacent energy eigenstates of the oscillator, written as

$$|\phi\rangle = c_{n-1}|\phi_{n-1}\rangle + c_n|\phi_n\rangle, \tag{20}$$

where c_{n-1} and c_n are complex numbers [9]. With the use of Equaitons (8), (9), (18) and (19), we obtain the time dependence for the expectation values of the operators $\hat{r}(t)$ and $\hat{p}(t)$ written as

$$\hat{r}(t) = \left(\frac{\hbar}{m\omega}\right)^{\frac{1}{2}} \left[c_{n-1}^{*} c_n \langle \phi_{n-1} | \hat{a}(0) | \phi_n \rangle e^{-i\omega t} + c_n^{*} c_{n-1} \langle \phi_n | \hat{a}^{\dagger}(0) | \phi_{n-1} \rangle e^{i\omega t} \right] \tag{21}$$

$$\hat{p}(t) = (m\omega\hbar)^{\frac{1}{2}} \left[c_{n-1}^{*} c_n \langle \phi_{n-1} | \hat{a}(0) | \phi_n \rangle e^{-i\omega t} - c_n^{*} c_{n-1} \langle \phi_n | \hat{a}^{\dagger}(0) | \phi_{n-1} \rangle e^{i\omega t} \right], \tag{22}$$

where, according to Equations (6) and (7), the first braket in Equation (21) represents the probability amplitude for a transition from the eigenstate $|\phi_n\rangle$, with energy E_n to the eigenstate $|\phi_{n-1}\rangle$, and energy E_{n-1} with the annihilation of a photon. The second braket represents the probability amplitude for a transition from the eigenstate $|\phi_{n-1}\rangle$ with energy E_{n-1} to the eigenstate $|\phi_n\rangle$ with energy E_n creating a photon with the same energy. The same interpretation is given to Equation (22). These two equations tell us that for an isolated atom, the electron is exchanging energy with the particle with which it is interacting, the nucleus of the atom, losing and receiving a quantum unit of energy $\hbar\omega$, with magnitude as calculated previously. The constant exchange of energy between the proton and electron explain the stability of the atom. Besides its small energy magnitude of the photon, the exchange of energy between the two particles cannot be observed because it is an internal affair of the atom. According to the interpretation given above, the time required for the electron to make one cycle oscillation between the classical turning points R_1 and R_2 is 4 times larger than the time required for the electron to span n eigenstates $|\phi_m\rangle$ between highest and the lowest energy levels allowed for the quantum oscillator, emitting and receiving a quantum unit of energy $\hbar\omega$.

Another way to see the time evolution of the electron in one cycle of oscillation as described above, is to make use of the Equations (4) and (5), to obtain [9]

$$r(t) = r(0)\cos\omega t + \left(\frac{1}{m\omega}\right)p(0)\sin\omega t \tag{23}$$

$$p(t) = -m\omega r(0)\sin\omega t + p(0)\cos\omega t. \tag{24}$$

These two equations present a close resemblance to the classical equations for a simple harmonic oscillator. We can see that the operators $r(t)$ and $p(t)$ oscillate just like their classic analogues where, in a cycle of oscillation, the energy of the electron is exchanged between the kinetic energy $E_k = p^2/2m$ and the potential energy $E_p = (1/2)m\omega^2 r^2$, according to the sequence $E_k(r = r_1) \rightarrow E_p(r = R_1) \rightarrow E_k(r = r_1) \rightarrow E_p(r = R_2) \rightarrow E_k(r = r_1)$, making one complete cycle of oscillation. Thus, according to this model, we argue that the mechanism responsible for the stability of the atom is the exchange of energy between the electron and the proton, mathematically represented by the action of the operators of creation $\hat{a}^\dagger$ and the operator o annihilation $\hat{a}$ acting on the eigenstates, creating and annihilating a photon of energy $\hbar\omega$, in such a way that the loss of energy by the atom is zero. Base on the Quantum Oscillatory Modulated Potential, for the first time, as far as we know, we have an idea of the energy magnitude of the empty space (the so-called vacuum state energy), and how the electron and the proton in the hydrogen atom are exchanging energy, to keep the atom a stable system.

4. CONCLUSION

——The Stability of the Hydrogen Atom The use of Quantum Oscillatory Modulate Potential for describing the interaction between the proton and the electron in the hydrogen atom, lead us to conclude that inside potential well created around the nucleus of the atom, the electron behaves approximately as a quantum harmonic oscillator, oscillating radially bouncing back and forth, confined by two potential barriers. Assuming that in the ground state of the atom, the amount of energy absorbed or emitted by the electron per unit of time, at absolute temperature, $T = 0$, can be calculated by a formula obtained with the base in the stochastic electrodynamics, it was possible to determine the zero point energy of the of the oscillator, and consequently, the quantum unit of energy, a photon of very small magnitude, that the electron is exchanging with the nucleus of the atom during one cycle of oscillation. We are concluding that the constant exchange of energy between the proton and the electron through the emission and absorption of photon, mathematically represented by the action of the operators of creation and annihilation, is the mechanism responsible for the stability of the atoms and consequently, for its own existence.

REFERENCES

1. D. A. McQuarrie, "Quantum Chemistry," Universe Science Books, Margate, 1983.
2. W. W. Filho, "Mecânica Quântica," Editora da Universidade Federal de Goiás, Goiânia, 2002.
3. J. S. Towsend, "A Modern Approach to Quantum Mechanics," McGraw Hill International Edition, 1992.
4. W. Hauser, "Introduction to the Principles of Electromagnetism," Addison Wesley Publishing Company Inc., Boston, 1986.
5. P. A. Tipler, "Modern Physics," Worh Publishers Inc., New York, 1977.
6. R. M. Eisberg, "Fundamentals of Modern Physics," John Wesley & Sons Inc., New York, 1961.
7. R. da Silva and H. Prança, "A Estabilidade do Átomo de Hidrogênio Segundo a Eletrodinâmica Estocástica," Revista Brasileira de Ensino de Física, Vol. 24, No. 1, 2002, pp. 1-13.
8. E. Merzbacher, "Quantum Mechanics," John Willey & Sons Inc., New York, 1976.
9. J. J. Sakurai, "Modern Quantum Mechanics," The Benjamin Cummings, Publishing Company Inc., Menlo Park, 1985.

CHAPTER 14

Determination of the Structural Constant of the Atom

Milan Perkovac*

The First Technical School TESLA, Zagreb, Croatia

ABSTRACT

The equations for energy, momentum, frequency, wavelength and also Schrödinger equation of the electromagnetic wave in the atom are derived using the model of atom by analogy with the transmission line. The action constant $A_0 = (\mu_0/\varepsilon_0)^{1/2} s_0^2 e^2$ is a key term in the above mentioned equations. Besides the other well-known quantities, the only one unknown quantity in the last expression is a structural constant s_0. Therefore, this article is dedicated to the calculation of the structural constant of the atoms on the basis of the above mentioned model. The structural constant of the atoms $s_0 = 8.277\ 56$ shows up as a link between macroscopic and atomic world. After calculating this constant we get the theory of atoms based on Maxwell's and Lorentz equations only. This theory does not require Planck constant h, which once was introduced empirically. Replacement for h is the action constant A_0, which is here theoretically derived, while the replacement for fine structure constant α is $1/(2s_0^2)$. In this way, the structural constant s_0 replaces both constants, h and α. This paper also defines the stationary states of atoms and shows that the maximal atomic number is equal to $2s_0^2 = 137.036$, i.e., as integer should be $Z_{max}=137$. The presented model of the atoms covers three of the four fundamental interactions, namely the electromagnetic, weak and strong interactions.

Keywords: Action Constant; Fine Structure Constant; Lecher's Line; Phase Velocity; Planck's Constant; Stability of Atoms; Stationary States; Structural Coefficient; Structural Constant; Transmission Line

1. INTRODUCTION

Although for a long time classical physics and Maxwell's equations were pushed out of research in the field of atomic phenomena, the article [1] showed that the equations of atom theory can be derived with the help of Maxwell's and Lorentz equations in the classical way. Thus, we derived equations for energy, momentum, frequency and wavelength of an electromagnetic wave in an atom, which coincide with the corresponding equations in quantum mechanics. Moreover, we can derive Schrödinger wave equation as well. To achieve that, a new approach based on the classical theory is used. In that approach, an analogy between the electromagnetic wave in the atom and the wave of voltage and electric current in the transmission line plays a crucial role. The basic novelty of this approach is the introduction of a new concept of the so-called structural constant s_0, which brings together both the parameters of transmission line (macro world) and the charge of the atoms (micro world). This article focuses on the calculation of a structural constant of the atom s_0.

2. MODEL OF AN ATOM

2.1. Atom and Transmission Line

According to [1], using Maxwell equations, electromagnetic wave in an atom is described by two partial differential equations of second order, the so-called wave equations [2]:

$$\nabla^2 \boldsymbol{E} - \frac{1}{u_{\text{em}}^2}\frac{\partial^2 \boldsymbol{E}}{\partial t^2} = 0; \quad \nabla^2 \boldsymbol{H} - \frac{1}{u_{\text{em}}^2}\frac{\partial^2 \boldsymbol{H}}{\partial t^2} = 0, \tag{1}$$

where vector E is the electric field strength, vector H is the magnetic field strength, ∇^2 is del-squared (i.e., Laplace operator), t is time, and $u_{em} = \lambda\nu$ is phase velocity of the electromagnetic wave in the atom, where λ is the wavelength and ν is the frequency of electromagnetic wave in the atom.

The same form of wave equations as (1), but only in one dimension, is also present on the parallel-wire transmission line (Lecher's line), consisting of a pair of ideal conductive nonmagnetic parallel wires of radius ρ, separated by δ, where the ratio $\delta/\rho = \chi$, [3]:

$$\frac{\partial^2 u}{\partial z^2} - L'C'\frac{\partial^2 u}{\partial t^2} = 0; \quad \frac{\partial^2 i}{\partial z^2} - L'C'\frac{\partial^2 i}{\partial t^2} = 0, \tag{2}$$

where u is voltage at the entrance of the element dz of Lecher's line, i is electric current at the entrance of the element dz of Lecher's line, $C' = \pi\varepsilon \Big/ \ln\left[\chi/2 + \left(\chi^2/4 - 1\right)^{1/2}\right]$ is capacitance of Lecher's line per unit length,

$L' = \mu\left(\ln\chi + 1/4\right)/\pi$ is inductance of Lecher's line per unit length, [4], ε is the permittivity of space in the atom, or in space of Lecher's line, and μ is the permeability of the space in the atom, or in the space of Lecher's line.

Systems with the same differential equations behave equally. Hence, in [1] it was concluded that the electric field in the atom behaves as the voltage on Lecher's line, and the magnetic field in the atom behaves as the current in Lecher's line. Furthermore, [1] shows that the Lecher's line may be represented as the inductive-capacitive network (so-called LC network), which finally makes the oscillatory circuit (LC circuit). The natural frequency of LC circuit $\nu = \left[2\pi\left(LC\right)^{1/2}\right]^{-1}$, where C is the sum of all small capacitors of the LC network on the open end of the network, and L is sum of all small inductances of the LC network on the shortcircuited of the network [5].

2.2. Electromagnetic Energy in the Atom

According to the model presented in [1], electromagnetic energy in the atom E_{em} can be described as the electromagnetic energy of LC circuit, i.e. $E_{em} = \Theta^2/(2C)$, where Θ is maximal charge on the said capacitor C, [6]. On the other hand, from the balance of forces in the atom follows [6]:

$$\frac{mv^2}{r\sqrt{1-\beta^2}} = \frac{|qQ|}{4\pi\varepsilon r^2}; \; r = \frac{|qQ|}{4\pi\varepsilon mc^2}\frac{\sqrt{1-\beta^2}}{\beta^2}$$

,(3)

where r is the radius of the circular orbit of the electron, q is the charge of the electron (q = −e), e is elementary charge, Q is the charge of the nucleus (Q = Ze), Z is atomic number (which theoretically is not in integral domain), m is the electron rest mass, $c = 1/\left(\mu_0 \varepsilon_0\right)^{1/2}$ is the speed of light in vacuum, $\beta = v/c$, where v is the velocity of the electron, $\varepsilon = \varepsilon_r\varepsilon_0$, as stated above is permittivity, ε_r is relative permittivity, ε_0 is permittivity of free space, $\mu = \mu_r\mu_0$, as stated above, is permeability, μ_r is relative permeability, μ_0 is permeability of free space, the transverse mass of the electron is $m/(1-\beta^2)^{1/2}$, [7]. Increase of transverse mass of the electron is

$$\Delta m = m/\left(1-\beta^2\right)^{1/2} - m.$$

The kinetic energy of the electron, [6], is K = Dmc², Figure 1. Using Equation (3) and noting that an electron holds an opposite charge to the nucleus, the potential energy of electron, [3,6], is $U = qQ/(4\pi\varepsilon r) = -mc^2\beta^2/\left(1-\beta^2\right)^{1/2}$.
The total mechanical energy of an electron, [6], is the sum of its kinetic and potential energies, $W = K + U = -mc^2\left[1-\left(1-\beta^2\right)^{1/2}\right]$. According to the law of conservation of energy, this energy is equal to the negative emitted electromagnetic energy of the atom, $E_{em} = -W = eV$, where V is the potential difference through which the electron passes to get an equal energy as electromagnetic energy E_{em}. Out of the equation $E_{em} = mc^2\left[1-\left(1-\beta^2\right)^{1/2}\right]$ comes $\left(1-\beta^2\right)^{1/2} = 1 - E_{em}/mc^2$ and $\beta^2 = 2E_{em}\left(1-E_{em}/2mc^2\right)/mc^2$ so the radius r in Equation (3) through arranging leads to:

$$r = \frac{|qQ|}{8\pi\varepsilon E_{em}} \frac{1-E_{em}/mc^2}{1-E_{em}/2mc^2} \tag{4}$$

or

$$E_{em} = \frac{1}{2}\frac{|qQ|}{4\pi\varepsilon r}\frac{1-E_{em}/mc^2}{1-E_{em}/2mc^2} = \frac{1}{2}|U|\frac{1-E_{em}/mc^2}{1-E_{em}/2mc^2}. \tag{5}$$

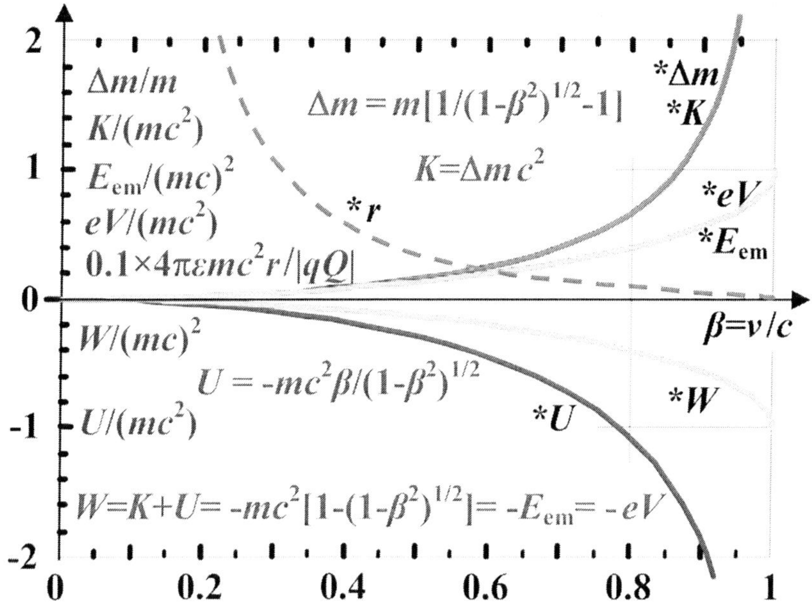

Figure 1. Energy relations in an atom (these relationships are valid for both classical and quantum physics; please note that the normalized quantities are marked with *). Radius of the circular orbit of the electron *r = r/[10|qQ|/(4pεmc²)], kinetic energy of the electron *K = K/(mc²), potential energy of the electron *U = U/(mc²), total mechanical energy of the electron *W = *K + *U, electromagnetic energy of the atom *E$_{em}$ = −*W = *eV; all versus normalized velocity of the electron β = v/c.

On the other hand, electromagnetic energy is $E_{em} = \Theta^2/(2C)$, which means that Equation (5) we can write as:

$$\frac{1}{2}\frac{|qQ|}{4\pi\varepsilon r}\frac{1-E_{em}/mc^2}{1-E_{em}/2mc^2} = \frac{1}{2}\frac{\Theta^2}{C}$$.(6)

The single Equation (6) has two unknowns, i.e., parameter C and variable Θ. By using Diophantine equations we get one of the many solutions: C = 4πεr, and $\Theta^2 = |qQ|\left(1-E_{em}/mc^2\right)/\left(1-E_{em}/2mc^2\right)$ [8].

2.3. Natural Frequency of the Electromagnetic Wave in the Atom

Equation (6), which represents the electromagnetic energy in an atom, can be written like this [1]:

$$E_{\mathrm{em}} = \frac{1}{2}\frac{\Theta^2}{C} = \frac{1}{2}\frac{\pi}{\pi}\frac{\Theta^2}{\sqrt{C}\sqrt{C}}\frac{\sqrt{L}}{\sqrt{L}} = \pi\sqrt{\frac{L}{C}}\Theta^2\frac{1}{2\pi\sqrt{LC}} = \pi Z_{\mathrm{LC}}\Theta^2 \nu = \pi Z_{\mathrm{LC}}\left|qQ\right|\frac{1-E_{\mathrm{em}}/mc^2}{1-E_{\mathrm{em}}/2mc^2}\nu = A\nu, \tag{7}$$

where

$$A = \pi Z_{\mathrm{LC}}\left|qQ\right|\frac{1-E_{\mathrm{em}}/mc^2}{1-E_{\mathrm{em}}/2mc^2} \tag{8}$$

is the action of the electromagnetic oscillator, and

$$Z_{\mathrm{LC}} = \sqrt{\frac{L}{C}} = \sqrt{\frac{L'\mathrm{d}z}{C'\mathrm{d}z}} = \sqrt{\frac{L'}{C'}} = \sqrt{\frac{\mu}{\varepsilon}}\frac{\sqrt{\left[\ln\left(\chi/2+\sqrt{\chi^2/4-1}\right)\right]\left(\ln\chi+1/4\right)}}{\pi} = \sqrt{\frac{\mu}{\varepsilon}}\frac{\sigma(\chi)}{\pi} \tag{9}$$

is the characteristic impedance of Lecher's line, while

$$\sigma(\chi) = \sqrt{\left[\ln\left(\chi/2+\sqrt{\chi^2/4-1}\right)\right]\left(\ln\chi+1/4\right)} \tag{10}$$

is the structural coefficient of Lecher's line.

From Equation (7), in fact from $E_{\mathrm{em}} = \Theta^2/(2C) = \pi Z_{\mathrm{LC}}\Theta^2\nu$, furthermore from Equations (9) and C = 4πεr, follows the natural frequency of Lecher's line, i.e., follows the natural frequency of electromagnetic wave in an atom connected with an electron which is located on the orbit of radius r:

$$\nu = \frac{1}{2\pi Z_{\mathrm{LC}}C} = \frac{1}{8\pi\sqrt{\mu\varepsilon}\,\sigma(\chi)\,r} \tag{.(11)}$$

Therefore in an atom with multiple electrons there are multiple natural frequencies ν. It should be noted that the same expression as (11) came from simultaneously multiplying and dividing the right side of the original expression $\nu = \left[2\pi(LC)^{1/2}\right]^{-1}$ with $C^{1/2}$, and then using the expression C = 4πεr and Equation (9).

Equation (11) shows that the natural frequency ν does not depend on the amount of charge in the atom, but depends on the properties of the space (μ, ε), structural coefficient of Lecher's line σ(χ), (which for its part depends only on the parameters of Lecher's line d and r), and the radius r of the circular orbit. Indeed, electromagnetic oscillations in the atom require a charge, but amount of the charge does not affect the amount of the frequency ν.

If Equation (4) is inserted in Equation (11) we obtain natural frequency of electromagnetic wave in an atom:

$$\nu = \frac{E_{em}}{\sqrt{\mu/\varepsilon}\,\sigma(\chi)|qQ|}\,\frac{1-E_{em}/2mc^2}{1-E_{em}/mc^2}$$.(12)

2.4. Structural Constant of the Atom

In the Equation (12) charges q and Q appear in the form of the product |qQ|. In order to satisfy the condition that the frequency ν is independent of the charge, and to avoid direct or indirect involvement of the charge, the total product $(\mu/\varepsilon)^{1/2}\sigma(\chi)|qQ| = (\mu/\varepsilon)^{1/2}\sigma(\chi)Ze^2$ in Equation (12) must be a constant, i.e., only a product σ(χ)Z must be constant, because $\mu/\varepsilon = \mu_0/\varepsilon_0$ and e are already considered as constants. This product, in the form

$$s_0 = \sqrt{\sigma(\chi)Z} = const.,$$ (13)

is called the structural constant of the atom. In the article [1] the value of structural constant has already been estimated (s_0 = 8.277 56). However, the present paper shows accurate procedure for calculating this constant.

2.5. Action Constant of the Atom

From Equations (8), (9) and (13) follows

$$A = \sqrt{\frac{\mu}{\varepsilon}}s_0^2e^2\,\frac{1-E_{em}/mc^2}{1-E_{em}/2mc^2} = A_0\,\frac{1-E_{em}/mc^2}{1-E_{em}/2mc^2}, \qquad \frac{A}{A_0} = \frac{1-E_{em}/mc^2}{1-E_{em}/2mc^2},$$ (14)

where

$$A_0 = \pi Z_{\mathrm{LC}} |qQ| = \sqrt{\frac{\mu}{\varepsilon}} s_0^2 e^2 = \sqrt{\frac{\mu_r \mu_0}{\varepsilon_r \varepsilon_0}} s_0^2 e^2 = \sqrt{\frac{\mu_0}{\varepsilon_0}} s_0^2 e^2, \tag{15}$$

is action constant, which implies that $\mu_r = \varepsilon_r$. Electromagnetic energy in Equation (7), i.e., $E_{em} = A\nu$, using Equation (14), we can now write as:

$$E_{\mathrm{em}} = A_0 \nu \frac{1 - E_{\mathrm{em}} / mc^2}{1 - E_{\mathrm{em}} / 2mc^2} \tag{16}$$

By solving this equation we obtain:

$$E_{\mathrm{em}} = A_0 \nu + mc^2 - \sqrt{(A_0 \nu)^2 + (mc^2)^2},$$
$$A = A_0 + mc^2 / \nu - \sqrt{A_0^2 + (mc^2 / \nu)^2}. \tag{17}$$

Taking into account that $E_{em} = eV$ it is as follows:

$$eV / mc^2 = A_0 \nu / mc^2 + 1 - \sqrt{(A_0 \nu / mc^2)^2 + 1} \tag{18}$$

The extended Duane–Hunt's law we get from Equations (14) and (16) using $E_{em}=eV$ [8], and using

$(1 - \beta^2)^{1/2} = 1 - E_{\mathrm{em}} / mc^2$ and $\beta^2 = 2E_{\mathrm{em}} (1 - E_{\mathrm{em}} / 2mc^2) / mc^2$ we get also a portion $A_0\nu$:

$$\nu = \frac{eV}{A_0} \frac{1 - eV / 2mc^2}{1 - eV / mc^2}; \quad \nu = \frac{eV}{A}; \quad A_0 \nu = \frac{1}{2} \frac{m}{\sqrt{1 - \beta^2}} v^2. \tag{19}$$

From Equations (5) and (16) follows:

$$\nu = \frac{|U|}{2A_0} \tag{20}$$

2.6. Phase Velocity of the Electromagnetic Wave in the Atom

Thanks to the analogy between the electromagnetic wave in the atom and the wave of voltage and current at the Lecher's line, according to Equations (1) and (2), and $C' = \pi\varepsilon / \ln\left[\chi/2 + \left(\chi^2/4 - 1\right)^{1/2}\right]$, $L' = \mu\left(\ln\chi + 1/4\right)/\pi$ the phase velocity is, [3]:

$$u_{\text{em}} = \lambda\nu = \frac{1}{\sqrt{L'C'}}, \tag{21}$$

i.e., in normalized form [9]:

$$u_{\text{em}}\sqrt{\varepsilon\mu} = \sqrt{\ln\left(\frac{\chi}{2} + \sqrt{\frac{\chi^2}{4} - 1}\right) \Big/ \left(\ln\chi + \frac{1}{4}\right)} = F(\chi), \tag{22}$$

where $(\varepsilon\mu)^{1/2} = (\varepsilon_r\mu_r)^{1/2}/c$. In case χ □ 1 expression under the square root in Equation (22) then it is approximately equal to one as described in [1]. In this paper, however, we use the integral Equation (22) and analyze it in the area around χ=2, which is suitable for calculating the structural constant s_0. Now we have a system of two equations with two unknowns, ε and μ,

$$\sqrt{\mu/\varepsilon} = \sqrt{\mu_0/\varepsilon_0}, \quad u_{\text{em}}\sqrt{\mu\varepsilon} = F(\chi), \tag{23}$$

for which the solutions are (Figure 2):

$$\varepsilon = \varepsilon_r\varepsilon_0 = \sqrt{\frac{\varepsilon_0}{\mu_0}}\frac{F(\chi)}{u_{\text{em}}}, \quad \mu = \mu_r\mu_0 = \sqrt{\frac{\mu_0}{\varepsilon_0}}\frac{F(\chi)}{u_{\text{em}}}, \quad \varepsilon_r = \mu_r = \frac{1}{\sqrt{\varepsilon_0\mu_0}}\frac{F(\chi)}{u_{\text{em}}} = \frac{c}{u_{\text{em}}}F(\chi). \tag{24}$$

2.7. Wavelength and Momentum of Electromagnetic Wave in the Atom

The momentum of the electromagnetic wave in the atom is equal to the momentum of a photon [1], $p_{\text{em}} = E_{\text{em}}/u_{\text{em}} = E_{\text{em}}/(\lambda\nu)$, Figure 2,

$$p_{em} = \frac{E_{em}}{\lambda \nu} = \frac{A_0}{\lambda} \frac{1 - eV/mc^2}{1 - eV/2mc^2} = \frac{A}{\lambda}. \quad (25)$$

In accordance to the law of conservation of momentum, this momentum is equal to the linear momentum of the electron, [1],

$$\frac{mv}{\sqrt{1-\beta^2}} = \frac{mc\beta}{\sqrt{1-\beta^2}} = \frac{A}{\lambda}. \quad (26)$$

By applying the expressions $\left(1-\beta^2\right)^{1/2} = 1 - E_{em}/mc^2$ and $\beta^2 = 2E_{em}\left(1 - E_{em}/2mc^2\right)/mc^2$ to Equation (26) we obtain

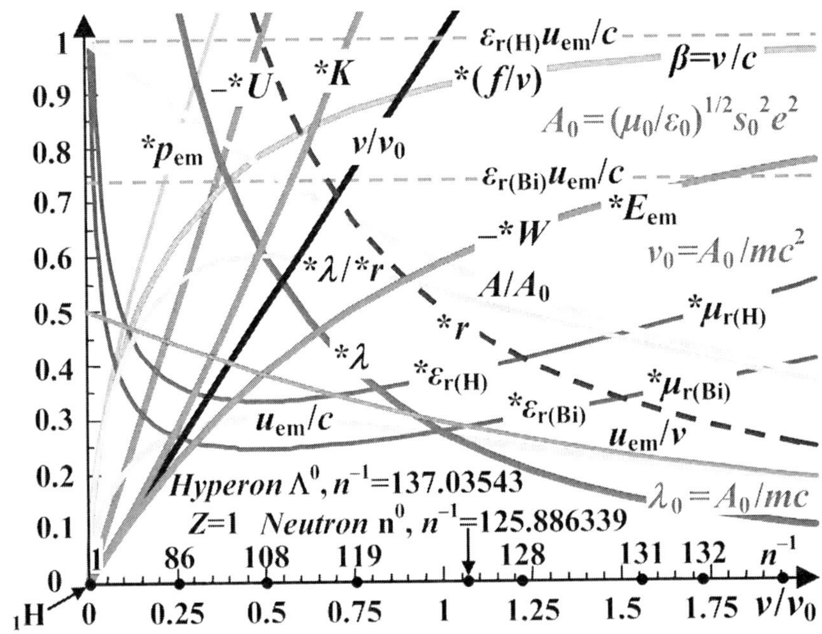

Figure 2. (Please note that the normalized quantities are marked with *). Radius of the circular orbit of the electron *r

= r/[|qQ|/(4pεmc^2)], normalized velocity of the electron β = v/c, kinetic energy of the electron $^*K = K/(mc^2)$, potential energy of the electron $^*U = U/(mc^2)$, total mechanical energy of the electron $^*W = {}^*K + {}^*U$, electromagnetic energy of the atom $^*E_{em} = -{}^*W$, momentum of electromagnetic wave in the atom $^*p_{em} = p_{em}/(mc)$, wavelength of the electromagnetic wave in the atom $^*l = l/[A_0/(mc^2)]$, normalized phase velocity of the electromagnetic wave in the atom u_{em}/c and u_{em}/v, normalized action of the electromagnetic oscillator A/A_0, relative permittivity and relative permeability of the space in the hydrogen atom $^*\varepsilon_{r(H)} = {}^*\mu_{r(H)} = \varepsilon_{r(H)}/10 = \mu_{r(H)}/10$ and also relative permeability and relative permeability of the space within atom of bismuth $^*\varepsilon_{r(Bi)} = {}^*\mu_{r(Bi)} = \varepsilon_{r(Bi)}/10 = \mu_{r(Bi)}/10$, product of relative permittivity and normalized phase velocity $\varepsilon_{r(H)}u_{em}/c = \mu_{r(H)}u_{em}/c$ in the case of hydrogen, product of relative permittivity and normalized phase velocity $\varepsilon_{r(Bi)}u_{em}/c = \mu_{r(Bi)}u_{em}/c$ in the case of bismuth, the ratio of wavelength and radius $^*l/{}^*r = (l/r)/(4p\varepsilon cA_0/|qQ|) = 2u_{em}/c$, the ratio of the frequency f of rotation of the electron orbiting atom and the frequency ν of the electromagnetic wave $^*(f/\nu) = (f/\nu)/(4\varepsilon cA_0/|qQ|)$, all versus normalized frequency of the electromagnetic wave in the atom ν/ν_0, where $\nu_0 = A_0/mc^2$; $n^{+1} = 1, 2, 3, \ldots$ is ordinal number of stationary orbits in the atom, $n^{-1} = 1/1, 1/2, 1/3, \ldots 1/n$, for hyperon X^0 $n^{-1} = 137.03587$.

$$\lambda = \frac{A_0}{\sqrt{2meV}} \frac{\left(1 - eV/mc^2\right)^2}{\sqrt{\left(1 - eV/2mc^2\right)^3}} \tag{27}$$

and using Equation (19) it becomes

$$\lambda = \sqrt{\frac{A_0}{2m\nu} \frac{\sqrt{\left(1 - eV/mc^2\right)^3}}{1 - eV/2mc^2}} \tag{28}$$

Phase velocity of the electromagnetic wave in an atom is obtained by multiplying Equations (19) and (27):

$$u_{em} = \lambda\nu = \sqrt{\frac{eV}{2m}} \frac{1 - eV/mc^2}{\sqrt{1 - eV/2mc^2}}. \tag{29}$$

From (25) and (27) follows

$$p_{em} = \sqrt{2meV}\,\frac{\sqrt{1-eV/2mc^2}}{1-eV/mc^2} \quad (30)$$

The ratio of the wavelength of the electromagnetic wave in the atom and the atom radius are obtained from Equations (4) and (27) using E_{em}=eV and Equation (29), Figure 2:

$$\frac{\lambda}{r} = \frac{8\pi\varepsilon A_0}{|qQ|}\sqrt{\frac{eV}{2m}}\,\frac{1-eV/mc^2}{1-eV/2mc^2} = \frac{8\pi\varepsilon A_0}{|qQ|}u_{em} \quad (31)$$

This expression, with $|U| = |qQ|/(4\pi\varepsilon r)$, leads to:

$$\lambda = 2A_0\frac{u_{em}}{|U|} \quad (32)$$

A minimum of two separate oscillating processes are performed simultaneously within an atom, i.e., the circular motion of electrons around the nucleus and oscillation of electromagnetic wave energy [10]. The time period of one circular tour of electrons around the nucleus is T_e = 2rp/v = 1/f, where f is the frequency of circulation of electrons around the nucleus. The duration of the period of the electromagnetic wave is T_{em} = 1/ν. Hence, ν/f = 2pνr/v. Using Equation (31), as well as v/c = β and lν = u_{em} follows, Figure 2:

$$\frac{T_e}{T_{em}} = \frac{\nu}{f} = \frac{|qQ|}{4\varepsilon A_0 v}, \quad \frac{f}{\nu} = \frac{4\varepsilon A_0 c}{|qQ|}\beta. \quad (33)$$

2.8. Stationary States of the Atoms

Long term existence of the rotation of electrons and long term existence of the electromagnetic wave in the atom (stationary state) is only possible if there is synchronism between them (synchronously stationary state) [10,11]. To be coherent with the active power of the electromagnetic wave in an atom, the electron needs to oscillate (i.e., rotate) with dual frequency of the wave, because the active power of wave oscillates with dual frequency 2w = 2(2pν), (this will be further discussed in Sub–Heading 2.9). This means that in the synchronously stationary state of the atom, the time period of electron rotation T_e is a half period of T_{em} (or, for reasons of synchronism, is $n^{\pm1}$-multiple of a half period of T_{em}), i.e., T_e = $n^{\pm1}T_{em}/2$, where n = 1, 2, 3, ... is ordinal number of stationary orbits in the atom (or n^{-1} = 1/1, 1/2, 1/3, ... 1/n). Equation (33) gives the speed of electron in a synchronously stationary state ([11], compare with [6]):

$$v_n = \frac{1}{n^{\pm 1}} \frac{|qQ|}{2\varepsilon A_0} \quad (34)$$

The Equations (3) and (34) give the radius of the electron orbits in the synchronously stationary states:

$$r_n = \left(n^{\pm 1}\right)^2 \frac{\varepsilon A_0^2 \sqrt{1-\beta^2}}{\pi m |qQ|} \quad (35)$$

From Equations (33), (34) and (35) follows [11]:

$$f_n = \frac{1}{\left(n^{\pm 1}\right)^3} \frac{|qQ|^2 m}{4\varepsilon^2 A_0^3 \sqrt{1-\beta^2}}, \quad (36)$$

$$v_n = \frac{1}{\left(n^{\pm 1}\right)^2} \frac{|qQ|^2 m}{8\varepsilon^2 A_0^3 \sqrt{1-\beta^2}}, \quad (37)$$

and, [10],

$$f_n = 2\left(\frac{v_n}{n^{\pm 1}}\right). \quad (38)$$

The total mechanical energy of an electron $W_n = -E_{em(n)}$ follows from Equations (17) and (37):

$$W_n = -\frac{1}{\left(n^{\pm 1}\right)^2} \frac{|qQ|^2 m}{8\varepsilon^2 A_0^2 \sqrt{1-\beta^2}} - mc^2 + \sqrt{\left(\frac{1}{\left(n^{\pm 1}\right)^2} \frac{|qQ|^2 m}{8\varepsilon^2 A_0^2 \sqrt{1-\beta^2}}\right)^2 + \left(mc^2\right)^2}. \quad (39)$$

For energies much smaller than mc^2:

$$W_n \approx -\frac{1}{\left(n^{\pm 1}\right)^2} \frac{|qQ|^2 m}{8\varepsilon^2 A_0^2 \sqrt{1-\beta^2}} \qquad (40)$$

If assume the maximum speed of electron is equal to the speed of light in a given medium, i.e., according to Equation (22) $v^{max} = u_{em} = F(\chi)/(\mu\varepsilon)^{1/2}$ (to increase the speed of electron should be $n^{\pm 1} = 1/n_{max}$) from Equations (15) and (34) we get:

$$n_{max} = \frac{2s_0^2}{Z} F(\chi) \qquad (41)$$

From Equation (41) follows the greatest possible atomic number Z_{max} when n_{max} is minimal and $F(\chi)$ is maximal, really when $n_{max} = 1$ and $F(\chi) = 1$, i.e.,

$$Z_{max} = \frac{2s_0^2}{n_{max}} F(\chi) = 2s_0^2 \qquad (42)$$

2.9. Wave Equations of the Electromagnetic Wave in the Atom

Wave equations of electromagnetic wave in an atom are expressed by Equation (1). If we insert phase velocity u_{em} in these equation from Equation (29), i.e.,

$$u_{em}^2 = \frac{eV}{2m} \frac{\left(1 - eV/mc^2\right)^2}{1 - eV/2mc^2}, \qquad (43)$$

we obtain

$$\nabla^2 \boldsymbol{E} - \frac{2m}{eV} \frac{1 - eV/2mc^2}{\left(1 - eV/mc^2\right)^2} \frac{\partial^2 \boldsymbol{E}}{\partial t^2} = 0, \quad \nabla^2 \boldsymbol{H} - \frac{2m}{eV} \frac{1 - eV/2mc^2}{\left(1 - eV/mc^2\right)^2} \frac{\partial^2 \boldsymbol{H}}{\partial t^2} = 0. \qquad (44)$$

Wave equations (1) and (44) have a lot of solutions. We will apply the solutions that correspond to the transmission line, i.e., to the LC network. These solutions are standing waves [12,13]:

$$E_x(z,t) = E_0 \sin\left(\frac{2\pi z}{\lambda}\right)\cos(\omega t), \quad H_y(z,t) = -\frac{E_0}{\sqrt{\mu/\varepsilon}}\cos\left(\frac{2\pi z}{\lambda}\right)\sin(\omega t), \tag{45}$$

where E_0 is the maximum value, i.e., the amplitude of electric field strength E, $E_x(z,t)$ is the x-component of the electric field strength dependent on the z-axis and the time t, and $H_y(z,t)$ is the y-component of the magnetic field strength H dependent on the z-axis and the time t, w = 2pν. All mathematical operations we perform for the y-component of the magnetic field $H_y(z,t)$ can be performed for the x-component of the electric field $E_x(z,t)$ in the same way.

In the standing waves (45) the energy oscillates between the electric and magnetic form. The electrical energy is maximum when the magnetic energy is zero, and vice versa. Furthermore, the standing wave transfers no energy through the space because the average active power of the wave is equal to zero. The current value of the active power oscillates in both directions, + and – of z axis, with dual frequency 2w from point to point of z axis [12]. As already mentioned, this is why (for the maintenance of stationary state of the atom) the electron has to rotate twice as fast compared to the lower harmonics (n^{+1}), or twice as fast compared to the upper harmonics (n^{-1}), i.e., f = 2(ν/$n^{\pm1}$) in accordance with Equation (38).

If we use the second derivative with respect to z of the y-component $H_y(z,t)$ of the magnetic field strength in Equation (45), we get: ¶$^2H_y(z,t)$/¶z^2 + (2p/l)$^2H_y(z,t)$ = 0. After inclusion of the wavelength l from Equation (27) we obtain:

$$\frac{\partial^2 H_y(z,t)}{\partial z^2} + \frac{8\pi^2 meV}{A_0^2}\frac{\left(1 - eV/2mc^2\right)^3}{\left(1 - eV/mc^2\right)^4}H_y(z,t) = 0 \tag{46}$$

If eV/mc^2 << 1, then eV » K = W – U, and Equation (46) becomes

$$\frac{\partial^2 H_y(z,t)}{\partial z^2} + \frac{8\pi^2 m}{A_0^2}(W - U)H_y(z,t) = 0. \tag{47}$$

The second derivative of $H_y(z,t)$ with respect to t gives:

¶$^2H_y(z,t)$/¶t^2 + w$^2H_y(z,t)$ = 0. After inclusion of the angular frequency w = 2pν from Equation (19) we obtain:

$$\frac{\partial^2 H_y(z,t)}{\partial t^2}+\left(\frac{2\pi}{A_0}eV\frac{1-eV/2mc^2}{1-eV/mc^2}\right)^2 H_y(z,t)=0 \quad .(48)$$

If $eV/mc^2 << 1$, then $eV \gg K = W - U$, and equation (48) becomes

$$\frac{\partial^2 H_y(z,t)}{\partial t^2}+\frac{4\pi^2}{A_0^2}(W-U)^2 H_y(z,t)=0 \quad .(49)$$

3. CALCULATION OF THE STRUCTURAL CONSTANT S_0

Structural constant of the atom s_0 can be determined in several ways, e.g., by measuring two quantities, the voltage V and frequency ν and calculating the action constant A_0 by Duane–Hunt's law, i.e., using Equations (15) and (19), [8]. However, here we will use a more direct theoretical calculation, with only one empirical item necessary.

Namely, the increase of the nuclear charge in the atom increases atomic number Z. In accordance with Equation (13), the value of structural coefficient $\sigma(\chi) = s_0^2/Z$ is assigned to each atom. So, greater atomic number means a lower structural coefficient $\sigma(\chi)$.

On the other hand, there is a critical nuclear charge which ensures stability of the atom [2,14]. In other words, to reduce $\sigma(\chi)$ means to grow instability of the atom. In general, the higher atomic number means the less stability (i.e., the less half-life, or $t_{1/2}$) of the atom, starting from bismuth $_{83}$Bi (Z = 83, $t_{1/2} = 6 \times 10^{26}$ s, [15]) to ununoctium $_{118}$Uuo (Z = 118, $t_{1/2}$ = 5 ms), [exceptions are atoms of technetium ($_{43}$Tc, Z = 43, $t_{1/2} = 1.3 \times 10^{14}$ s) and promethium ($_{63}$Pm, Z = 63, $t_{1/2} = 5.6 \times 10^{8}$ s)].

For the calculation of structural constant s_0, it is enough to find only one associated pair of $\sigma(\chi)$ and Z. The curve $\sigma(\chi)$ has no extremes, Figure 3. Thus it is not easy to find a mentioned pair of $\sigma(\chi)$ and Z. In that sense, a better situation is with the phase velocity u_{em}, specifically with the normalized phase velocity $u_{em}(\mu\varepsilon)^{1/2} = F(\chi)$ of electromagnetic wave in the atom, Figure 3, [9]. Neither of these two curves have extremes, but there is a sharp knee on $F(\chi)$ which can be used to determine the structural constant s_0.

Although there is no theory about the connection between the phase velocity of electromagnetic waves in the atom and the stability of the atoms, it is still possible to use this mathematical benefit of sharp knee for those atoms, in which there is the lower phase velocity of electromagnetic waves that exhibit greater instability. Use of this result will be discussed just a little bit later.

The nuclear binding energy per nucleon slightly decreases with increase the atomic number (starting from the first radioactive element bismuth, $_{83}$Bi, 7.848 MeV, to the unoctium, $_{118}$Uuo, 7.074 MeV, about 0.31% decrease for each 35 atoms in that area [15]). Physically this means that the boundary between stable and unstable areas is not emphasized. Mathematically it allows that between the two areas set up so-called transition area, Figure 3. This is, at the moment, the most accurate way to determine the boundary between stable atoms and the others. Indeed, the first unstable atom must be located on that border. This is a key fact to determine the structural constant s_0 of the atom.

Before calculating, we observe the first derivative $F'(\chi)$ of the curves of normalized phase velocity $F(\chi)$ of electromagnetic waves in an atom (Figure 3). When this derivative is greater than 1, it means that the phase velocity rapidly decline, it is a zone of unstable atoms ($2 < \chi < 2.129\ 531\ 7$). It should be noted that the situation $\chi \leq 2$ is theoretically impossible because then there is no Lecher's line.

When the second derivative $F''(\chi)$ of the normalized phase velocity $F(\chi)$ is greater than 1, it means that the phase velocity starts to rapidly decline (Figure 3), this is a transition zone ($2.129\ 531\ 7 < \chi < 2.382\ 788$).

The border crossing from the transition zone to the stable zone (i.e., $\chi_0 = 2.382\ 788$), in accordance with the experiments, [15], is closest to the bismuth atom. Bismuth atom ($_{83}$Bi) is the first unstable atom, in the entire chain of stable atoms, which ends with lead ($_{82}$Pb). The corresponding value of the structural coefficient in that place is $\sigma(\chi_0) = 0.825\ 402$, Figure 3. Bismuth is a chemical element with atomic number Z = 83, with half-life more than a billion times the estimated age of the universe. Even though charges in reality take discrete values (e, 2e, 3e, ... Ze), theoretical value of Z in Equation (13) can be within the range $Z_{th} = 83 \pm 1/2$. Thus, according to Equation (13) we get the structural constant of the atom $[0.825\ 402 \times (83 \pm 1/2)]^{1/2}$, i.e., $8.252 < s_0 < 8.302$, with a mean value 8.277 and with sample standard deviation ±0.035 355 or as a percentage $s_0 = 8.277 \pm 0.43\%$.

4. CONCLUSION

In classical physics, the use of Maxwell's equations and Lorentz theory of electrons gives us equations that describe the behavior of atom. For this description, usual classical physical constants are sufficient. The only new

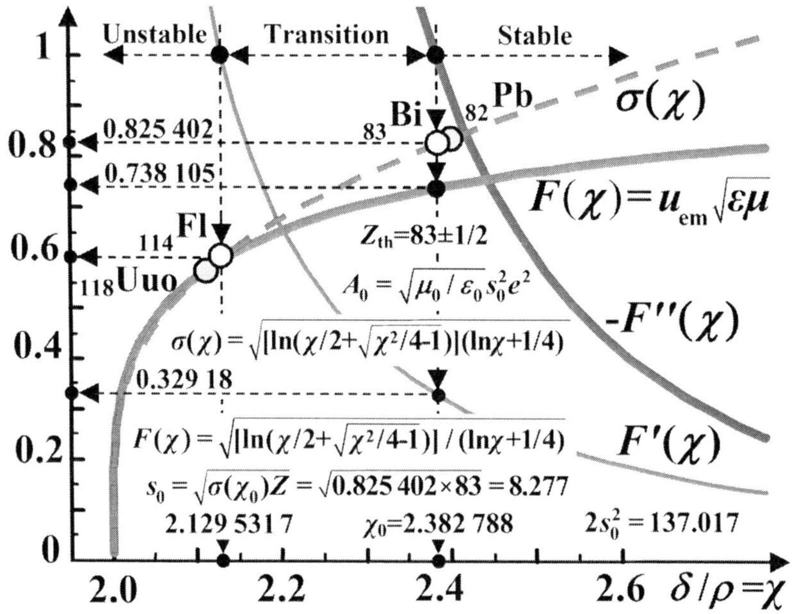

Figure 3. Structural coefficient of Lecher's line $\sigma(\chi)$, normalized phase velocity of the electromagnetic wave in the atom $F(\chi) = u_{em}(\mu\varepsilon)^{1/2}$, the first derivative of the normalized phase velocity $F'(\chi)$, inverted second derivative of the normalized phase velocity $F''(\chi)$, all versus ratio $\delta/\rho = \chi$ of the transmission Lecher's line, consisting of a pair of ideal conducting nonmagnetic parallel wires of radius ρ separated by δ, which represents a model of an atom.

necessary physical constant is the structural constant of the atom s_0. This new physical constant was calculated in the framework of a unified theory based on the model of the atom by analogy with the transmission line. To calculate it, we need only one empirical item, i.e., the atomic number of the first unstable atom in periodic table of elements. It is known, that it is bismuth, with atomic number Z = 83. From this data, the structural constant s_0 is calculated. Thus, we get $s_0 = 8.277 \pm 0.43\%$. Comparing with the fine structure constant we get $s_0 = 8.27756$, which is consistent with the calculation performed here (the relative difference is less than 0.0068%). This constant, thanks to the expressions $(\mu_0/\varepsilon_0)^{1/2}s_0^2e^2$ and $(2s_0^2)^{-1}$, can replace two existing constants, Planck's constant h and the fine structure constant α. We get also the stationary states of the atoms and the maximal atomic number $Z = 2s_0^2 = 137.036$, i.e., as integer should be $Z_{max} = 137$. The difference between $2s_0^2$ and Z_{max} we call the slipping. This indicates that, in addition to integer synchronously stationary states, non-integer (asynchronously) stationary states in atoms are also possible. Now, just a mention, as it seems to asynchronously stationary states can produce a continuous spectrum of atoms. This, however, requires more detailed research. This model of the atoms covers the energies of the electromagnetic, weak and strong fundamental interactions [16].

ACKNOWLEDGEMENTS

Wolfram Research, Inc. Mathematica software is used by courtesy of Systemcom Ltd., Zagreb, Croatia. The author thanks Ms. Erica Vesic for editing this article in English, Mr. Damir Vuk for the useful discussions and Prvomajska TZR Ltd., Zagreb, Croatia, www.prvomajska-tzr.hr, for the financial support.

REFERENCES

1. M. Perkovac, "Model of an Atom by Analogy with the Transmission Line," Journal of Modern Physics, Vol. 4, No. 7, 2013, pp. 899-903.
2. H. Hänsel and W. Neumann, "Physik," Spektrum Akademischer Verlag, Heidelberg, 1995.
3. Czichos and Association Hütte, "Die Grundlagen der Ingenieurwissenschaften," Springer-Verlag, Berlin, 1989.
4. T. Bosanac, "Teoretska elektrotehnika 1,"Tehnicka knjiga, Zagreb, 1973.
5. R. Rüdenberg, "Elektrische Schaltvorgänge," Julius Springer, Berlin, 1923.
6. D. C. Giancolli, "Physics for Scientists and Engineers," Prentice Hall, Englewood Cliffs, 1988.
7. L. Page and N. I. Adams, "Electrodynamics," D. Van Nostrand Company, Inc., New York, 1940.
8. M. Perkovac, "Statistical Test of Duane—Hunt's Law and Its Comparison with an Alternative Law," 2010.
9. M. Perkovac, "Maxwell Equations for Nanotechnology," Proceedings of the 35th International Convention of the IEEE MIPRO, Opatija, 21-25 May 2012, pp. 429-436.http://ieeexplore.ieee.org/xpl/freeabs_all.jsp?arnumber=6240683.
10. M. Perkovac, "Quantization in Classical Electrodynamics," Physics Essays, Vol. 15, No. 1, 2002, pp. 41-60.
11. M. Perkovac, "Absorption and Emission of Radiation by an Atomic Oscillator," Physics Essays, Vol. 16, 2003, pp. 162-173.
12. Z. Haznadar and Z. Stih, "Elektromagnetizam," Skolska knjiga, Zagreb, 1997.
13. J. D. Jackson, "Classical Electrodynamics," John Wiley & Sons, Inc., New York, 1999.
14. R. D. Benguria, M. Loss and H. Siedentop, "Stability of Atoms and Molecules in an Ultrarelativistic Thomas-FermiWeizsäcker Model," Journal of Mathematical Physics, Vol. 49, 2007, Article ID: 012302.

CHAPTER 15

The Fractional Hydrogen Atom: A Paradigm for Astrophysical Phenomena

A. I. Arbab

Department of Physics, Faculty of Science, University of Khartoum, Khartoum, Sudan

ABSTRACT

We have found that fractional principal quantum numbers are permitted in hydrogen atom which yield the conditions for neutron and white dwarf stars evolution. The number densities of neutron and white dwarf stars reveal that these systems have the maximal conductivity of $1.37\times10^{10}\,\Omega^{-1}\cdot\mathrm{m}^{-1}$. They are giant perfect conductors at very high temperature and magnetic field.

Keywords: Hydrogen Atom; Fractional Quantum Number; Neutron Stars; White Dwarfs; Perfect Conductors

1. INTRODUCTION

Bohr put forward the postulates that lead to the emergence of the hydrogen atom. The orbital angular momentum is quantized and that atomic emission is possible only when an electron makes a transition form a higher level to a lower one. Moreover, the orbit, velocity and energy of the electron are quantized in terms of the principal quantum number n that takes integer values. Under high pressure the hydrogen atom is transformed in a system of degenerate electron gas. This state is identified astrophysically as a white dwarf. White dwarfs are studied by Chandreshekar [1] who set a limit to their existence as stable stars. Investigation of white dwarfs provides important constraints on the theory of stellar evolution. If more pressure is applied on the hydrogen atom, a new state of neutrons results in. This new state of hydrogen is known as neutron stars. If more pressure is applied a final state of black hole is reach that not even light can escape from. The electron and neutron gases are described statistically by Fermi-Dirac distribution. However, the neutron gas provides a large pressure than the electron gas. The neutron gas can withstand more gravity than electron gas does.

The state of hydrogen atom under this high compression has not been provided by Bohr theory. In Bohr theory, the electron can only exist in certain orbits from the nucleus. The closest (minimum) one from the nucleus is the Bohr orbit. It amounts to a_0 = 0.53 Å. Orbit quantization does not permit a lower orbit than this value. The question posed is that how can electron exist under high pressure? Bohr theory does not provide an answer to this question.

Recently, Arbab [2] has considered the effect of the internal magnetic field resulting from the electronic orbital current, and shown that this field is equivalent to considering a Bohr orbit with a fractional value of n. In particular, this corresponds to, $n = \alpha$, where α is the fine structure constant. Fractional values of the quantum number may result from existence of other form of interactions that we may not now know.

This consideration leads to the existence of a new state of hydrogen with an electron orbiting close to the proton at a distance equals to the electron classical radius. We call this electron-proton state a protonium. The physical properties of this state is similar to that of the neutron. However, the anomalous magnetic moment under this consideration is convincingly explained.

Fractional quantum numbers are observed in the quantum Hall effect (FQHE) by Tsui et al. [3], where the Hall conductance of 2D electrons shows precisely quantized plateaus at fractional values of α. In this effect the electrons bind magnetic flux lines to make new quasiparticles. We argue that such a phenomenon is also reflected by hydrogen atom, when subject to external compression. In Bohr theory, an electron can't exist at a distance less than Bohr radius.

We propose here a fractional hydrogen atom whose implication will be prominent in astrophysical objects, e.g., Neutron stars and White Dwarfs, in which gravitational compression is so immense. The electron in the atoms of these objects does not fall into the nucleus, but rather jumps into a closer stable orbits to the nucleus with $n = \alpha$ and $n = \sqrt{\alpha}$, respectively. A strong magnetic field can transform a boson into a fermion or vice versa. Hence, we argue that for magnetic white dwarfs the magnetic field may alter the statistics of the white dwarf's electrons from fermionic to bosonic. This would affect the stars structure, giving it a smaller than expected radius, and a lower than expected temperature. In some extreme cases one could imagine that this effect could lead to the collapse of the white dwarf into a neutron star.

This hypothesis is inspired by our recent model for the neutron. In this model, the electron exists in an orbit with a principal quantum number n equals to the fine structure constant, α. This state is reached when an atom is under a very large compression. Such a state yields very huge magnetic field that is of the same order of magnitude of the neutron star. Fowler [4] showed that a system of many electrons (stars) must be governed by quantum statistical physics besides quantum mechanics, since electrons are fermions. Under high pressure electrons behave as a condensed matter system. They are governed by the laws of many-body system.

In this paper, we would like to investigate other possible fractional values of n and seek their astrophysical consequences. This paper is structured as follows: we study in Section II the implication of our hypothesis for the formation and characterization of neutron stars. The huge magnetic field observed in neutron stars is a result of the current developed by electrons forming our neutrons model. The surface area of a neutron star is directly proportional to its mass. The pressure of the star is related to the square of its magnetic field. We provide in Section III how one can obtain all white dwarfs stars by applying our hypothesis to hydrogen atom where the principal quantum number is a fraction. This gives exactly are characterizations (pressure, magnetic field, electric field and conductivity) of white dwarfs now known. We further apply our hypothesis of the maximal conductivity and found that white dwarfs as well as neutron stars are perfect conductors. We finally provide in Section IV the implications of our hypothesis to ordinary stars, like our sun. These stars systems exhibit the Hall effect observed in conductors under application of electric and magnetic fields. A table showing the relations between different quantities describing neutron stars, white dwarfs and ordinary stars is provided. These quantities are shown to be multiples of some powers of the fine structure constant, α. This model can be compared with that standard model of formation of neutron stars and white dwarfs stars where Fermi-Dirac statistics is employed.

2. NEUTRON STAR

Recall that in Bohr theory, the orbit, velocity and energy of the electron are given by

$$r_n = \left(\frac{\hbar^2}{mke^2}\right)n^2,\ v_n = \left(\frac{ke^2}{\hbar}\right)\frac{1}{n},$$

$$\varepsilon_q = -\frac{mk^2e^4}{2\hbar^2 n^2},\qquad (1)$$

where $n = 1, 2, 3, \cdots$. We have found that when the magnetic force inside the atom is considered, a new stable orbit exists. This stable orbit is equivalent to considering a fractional quantum number, $n = \alpha$ in Bohr's model. Hence, Equation (1) yields

$$r_q = \left(\frac{ke^2}{mc^2}\right), v_q = c, E_n = -\frac{1}{2}mc^2.\qquad (2)$$

Interestingly, the radius of this orbit is equal to the classical electron radius, r_e.

One can calculate the current resulting from the motion of the electron in this state as

$$I_q = \frac{ec}{2\pi r_q} = \frac{mc^3}{2\pi k e} = 2.7\times 10^3\ \text{A}\qquad (3)$$

and the resulting magnetic field is that of a current loop, viz.,

$$B_q = \frac{\mu_0 I_q}{2r_q} = \frac{\mu_0 m^2c^5}{4\pi k^2 e^3} = \frac{m^2c^3}{ke^3}$$
$$= 6.06\times 10^{11}\,\text{T},\qquad (4)$$

where $\frac{\mu_0 c^2}{4\pi} = k.$ The magnetic field predicted by Equation (4) coincides with the magnetic field of neutron stars obtained by Wunner [5] & Leiby

[6]. Hence, the magnetic field observed in neutron stars results from the magnetic field produced by the electron motion inside the hydrogen atom. The magnetic flux is defined by

$$\phi_q = B_q A = 4\pi \frac{ke}{c} = \frac{e}{\varepsilon_0 c} = 6.03\times 10^{-17}\,\mathrm{W}b.$$

This can be compared with the quantum of magnetic flux that passes through the superconductor that is defined by $\phi_0 = \frac{h}{2e} = 2.067\times 10^{-15}\,\mathrm{W}b$. Does this mean that the neutron star is a giant superconductor? If we assume that the above magnetic flux is quantized in terms of s, then one can write $\left(\frac{h}{2e}\right)s = \frac{e}{\varepsilon_0 c}$, hence

$s = 4\frac{ke^2}{hc} = 4\alpha$. This means the flux of the neutron star is quantized in units of α, in the same way as the principal quantum numbers.

The capacitance and inductance of the protonium are defined by

$$L_q = \frac{\phi_q}{I_q} = \frac{\mu_0^2 e^2}{2m} = 2.24\times 10^{-20}\,\mathrm{H},$$

$$C_q = \frac{e}{ke/r_q} = \frac{e^2}{mc^2}. \tag{5}$$

Let us now consider the magnetic moment due to the electron in the protonium state. This is given by

$$\mu_q = -\left(\frac{e}{2m}\right)L = -\frac{e}{2m}(h\alpha) = -\frac{ke^3}{2mc}. \tag{6}$$

This can be compared with the measured value of the neutron and protons magnetic moments, viz. $\mu_n = -9.66\times 10^{-27}\,\mathrm{J/T}$ and $\mu_p = 1.41\times 10^{-26}\,\mathrm{J/T}$. We can get the exact value by making some linear combinations of Equation (6) and μ_p. Using Equations (4) and (6), the Zeeman energy due to the presence of the magnetic field B_q can be written as

$$E_z = -\mu_q B_q = \frac{1}{2}mc^2.$$

It is interesting that this Zeeman energy is equal to the orbital energy of the protonium. Hence, the total energy of the system is zero. Therefore, the system is at $T = 0$ temperature. This is the case when a Fermi-Dirac statistic is considered to describe such a system. One can further find the electric field due to protonium state as

$$E_q = \frac{ke}{r_e^2} = \frac{m^2c^4}{ke^3} = 1.8 \times 10^{20}\ \text{V/m}. \tag{7}$$

The electromagnetic energy density (or pressure) inside the protonium is

$$u_q = \frac{1}{2}\varepsilon_0 E_q^2 + \frac{B_q^2}{2\mu_0} = \frac{m^4c^8}{4\pi k^3 e^6} = 2.93 \times 10^{29}\ \text{J/m}^3. \tag{8}$$

If we consider the protonium as a dipole, we can calculate the power radiated due to Larmor dipole radiation as shown in Jackson [7]

$$P_q = \frac{2}{3}\frac{ke^2 a_q^2}{c^3} = \frac{2}{3}\frac{m^2c^5}{ke^2} = 5.8 \times 10^9\ \text{W}. \tag{9}$$

Stars (mainly hydrogen) are affected by their selfgravitating energy and the gas energy. At equilibrium (balance) the gravity and gas (neutron Fermi gas) pressures are equal. The gravitational pressure is defined as[1]

$$P_g = f\frac{GM^2}{R^4}, \tag{10}$$

where f is some fixed number related to the equation of state of the matter under consideration. According to Newtonian gravity, this occurs when the gas radius (R) is related to the star mass M as shown by Phillips [8]

$$R_N = \frac{3}{8}\left(\frac{3}{2\pi^4}\right)^{1/3}\left(\frac{h^2}{Gm_n^{8/3}}\right)M^{-1/3}, \tag{11}$$

where $f_q = 3/(20\pi)$. Owing to Equations (10) and (11), a neutron star of mass $M = 1.5M_e$ will produce a radius of $10.75\ \text{km}$ and a pressure of $2\times10^{33}\ \text{Pa}$. In our present model, a pressure of ~10^{33} Pa will produce a magnetic field of $10^{12}\ \text{T}$. This is also apparent from substituting Equations (4) and (8) to yield

$$P_q = \left(\frac{4\pi}{\mu_0}\right) B_q^2. \tag{12}$$

Using Equations (8) and (10) the radius mass relation is

$$R_q = \left(\frac{3k^3e^6G}{5m^2c^8}\right)^{1/4}\left(\frac{M}{m}\right)^{1/2}. \tag{13}$$

This implies that the pressure is constant. It is thus different from the standard relation, i.e., $R \propto M^{-1/3}$. The radius in Equation (13) can be made to give a different value if we take into consideration the relativistic mass resulting from Equation (2) that can be defined to yield the correct value of the magnetic moment of the neutron defined in Equation (6). To resolve this, the mass of the electron should be replaced by 7 m. This replacement changes the pressure (energy density) in Equation (8) to $7\times10^{32}\ \text{Pa}$, and the magnetic field in Equation (4) to $2.96\times10^{13}\ \text{T}$. Equations (2), (5) and (9) will also change. Hence, the radius in Equation (13) will be reduced by 1/7.

Using Equation (1), one can estimate the internal temperature of this state using the energy $\varepsilon_q = k_B T_q$. This yields a value of $T_q = 1.28\times10^9\ \text{K}$. However, if we consider the protonium as a black body, it will radiate at a temperature of $T = 0.88\times10^9\ \text{K}$. This is very close to T_q. This may permit us to assume that fermions in this state undergo a Bose-Einstein state of condensation. The energy and number densities of a Fermi-Dirac electron/neutron gas are given by

$$u = \frac{\pi^2}{15}\frac{(k_B T)^4}{(hc)^3},\ n = \frac{2.4}{\pi^2}\left(\frac{k_B T}{hc}\right)^3.$$

Let us now calculate the electrical conductivity of the protonium state as

$$J = \sigma E, J = \text{nev}. \tag{14}$$

Using Equations (2) and (7), one obtains

$$\sigma_q = \frac{nke^4}{m^2 c^3}. \tag{15}$$

This can be set to the maximum conductively, $\sigma_m = \frac{2m}{\mu_0 h}$, hypothesized by Arbab [9] which yields a number density

$$n_q = \frac{m^3 c^5}{k^2 e^4 h} = 3.4 \times 10^{44} \text{ neutron/m}^3. \tag{16}$$

This is a typical number density of nuclear matter. It is a number density of perfect conductor. Hence, neutron stars are perfect conductor. This supports the recent discovery by Page et al. [10] who have provided the first direct evidence for a superconductor state at the core of neutron stars. The current inside lasts forever as far as electrons keep circulating. The number density in Equation (16) agrees with the number density found for neutron stars by considering the Fermi statistics of the degenerate neutrons [4,11]. It is an interesting agreement. We conclude here that the above maximal conductivity hypothesis due to Arbab [9] is then verified.

Let us now calculate the impedance of the protonium state using Equation (5)

$$Z_q = \sqrt{\frac{L_q}{C_q}} = \frac{1}{\varepsilon_0 c}. \tag{17}$$

This is equal to the impedance of the vacuum. Thus, the space inside the protonium behaves as vacuum.

3. THE WHITE DWARFS

Let us now consider the case when,

$$n = \sqrt{\alpha}. \tag{18}$$

Apply this in Equation (1) to obtain

$$r_d = \frac{\alpha h^2}{mke^2} = \frac{h}{mc},$$
$$v_d = \frac{ke^2}{\sqrt{\alpha}\, h} = \sqrt{\alpha}\, c, \quad \varepsilon_d = -\frac{1}{2} mc^2 \alpha. \tag{19}$$

Such a case will lead to an interesting physical situation. It is interesting to see that the radius r_d in Equation (19) is equal to the Compton wavelength of the electron. We, consequently, propose that an electron under high pressure jumps to this quantum state. The physical properties of this state is described in the subsequent equations.

One can also calculate the electronic current resulting from orbital motion of the electron in the quantum state described by Equation (19).

$$I_d = \frac{ev}{2\pi r_d} = \frac{\sqrt{\alpha}\, emc^2}{h} = 1.69\ \text{A} \tag{20}$$

and the resulting magnetic field is that of a current-loop, i.e.,

$$B_d = \frac{\mu_0 I_d}{2r_d} = \frac{\sqrt{\alpha}\, e\mu_0 m^2 c^3}{4\pi h^2} = 6.9\times 10^5\ \text{T}. \tag{21}$$

It is interesting to note that the above magnetic field is a typical field of white dwarfs ($\sim 10^9$ G) as obtained by Jordan [11,12]. The magnetic flux is defined by

$$\phi_d = B_d A = \sqrt{\alpha}\, e\mu_0 c = 5.1\times 10^{-18}\ Wb.$$

The capacitance and inductance of the state in Equation (19) are defined by

$$L_d = \frac{\phi_d}{I_d} = \frac{\mu_0 h}{mc} = 3.1\times 10^{-18}\ \text{H},$$
$$C_d = \frac{2\varepsilon_0 h}{mc} = 4.3\times 10^{-23}\ \text{F}. \tag{22}$$

The above equation suggests that the quantum of inductance is $\frac{\mu_0 h}{mc}$. This implies that the neutron star has a quantum of inductance that is α times that of the white dwarf. One can relate this to the maximum conductivity by $L = \frac{4\pi}{c}\sigma$. Moreover, the quantum of resistivity is $\frac{\mu_0 h}{m}$. Hence, the quantum of resistivity is equal to c

times the quantum of inductance. The quantum of conductivity represent the maximum conductivity that a system can have. Metals will have a conductivity that is equal to α times this quantum. Similarly, the quantum of capacitance is $\frac{2\varepsilon_0 h}{mc}$. These quanta are characteristic values of any quantum system.

We now calculate the electric field of the state defined by Equation (4) as

$$E_d = \frac{ke}{r_d^2} = \frac{kem^2c^2}{h^2} = 9.73\times10^{16}\ \mathrm{V/m}. \tag{23}$$

The electric energy densities are thus

$$u_d^e = \frac{1}{2}\varepsilon_0 E_d^2 = \frac{ke^2m^4c^4}{8\pi h^4}. \tag{24}$$

The pressure associated with this energy density amounts to $2.8\times10^{22}\ \mathrm{Pa}$.

Let us now consider the magnetic moment due to the electron in the above state. This is given by

$$\mu_d = -\left(\frac{e}{2m}\right)L = -\frac{eh}{2m}\sqrt{\alpha} = -7.91\times10^{-27}\ \mathrm{J/T} \tag{25}$$

Because of the magnetic field, as in Equation (21), the Zeeman energy is given by

$$E_z = -\mu_d B_d = \frac{mk^2e^4}{2h^2} = -E_1. \tag{26}$$

If we consider the state in Equation (19) as a dipole, we can calculate the power radiated due to Larmor dipole radiation as explained in Jackson [7]

$$P_d = \frac{2}{3}\frac{ke^2a_q^2}{c^3} = \frac{2}{3}\frac{k^3e^6m^2c}{h^4} = 16.5\ \mathrm{W}. \tag{27}$$

This exceedingly small power may explain the faintness of the white dwarfs.

Let us now calculate the conductivity of the state defined in Equation (14) using Equations (19) and (23) to obtain

$$\sigma_d = \frac{\sqrt{\alpha}\,h^2}{km^2c}n. \tag{28}$$

For a maximum conductivity $\left(\sigma = \frac{2m}{\mu_0 h}\right)$, one obtains

$$n_d = \frac{1}{2\pi\sqrt{\alpha}}\left(\frac{mc}{h}\right)^3 = 3.2\times10^{37}\ \mathrm{electron/m^3}. \tag{29}$$

The above value of n_d agrees with the number density found for white dwarfs by considering the Fermi statistics of the degenerate electrons as formulated by Fowler [4], Jordan [11] and Fassbinder [12]. Hence, white dwarfs can be thought of as perfect conductors. If we now equate the conductivity in Equations (15) and (28) to Drude conductivity of metals developed by Drude $\sigma = \frac{ne^2\tau}{m}$ [13], then the relaxation time, τ, for the neutrons in neutron star and the electrons in the white dwarfs, will be $\tau_q = \frac{r_q}{c}, \tau_d = \frac{a_0}{c}$, respectively. While $\tau_q = 9.3\times10^{-24}\ \mathrm{sec}$ and $\tau_d = 1.7\times10^{-19}\ \mathrm{sec}$, electrons in metal have $\tau \sim 10^{-14}\ \mathrm{sec}$, and a number density of $n \sim \times10^{28}\ \mathrm{electron/m^3}$. It is interesting to note that $r_q = \alpha r_d, r_d = \alpha a_0$.

The gravitational pressure of the white dwarfs, at equilibrium, is balanced by the electrons gas pressure. As seen from Equation (24), the gravitational energy of the white dwarfs is supported by their electrical energy. The magnetic energy of the white dwarf is negligible in comparison with the electric energy. Hence, under equilibrium one must have

$$\frac{1}{2}\varepsilon_0 E_d^2 = f\frac{GM^2}{R^4}, \tag{30}$$

where R is the radius of the star and M it's mass. Therefore, the equilibrium radius (R_d) is

$$R_d = \left(f_d \frac{8\pi h^4 G}{ke^2 m^2 c^4}\right)^{1/4} \left(\frac{M}{m}\right)^{1/2}, f_d = \frac{3}{20\pi}. \tag{31}$$

It is interesting to that the radius mass relation we have obtained, i.e., $R \propto M^{1/2}$, that implies the pressure is constant, is different from the standard formula, viz., $R \propto M^{-1/3}$. Despite this, we obtain similar result. Owing to Equation (31) a white dwarf of a mass $M = 1.0M_{\odot}$ will have a radius of $4155\ \text{km}$. This agrees with the observed radii of white dwarfs. This radius is close to the radius of the Earth (6378 km). Hence, we conclude that a white dwarf with mass of the Sun will have a radius of 4155 km and dominated by an electric field of $10^{16}\ \text{V/m}$ as obtained by Leiby [6] and Jordan [11]. These are typical values for the observed white dwarfs. Hence, hydrogen atom in white dwarfs are in state of a fractional quantum number, $n = \sqrt{\alpha}$. The electrons comprising the white dwarf are at a distance equals to the Compton wavelength of the electron in the hydrogen atom.

One can estimate the internal temperature of the electron by equating the energy in Equation (19) to $k_B T_d$, hence one finds $T_d = 2.16\times10^7\ \text{K}$.

4. ORDINARY STARS

An ordinary star (e.g., the Sun) is the one which is supported by hydrogen gas (with $n = 1$) under gravitational compression. The ground state of hydrogen atom is defined by

$$r_s = a_0, v_s = \alpha c, E_n = -\frac{mc^2}{2}\alpha. \tag{32}$$

The electric field is given by

$$E_s = \frac{ke}{a_0^2} = \frac{k^3 m^2 e^5}{h^4}. \tag{33}$$

This state is characterized by having an electric energy density of

$$u_s = \frac{1}{2}\varepsilon_0 E_s^2 = \frac{k^5 m^4 e^{10}}{8\pi h^8} = P_s. \tag{34}$$

This amounts to a value of 4.65×10^{12} Pa. We see from Equations (7), (23) and (33) that the electric fields of the three systems are

$$E_d = \alpha^2 E_q, E_s = \alpha^2 E_d. \tag{35}$$

The balance between the gravitational pressure and the internal hydrogen pressure will produce a stable star. Equating Equations (10) and (34) yields the radius of a stable star as

$$R_s = \left(f_s \frac{8\pi h^8 G}{k^5 e^{10} m^2}\right)^{1/4}\left(\frac{M}{m}\right)^{1/2}. \tag{36}$$

The above formula can be used as a benchmark to calculate the radii of stars from their respective masses or vice versa. To reproduce the Sun's radius,

$f_s = 0.0011 = \frac{3}{80\pi^3}$. Thus, Equation (36) can be written as

$$R_s = 4.707\times 10^{-22}\left(\frac{M}{m}\right)^{1/2} (\text{km}).$$

We remark here that the physical properties of ordinary metals are similar to ordinary stars (e.g., Sun). It is reported by Weir et al. that hydrogen becomes a metal at a pressure greater than 10^{11} Pa [14].

Using Equations (14) and (33), the conductivity in this state can be written as

$$\sigma_s = \frac{nh^3}{m^2 k^2 e^2} = \frac{2}{1}\left(\pi a_0^2 n\right)\frac{e^2}{h}. \tag{37}$$

For neutron and white dwarf stars, one finds

$$\sigma_q = \frac{2}{\alpha}\left(\pi r_q^2 n\right)\frac{e^2}{h}, \quad \sigma_d = \frac{2}{\sqrt{\alpha}}\left(\pi r_d^2 n\right)\frac{e^2}{h}. \tag{38}$$

Equations (37) and (38) can be combined to give

$$\sigma_q = \frac{2}{\gamma}\left(\pi r_q^2 n\right)\frac{e^2}{h}, \tag{39}$$

where $\gamma = 1$, α, $\sqrt{\alpha}$, correspond to the quantum number defining the hydrogen atom in the three states defined before (ordinary, neutron and white dwarf stars). These forms can be related to Ohm's law for a conductor of length L and cross-sectional area A, whose resistance is R, by

$$\sigma = \frac{L}{A}\frac{1}{R}, \frac{L}{A} \Rightarrow 2\pi a_0^2 n, \frac{1}{R} \Rightarrow \frac{1}{\gamma}\frac{e^2}{h}. \tag{40}$$

In FQHE the quantum of conductance is $\frac{e^2}{h}$. The FQHE is thought to result from many-electron interactions in 2-dimensional systems under low temperature and strong magnetic field. Here, the FQHE may take place in these systems due to the high compression of their matter as shown by Morawicz et al. [15].

If we consider the number density in hydrogen as,

$n = \frac{3}{4\pi a_0^3}$, then Equation (37) yields $n = 1.6\times 10^{30}\,\mathrm{m}^{-3}$ so that $\sigma_s = 1.09\times 10^6\,\Omega^{-1}\mathrm{m}^{-1}$. One can find the relaxation time for ordinary star medium (atoms) from equating the conductivity in Equation (36) to $\sigma = \frac{ne^2\tau}{m}$. This yields, $\tau_s = \frac{a_0}{\alpha c} = \frac{\tau_d}{\alpha}$.

Using Equations (4), (14) and (15), the Hall coefficient $\left(R_H = \frac{E}{JB}\right)$ for a neutron is

$$R_H = \frac{1}{ne}. \tag{41}$$

Using Equations (14), (21) and (28), the Hall coefficient for a white dwarf star is

$$R_H = \frac{1}{\alpha}\frac{1}{ne}. \tag{42}$$

Using Equations (14), (32) and (37), the Hall coefficient for an ordinary star

$$R_H = \frac{1}{\alpha^2}\frac{1}{ne}. \tag{43}$$

Equations (41), (42) and (43) clearly show that the Hall effect exists in astrophysical systems with quantized coefficient in powers of α. It is interesting to notice that the radii for ordinary, neutron and white dwarf stars can be related to each other. Therefore, using Equations (13), (31), (35) and (36), one obtains

$$R_q = \alpha\sqrt{\frac{M_q}{M_d}}R_d, R_d = \alpha\sqrt{\frac{M_d}{M_s}}R_s. \tag{44}$$

Owing to Equations (31) and (44), Sirius B of radius 5846 km will have a mass of $1.15M_e$ instead of $0.978M_e$ as obtained by Holberg et al. [16] and Liebert et al. [17]. However, Equations (31) and (44) agree with the Landsman et al. [18] findings of the hot white-dwarf companions.

We provide in **Table 1** the relation between different quantities describing neutron stars, white dwarfs and ordinary stars.

Table 1. A comparison between neutron, white dwarf and ordinary stars physical properties.

Quantity	Neutron star	White dwarf	Ordinary star
Magnetic field	B_q	$\alpha^{5/2}B_q$	$\alpha^5 B_q$
Electric field	E_q	$\alpha^2 E_q$	$\alpha^4 E_q$
Magnetic moment	μ_q	$\alpha^{-1/2}\mu_q$	$\alpha^{-1}\mu_q$
Radius	R_q	$\alpha^{-1}R_q$	$\alpha^{-2}R_q$
Hall coefficient	R_{Hq}	$\alpha^{-1}R_{Hq}$	$\alpha^{-2}R_{Hq}$

5. CONCLUSION

We have proposed in this paper that fractional quantum numbers exist in hydrogen atom. These are powers of the fine structure constant. The first number gives rise to the physical properties of the neutron star, and the second one to the white dwarf stars. These fractional quantum numbers were first shown to exist in quantum Hall effect. We propose a quantum of inductance and capacitance in addition to the quantum of flux and conductance. We have shown in this paper that neutron and white dwarf star got the maximum possible conductivity in nature. Hence, these stars are examples of giant perfect conductors at a relatively high temperature and magnetic field.

A. I. Arbab Moreover, these systems exhibit a fractional quantum effect. The Hall coefficient of these systems are quantized in terms of α . The radii of these systems are also intimately connected. It is important to point out that the other set of fractional values of n led to un physical situation.

REFERENCES

1. S. Chandrasekhar, "The Density of White Dwarf Stars," Philosophical Magazine, Vol. 11, No. 7, 1931, pp. 592- 596.
2. A. I. Arbab, "The Neutron Model," The Hadronic Journal, Vol. 33, 2011, p. 695.
3. D. C. Tsui, H. L. Stormer and A. C. Gossard, "Two-Dimensional Magnetotransport in the Extreme Quantum Limit," Physical Review Letters, Vol. 48, 1982, pp. 1559-1562.
4. R. H. Fowler, "On Dense Matter," Monthly Notices of the Royal Astronomical Society, Vol. 87, 1926, pp. 114-122.
5. G. Wunner, "Matter Under Extreme Conditions; Exotic QED Processes and Atoms in Strong External Fields," Lecture Notes in Physics, Vol. 440, 1994, pp. 101-151.
6. C. C. Leiby, "The Fields of Pulsars and Magnetic White Dwarf Stars," Astrophysics and Space Science, Vol. 122, No. 1, 1986, pp. 151.
7. J. D. Jackson, "Classical Electrodynamics," 2nd Edition, John Wiley & Sons, New York, 1975.
8. A. C. Phillips, "The Physics of Stars," 2nd Edition, Wiley, New York, 1999.
9. A. I. Arbab, "On the Electric and Magnetic Properties of Conductors," Advanced Studies in Theoretical Physics, Vol. 5, No. 12, 2011, pp. 595-607.
10. D. Page, M. Prakash, J. M. Lattimer and A. W. Steiner, "Rapid Cooling of the Neutron Star in Cassiopeia A Triggered by Neutron Superfluidity in Dense Matter," Physical Review Letters, Vol. 106, No. 8, 2011, pp. 081101-081104.
11. S. Jordan, "White Dwarfs," Proceedings of the Royal Society A, Vol. 97, 1920, p. 374.
12. P. Fassbinder and W. Schweizer, "Stationary Hydrogen Lines in Magnetic and Electric Fields of White Dwarf Stars," Astronomy & Astrophysics, Vol. 314, No. 2, 1996, pp. 700-706.
13. P. Drude, "Zur Elektronentheorie der Metalle," Annalen der Physik, Vol. 306, No. 3, 1900, pp. 566-613.
14. S. T. Weir, A. C. Mitchell and W. J. Nellis, "Metallization of Fluid Molecular Hydrogen at 140 GPa," Physical Review Letters, Vol. 76, No. 11, 1996, pp. 1860-1863.

15. N. G. Morawicz, K. W. J. Barnham and C. Zammit, "Observation of the Fractional Quantum Hall Effect under Hydrostatic Pressure," Physical Review B, Vol. 41, No. 18, 1990, pp. 12687-12692.
16. J. B. Holberg, M. A. Barstow, F. C. Bruhweiler, A. M. Cruise and A. J. Penny, "Sirius B: A New, More Accurate View," The Astrophysical Journal, Vol. 497, No. 2, 1998, pp. 935-942.
17. J. Liebert, P. A. Young, D. Arnett, J. B. Holberg and K. A. Williams, "The Age and Progenitor Mass of Sirius B," The Astrophysical Journal, Vol. 630, No. 1, 2005, pp. L69-L72.
18. W. Landsman, T. Simon and P. Bergeron, "The Hot White-Dwarf Companions of HR 1608, HR 8210, and HD 15638," Astronomical Society of the Pacific, Vol. 105, No. 690, 1999, pp. 841-847.

NOTES

[1]The gravitational energy density is $U = \frac{1}{8\pi}\frac{GM^2}{R^4}$. The pressure density is $\frac{2}{3}U$.

CHAPTER 16

Variation of the Structural Properties of IV Element Nano Clusters Due to Tensile Stress

Zainelabideen Y. Mijbil[1], Hamed I. Aboud[2], Ahmed M. Abdul-Lettif[2*]

[1]Basic Science Department, Veterinary Medicine College, University of Babylon, Hilla, Iraq; [2]Physics Department, College of Science, University of Babylon, Hilla, Iraq.

ABSTRACT

CNDO/2 method with 27 unit cells composed of 216 atoms, has been used for calculating and studying the lattice constant, density, total energy, cohesive energy, bulk modulus, microhardness, Young modulus, speed of sound, acoustic phonons energy, plasmon energy and vibrational energy, for diamond, silicon, germanium, and gray tin subjected to tensile stress. And to investigate the categories of these clusters; do these clusters have properties of nano materials or properties of bulk materials? It is found that most values of the investigated properties decrease with tensile stress except the lattice constant and the vibrational energy. In addition, two simple formulas have been established for both the bulk modulus and plasmon energy. It is found that the investigated clusters show bulk behavior.

Keywords: IV Elements; Structural Properties; Pressure; Nano Materials

1. INTRODUCTION

The general properties of IV materials make them useful in many fields [1-3], such as the field of electronics [4-6]. So these materials have been extensively investigated [7- 13]. The variation of material properties with pressure is an interesting field [14-15]. So, the first motive of the present work is to investigate the effect of pressure on some properties of diamond, silicon, germanium, and gray tin. The second motive is to answer the following question: Could the 216 atom clusters of these elements be regarded as nano materials or bulk materials concerning the structural properties?

A LUC (Large Unit Cell) composed of 27 unit cells (each of these units contains 8 atoms) has been used in the present analysis. The CNDO/2 method with STO (Slater Type Orbital) and LCAO (Linear Combination of Atomic Orbitals) approximation [16] was used to calculate the lattice constant, cohesive energy, total energy, and bulk modulus of the aforementioned elements. The Young modulus, microhardness, speed of sound, energy of phonons and plasmons have been also determined.

2. CALCULATIONS

The orbital exponent (ξ), bonding parameter (β), s-shell electronegativity (EES), and p-shell electronegativity (EEP), shown in **Table 1** for the four elements, are the pa rameters used in the CNDO/2 method [16] to calculate the lattice constant (a_o), cohesive energy (E_c) and total energy (E_o) in the ground state (zero pressure-temperature) [z-PT].

The calculated structural properties compared with other results are shown in **Table 2**. Our results were fitted according to Equation (1), the state equation of Murnaghan, in order to estimate the change of energy E(V) with pressure [30].

$$E(V) = E_o + \frac{B_o V}{B'_o}\left[\frac{(V_o/V)^{B'_o}}{B'_o - 1} + 1\right] - \frac{B_o V_o}{B'_o - 1} \tag{1}$$

Table 1. Parameters of CNDO/2 for C, Si, Ge and α-Sn.

			Diamond		
Par. (eV)	Pres.	Ref. [16]	Ref. [17]	Ref. [18]	Ref. [19]
ξ	1.76	1.652	1.83	1.765	1.820
$-\beta$	10.48	21	10	10.2	10.24
$-EES$	7.3	14.05	5.57	7	6.220
$-EEP$	5.55	5.572	4.39	5.5	4.352
			Silicon		
Par. (eV)	Pres.	Ref. [16]	Ref. [17]	Ref. [18]	Ref. [20]
ξ	1.56	2.075	1.635	1.54	1.59
$-\beta$	14.41	13.06	5.45	6.4	6.2
$-EES$	6.755	10.03	7.25	6.3	7.25
$-EEP$	4.479	4.133	4.95	4.5	5
			Germanium		
Par. (eV)	Pres.	Ref. [16]	Ref. [17]	Ref. [21]	
ξ	1.304	1.527	1.98	1.98	
$-\beta$	9.35		5.35	5.352	
$-EES$	9.35		9.35	9.357	
$-EEP$	5.4		5.4	5.404	
			Grey tin		
Par. (eV)	Pres.	Ref. [16]	Ref. [17]	Ref. [22]	
ξ	1.3	1.425	2.05	1.996	
$-\beta$	13.03		5.35	5.334	
$-EES$	9.35		9.46	9.452	
$-EEP$	4.4		4.69	5.632	

Table 2. The structural properties of C, Si, Ge, and α-Sn.

Element	a_o (Å)			B_o (GPa)		
	Pres	Theo	Exp	Pres	Theo	Exp
C	3.556	3.56 [19] 3.56 [17]	3.556 [23]	433.59	520 [19] 808 [17]	433 [23]
Si	5.4309	5.43 [17]	5.431 [23]	98.93	149.5 [17] 180.8 [20]	98 [23]
Ge	5.6579	5.65 [17] 5.65 [21]	5.658 [23]	86.096	137.8 [17] 103 [21]	75 [23]
Sn	6.4809	6.46 [17] 6.487 [22]	6.481 [23]	53.084	96.6 [17] 100 [22]	53 [23]

Element	E_{coh} (eV)			v_e m/s		
	Pres.	Theo.	Exp.	Pres.	Theo.	Exp.
C	7.3722	7.36 [19] 7.55 [17]	7.37 [24]	11089.155	15159 [17]	19161 [19]
Si	4.6319	4.68 [17]	4.63 [24]	6517.4732	8010 [17] 6600 [25]	9150 [25]
Ge	3.8501	3.85 [17] 3.83 [21]	3.85 [24]	4022.8669	5070 [17] 3830 [25]	5400 [25]
Sn	3.1447	3 [17] 3.14 [22]	3.14 [24]	3033.9389	4095 [17]	

Element	P_{max} (GPa)	$\rho_o \times 10^{-3}$ (Kg/m^3)	E_{ZPV} (eV)	E_{pl} (eV)		
				Pres.	Theo.	Obs.
C	1110 [26]	3.526 [27]	0.215 [24]	31.19739		
Si	11.7[h]	2.329 [27]	0.065 [24]	16.60106	16.4 - 16.9 [27]	16 [27]
Ge	10.6[h]	5.32 [27]	0.036 [24]	15.640944	16 - 16.4 [27]	16 [27]
Sn	1[i]	5.767 [27]	0.021 [24]	12.713636		

Element	B'_o	Θ (K)
C	3.553 [23]	2220 [24]
Si	4.153 [23]	674 [24]
Ge	4.678 [23]	374 [24]
Sn	4.804 [23]	220 [24]

This change, using Equation (2), determines the change of vibrational energy (F_{vib}) [26] as below:

$$P_T = -\frac{\partial E(V)}{\partial V} - \frac{\partial F_{vib}}{\partial V} \tag{2}$$

where P_T is the isothermal pressure, V_o and B_o are the volume and bulk modulus of the unit cell at [z-PT], and V is the volume at any pressure.

The bulk modulus at [z-PT] is calculated using Cohen empirical formula [31]:

$$B_o = \frac{1970 - 200I}{d^{3.5}} \tag{3}$$

where I is the ionicity factor which equals 0, 1 and 2 for IV, III-V and II-V groups respectively [31], and d (= 0.433 a) is the intratomic distance [23], where a is the lattice constant at any pressure.

Then both the isothermal bulk modulus at any pressure (B_T), Equation (4) [32], and the derivative of bulk modulus at zero pressure (B'_o), Equation (5) [23,33], have been calculated as:

$$B_T = -V\frac{\mathrm{d}P_T}{\mathrm{d}V} \tag{4}$$

$$B'_o = \left.\frac{\mathrm{d}B}{\mathrm{d}P}\right|_{P=0} \tag{5}$$

By solving Equations (4) and (5), one can get:

$$V = V_o \exp(-\varphi) \tag{6}$$

where

$$\varphi = \frac{P_T}{B_o + B'_o P_T} \tag{7}$$

The results of volume vs pressure are shown in Figure 1, in comparison with:

Ÿ Murnaghan equation of state EOS [34] of pressure (P_M);

$$P_M = \frac{B_o}{B'_o}\left[\left(\frac{V_o}{V}\right)^{B'_o} - 1\right] \tag{8}$$

Ÿ Birch-Murnaghan EOS of pressure (P_{BM});

$$P_{BM} = \frac{3}{2}B_o\left[\left(\frac{V_o}{V}\right)^{\frac{7}{3}} - \left(\frac{V_o}{V}\right)^{\frac{5}{3}}\right]\left\{1 - \frac{3}{4}(4 - B'_o)\left[\left(\frac{V_o}{V}\right)^{\frac{2}{3}} - 1\right]\right\} \tag{9}$$

Ÿ Vinet EOS of pressure (PBM) [35];

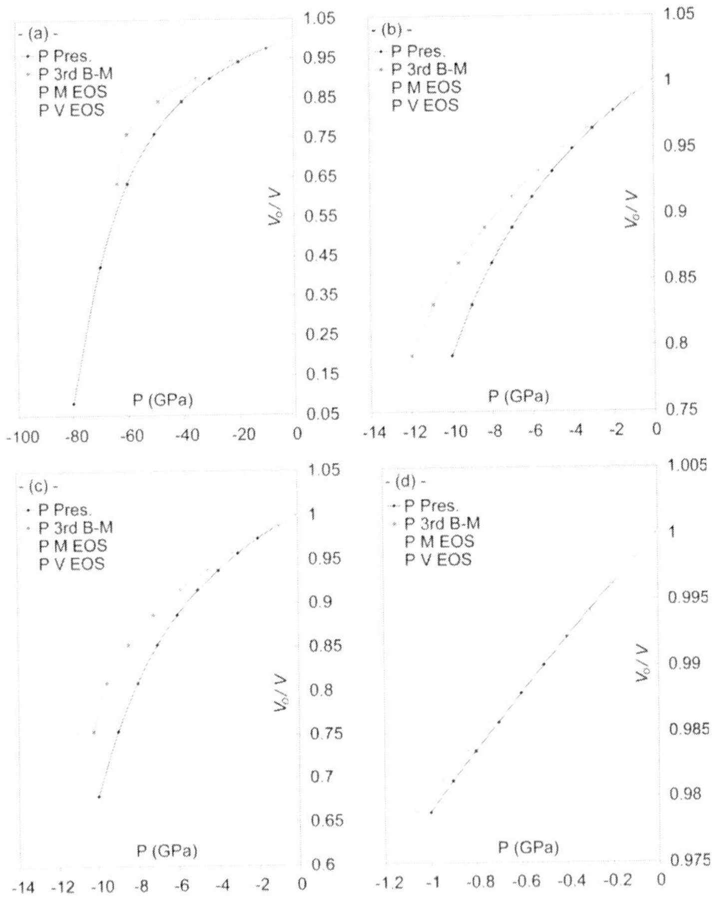

Figure 1. The dependence of volume upon the tensile stress, and a comparison between our general relation and each of Murnagham EOS, Birch-Murnagham EOS and Vinet EOS, for: (a) C; (b) Si; (c) Ge; and (d) α-Sn.

$$P_V = 3B_o\left[1-\left(\frac{V}{V_o}\right)^{\frac{1}{3}}\right]\left(\frac{V}{V_o}\right)^{-\frac{2}{3}}\exp\left\{\frac{3}{2}\left(B'_o-1\right)\left[1-\left(\frac{V}{V_o}\right)^{\frac{1}{3}}\right]\right\} \quad (10)$$

To get a wide understanding of the four elements properties, the microhardness (H) [36] and the Young modulus (E) [37] are also calculated as below:

$$H = (\zeta)^2 \frac{P_T}{3(1-\zeta)} \quad (11)$$

$$E = \frac{P_T}{\varepsilon} \quad (12)$$

where ζ is (a/a_o) and ε is the strain. Also the speed of sound (υ_s) is calculated as [29]:

$$\upsilon_s \approx \sqrt{\left(\upsilon_{os}^2 + \frac{B'_o P_T}{\rho_o}\right)\exp(-\varphi)} \quad (13)$$

where ρ_o and υ_{os} are the density and the velocity of sound at [z-PT], respectively. The velocity of sound equals the average of phonons velocity ($\bar{\upsilon}_{ph}$) [38]. And from [30], the phonon velocity, angular frequency ($\bar{w}_{ph}$), phonon energy ($\bar{E}_{ph}$) and the plasmon energy (E_{pl}) equal:

$$\bar{\upsilon}_{ph} = \upsilon_s \quad (14)$$

$$\bar{w}_{ph} = \left(\frac{6\pi^2 N}{V}\right)^{\frac{1}{3}} \upsilon_s \quad (15)$$

$$\bar{E}_{ph} = h\bar{w}_{ph} \quad (16)$$

$$E_{pl} = h\sqrt{\frac{n_e e^2}{\varepsilon_o m}} \quad (17)$$

where n_e is the volume of the bond which equals [39]:

$$n_e = \frac{3\sqrt{3}}{2d^3} \tag{18}$$

and N is the number of atoms in the unit cell which equals 8, e and m are the charge and the mass of electron respectively, and ε_0 is the permittivity of vacuum. By gathering Equations (15), (16) and (6), one can get:

$$\bar{E}_{ph}(eV) = \frac{hk}{e}\sqrt{\upsilon_{so}^2 + \frac{B_o' P_T}{\rho_o}\exp\left(-\frac{\varphi}{6}\right)} \tag{19}$$

$$k = \sqrt[3]{\frac{6\pi^2 N}{V_o}} \tag{20}$$

Gathering Equations (17), (18) and (6), one can get:

$$E_{pl}(eV) = 210.0563\sqrt{\frac{\exp(\varphi)}{V_o}} \tag{21}$$

3. RESULTS AND DISCUSSION

The results of the present work illustrated in **Table 2**, are in good agreement with experimental results of bulk materials. So we can assume that the structural properties of 216 atom clusters (1.21 nm, 4.32 nm, 4.89 nm, and 7.35 nm for C, Si, Ge, and Sn respectively) are properties of the corresponding bulk materials.

As shown in Figures 2 and 3, the increase of tensile stress leads to an increase of volume and a decrease in the density. Strain, caused by tensile stress, gives the following results:

1) Alterations of the orbitals [39], and an increase of the distance between atoms [40].

2) A decrease of the bond strength [41] which depends on the distance [42].

3) A decrease of the bulk modulus [43]. It is known that the bulk modulus represents the resistance of matter to the change in volume [44], and it can be considered as the sum of the bulk modulus of all atoms [45] according to the AIM (atoms in molecules) theory [46]. Because the bulk modulus depends directly upon the density [47] and the charge distribution between

atoms [48,49], i.e. the bonds [44], the bulk modulus decreases with the tensile stress.

4) A decrease of the Young modulus because it depends upon the bulk modulus [47].

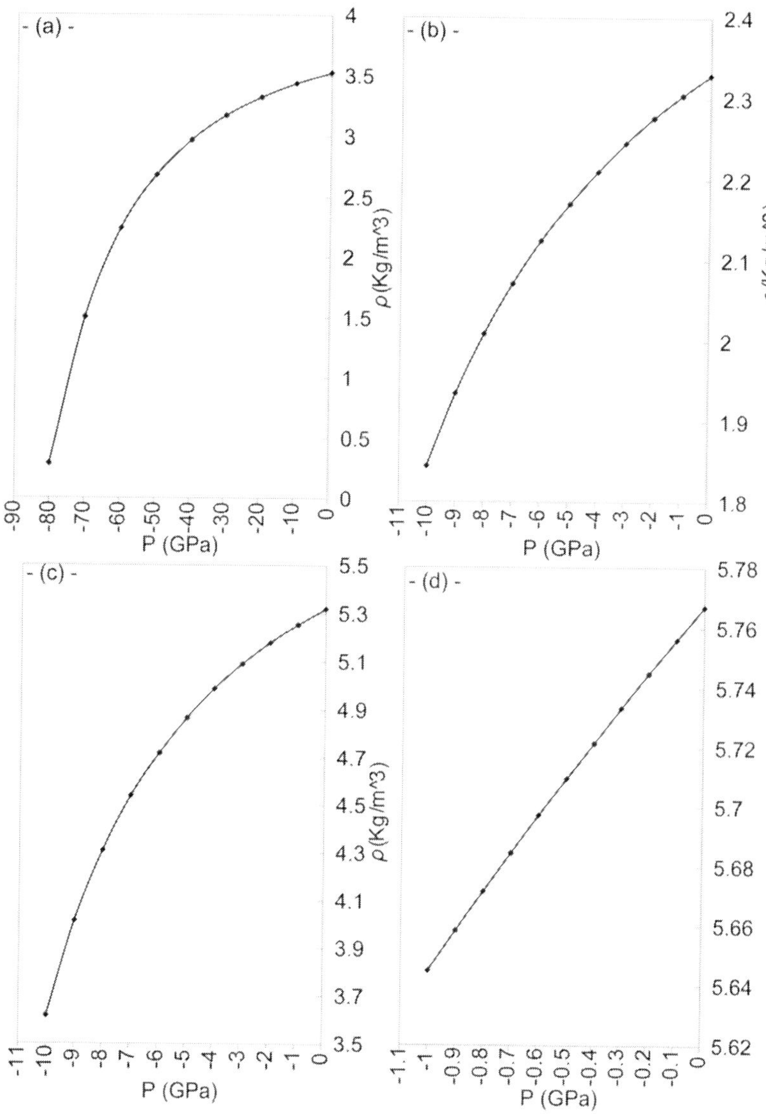

Figure 2. The change of density with tensile stress for: (a) C; (b) Si; (c) Ge; and (d) α-Sn.

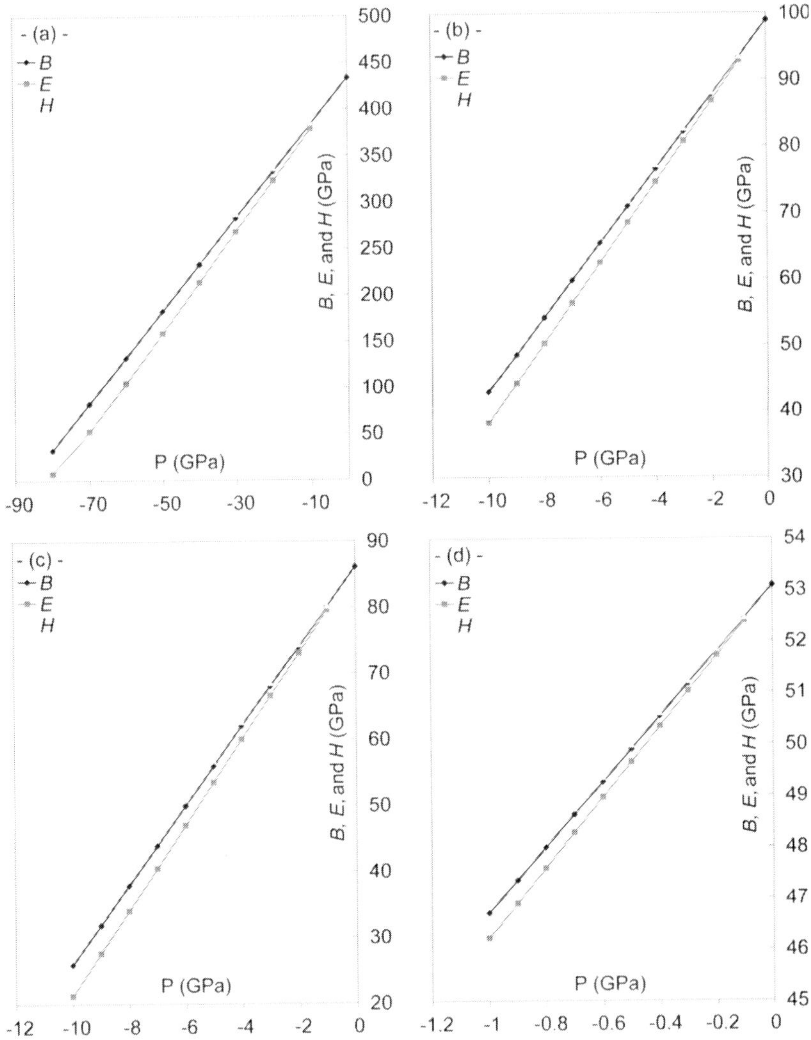

Figure 3. Effect of the tensile stress on the bulk modulus, Young's modulus and the microhardness for: (a) C; (b) Si; (c) Gr; and (d) α-Sn.

5) An increase of the rigidity because the tensile stress decreases the microhardness [50].

As shown in Figures 4-6, the speed of sound, the energy of phonon and the energy of plasmon decrease when the tensile stress increases.

The sound waves produce charge distribution, according to Kornyushin who has showed that the plasma affects the charge distribution and so the

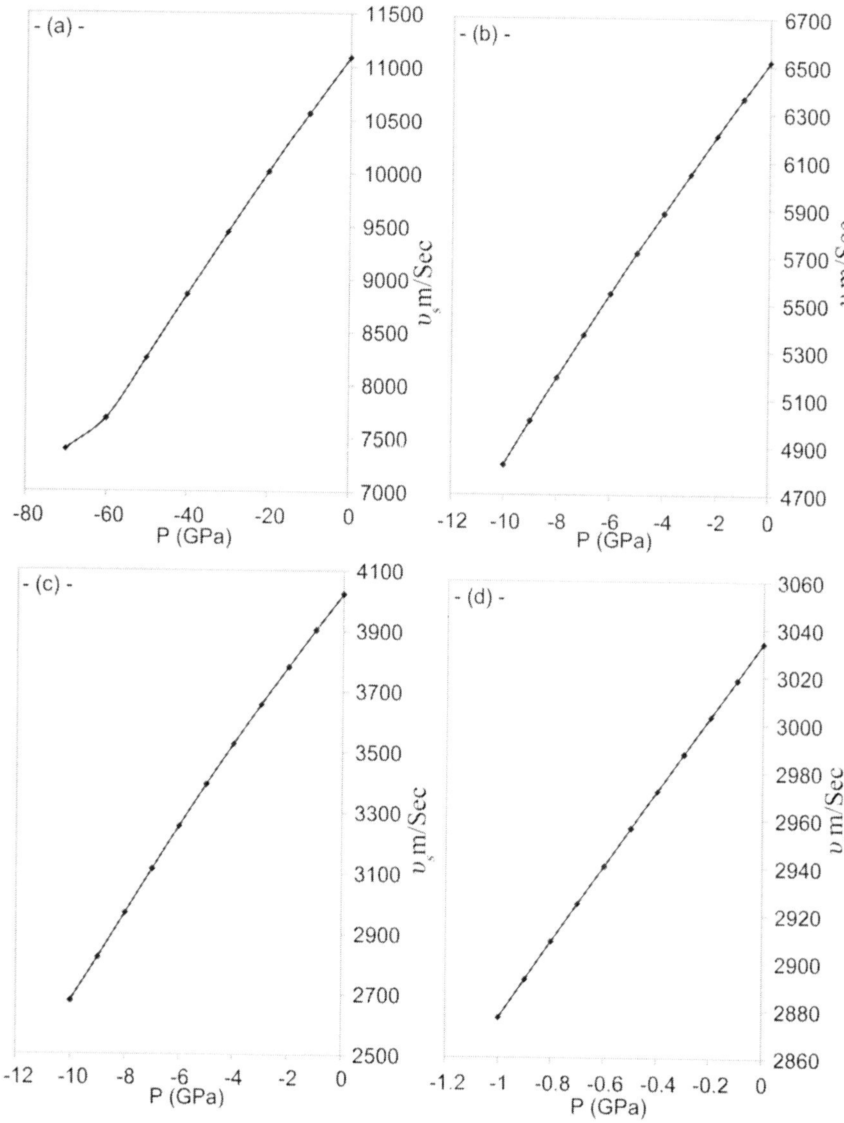

Figure 4. The change of speed of sound with tensile stress for: (a) C; (b) Si; (c) Gr; and (d) α-Sn.

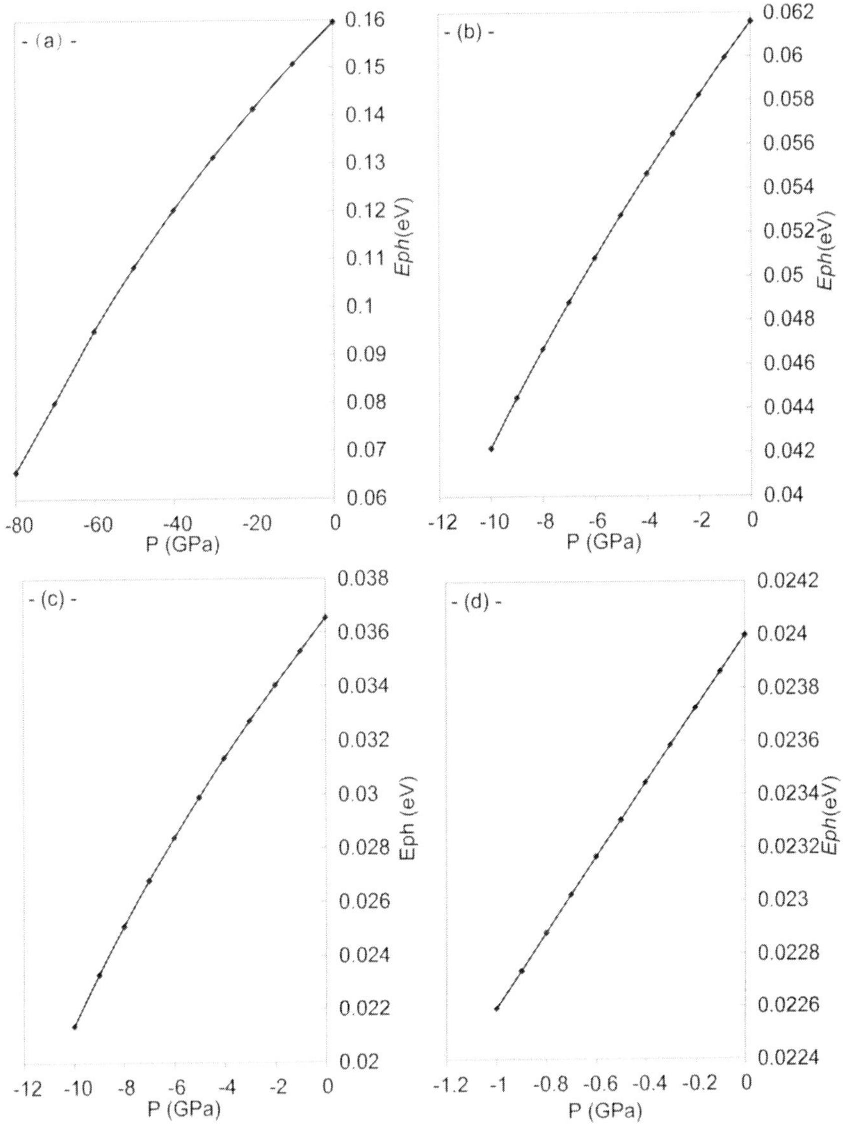

Figure 5. The change of speed of sound with tensile stress for: (a) C; (b) Si; (c) Gr; and (d) α-Sn.

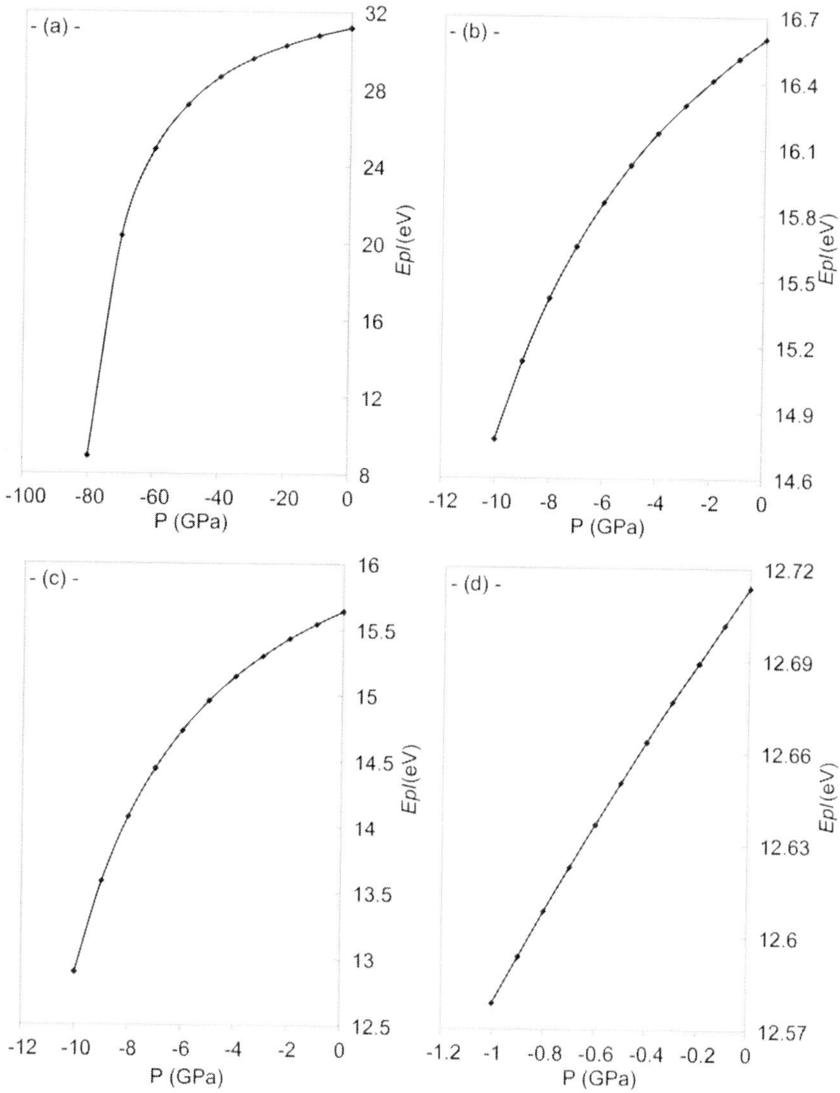

Figure 6. The change of plasmon energy with tensile stress for: (a) C; (b) Si; (c) Gr; and (d) α-Sn.

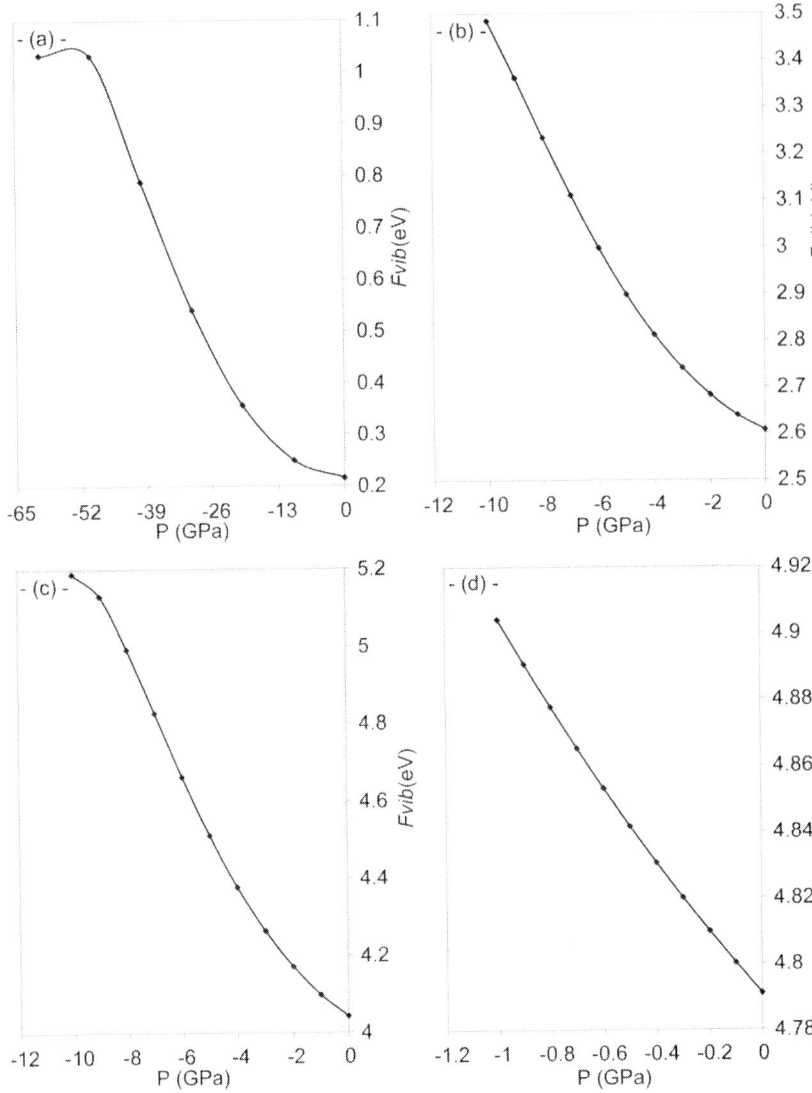

Figure 7. Effect of tensile stress on the vibrational energy for: (a) C; (b) Si; (c) Gr; and (d) α-Sn.

sound wave [51]. On the other hand, the phonon frequency affects the effective charge [52], because the effective charge depends upon the charge distribution and determines the plasma frequency [53]. According to Sanjurjo, the compressive pressure increases the effective charge for IV-IV materials and vice verse [52]. Meanwhile, depending on Burstein who has illustrated that the decrease in the effective charge decreases the plasmon frequency [53], one can find that our results of the behavior are in good agreement with experiment [52] for both the plasmon and phonon energies.

It is shown from **Figure 7** that the vibrational energy increases as the tensile stress increases. Many properties of solids depend on the phonons which determine the atoms vibration [54], where the velocity of sound depends on the atoms arrangement [55]. The crystal is considered as a group of harmonic oscillators [56], according to Bornvon Kármán theory, in order that the atoms in the lattice interplay with the nearest atoms only [57]. From all above and according to Polit who mentioned that the change of the lattice constant changes the charge distribution which affects the atoms' force field and these potentials modify the vibration energy [58].

4. CONCLUSIONS

From the results of the present work, it can be concluded that:

1) The 216 atom clusters of C, Si, Ge, and α-Sn can be considered as bulk materials concerning the structural properties.
2) The range of the elastic deformation for diamond is the smallest.
3) The higher the tensile stress, the higher the rigidity of the sample.
4) The inclusion of the bulk modulus in every change may be very important in some properties because it can limit their variation range caused by external factors.

Zainelabideen Y. Mijbil, Hamed I. Aboud, Ahmed M. Abdul-Lettif 5) The relation of phonons gives approximated results, however, it is simple and could be used for behavior study.

REFERENCES

1. F. Cataldo, "Carbon Allotropy and Carbon Black," Kautschuk Gummi Kunststoffe, Vol. 54, No. 1-2, 2001, pp. 22-28.
2. C. W. Myles, J. Dong and O. F. Sankey, "Structural and Electronic Properties of Tin Clathrate Materials," Physical Review B, Vol. 64, No. 16, 2001, pp. 165202-165212.
3. M. Gaith and I. Alhayek, "Correlation between Overall Elastic Stiffness, Bulk Modulus and Interatomic Distance in Anisotropic Materials: Semiconductors," Reviews on Advanced Materials Science, Vol. 21, No. 2, 2009, pp. 183-191.
4. M. Gabrysch, "Electronic Properties of Diamond," Uppsala University, 2008. http://urn.kb.se/resolve?urn=urn:nbn:se:uu:diva-8675

5. M. Virgilio, G. Pizzi and G. Grosso, "Optical Gain in Short Period Si/Ge Superlattices on [001]-SiGe Substrates," Journal of Applied Physics, Vol. 110, No. 8, 2011, pp. 083105-083112.
6. M. Upadhyay, S. Murugavel, M. Anbarasu and T. R. Ravindran, "Structural Study on Amorphous and Crystalline State of Phase Change Material," Journal of Applied Physics, Vol. 110, No. 8, 2011, pp. 083711-083716.
7. R. E. Stallcup II, L. M. Villarreal, S. C. Lim, I. Akwani, A. F. Aviles and J. M. Perez, "Atomic Structure of the Diamond (100) Surface Studied Using Scanning Tunneling Microscopy," Journal of Vacuum Science and Technology B, Vol. 14, No. 2, 1996, pp. 929-932.
8. B. K. Serdega, E. V. Nikitenko and Prikhodenko, "Effect of Surface Condition on Strain in Semiconductor Crystal Sample," Semiconductor Physics, Quantum Electronics and Optoelectronics, Vol. 4, No. 1, 2001, pp. 9-11.
9. J. Pollmann, P. Krüger, M. Rohlfing, M. Sabisch and D. Vogel, "Ab Initio Calculations of Structural and Electronic Properties of Prototype Surfaces of Group IV, III-V and II-VI Semiconductors," Applied Surface Science, Vol. 104-105, 1996, pp. 1-16.
10. A. Otero-de-la-Roza and V. Luaña, "Topological Characterization of the Electron Density Laplacian in Crystals. The Case of the Group IV Elements," Journal of Chemical Theory and Computation, Vol. 6, No. 12, 2010, pp. 3761-3979.
11. L. A. Woldering, L. Abelmann and M. C. Elwenspoek, "Predicted Photonic Band Gaps in Diamond-Lattice Crystals Built from Silicon Truncated Tetrahedrons," Journal of Applied Physics, Vol. 110, No. 4, 2011, pp. 043107- 043114.
12. Q. Zhou, X. Hu, K. Al-Hemyari, K. McCarthy, L. Domash and J. A. Hudgings, "High Spatial Resolution Characterization of Silicon Solar Cells Using Thermoreflectance Imaging," Journal of Applied Physics, Vol. 110, No. 5, 2011, pp. 053108-053113.
13. J. Kim and D. Ahn, "Effect of Indirect Interband Absorption in Ge/SiGe Quantum Wells," Journal of Applied Physics, Vol. 110, No. 8, 2011, pp. 083119-083128.
14. C. Q. Sun, "Size Dependence of Nanostructures: Impact of Bond Order Deficiency," Progress in Solid State Chemistry, Vol. 35, No. 1, 2007, pp. 1-159.
15. M. M. Shokrieh and R. Rafiee, "Prediction of Young's Modulus of Graphene Sheets and Carbon Nanotubes Using Nanoscale Continuum Mechanics Approach," Materials and Design, Vol. 31, No. 2, 2010, pp. 790-795.
16. J. A. Pople and D. L. Beveridge, "Approximate Molecular Orbital Theory," McGraw-Hill, New York, 1970.
17. M. A. Abdulsattar, "Self Consistent Field Calculations of Covalent Semiconductors," Ph.D. Thesis, University of Baghdad, Baghdad, 1997.
18. B. S. M. Baaee, "Theoretical Process to Study the Band Structure of Germanium Crystal by Use the Intermediate Neglect of Differential Overlap Method," M.Sc. Thesis, University of Babylon, Babylon, 2003.
19. I. O. R. Al-Jelawy, "Effect of Pressure and Temperature on Some Properties of Diamond Crystal Using HartreeFock Method," M.Sc. thesis, University of Babylon, Babylon, 2005.
20. M. M. Al-Hello, "Study of the Effect of Pressure on the Properties of Si-Crystal Using Semiempirical Method," M. Sc. Thesis, University of Babylon, Babylon, 2004.
21. A. H. Harker and F. P. Larkins, "A Large Unit Cell Semiempirical Molecular Orbital Approach to the Properties of Solids II. Covalent Materials: Diamond and Silicon," Journal of Physics C: Solid State Physics, Vol. 12, 1979, pp. 2497-2507.
22. M. G. Merdan, "Self-Consistent Field Calculations for the Effect of Pressure and Temperature on Some Properties of Grey Tin Crystal," M.Sc. Thesis, University of Babylon, Babylon, 2005.
23. S. Q. Wang and H. Q. Ye, "Plane-Wave Pseudopotential Study on Mechanical and Electronic Properties for IV and III-V Crystalline Phases with Zinc-Blende Structure," Physical Review B, Vol. 66, No. 23, 2002, pp. 235111- 235117.
24. S. H. Lee, J. H. Kang and M. H. Kang, "Structural Properties of Semiconductors in the Generalized Gradient Approximation," Journal of the Korean Physical Society, Vol. 31, No. 3, 1997, pp. 811-814.

25. J. S. Blakemore, "Solid State Physics," 2nd Edition, Cambridge University Press, Cambridge, 1985.
26. J. Xie, S. P. Chen, J. S. Tse, S. de Gironcoli and S. Baroni, "High-Pressure Thermal Expansion, Bulk Modulus, and Phonon Structure of Diamond," Physical Review B, Vol. 60, No. 13, 1999, pp. 9444-9449.
27. C. Kittel, "Introduction to Solid State Physics," 7th Edition, John Wiley and Sons, Hoboken, 1996.
28. A. Mujica, A. Rubio, A. Munoz and R. J. Need, "High Pressure Phaces of Group-IV, III-V, and II-VI Compounds," Reviews of Modern Physics, Vol. 75, No. 3, 2003, pp. 863-907.
29. B. Akdim, D. A. Papaconsttantopoulos and M. J. Mehl, "Tight-Binding Description of the Electronic Structure and Total Energy of Tin," Philosophy Magazine B, Vol. 82, No. 1, 2002, pp. 47-61.
30. C. A. Perottoni, A. S. Pereira, and J. A. H. da Jornada, "Periodic Hartree-Fock Linear Combination of Crystalline Orbitals Calculation of the Structure, Equation of State and Elastic Properties of Titanium Diboride," Journal of Physics: Condensed Matter, Vol. 12, No. 32, 2000, pp. 7205-7222.
31. J. Zheng, C. H. A. Haun, A. T. S. Wee, R. Wang and Y. Zheng, "Ground State Properties of Cubic C-BN Solid Solutions," Journal of Physics: Condensed Matter, Vol. 11, No. 3, 1999, pp. 927-935.
32. T. Faisst, "Temperature Dependence of the Thermal Expansion Coefficient, Bulk Modulus and Magnetic Grueneisen Constant of Nickel Near the Curie Point," Journal De Physique, Vol. 49, No. 12, 1988, pp. 65-66.
33. P. E. Van Camp, V. E. Van Doren, and J. T. Devreese, "Ground-State and Electronic Properties of Covalent Solids," Physical Review, Vol. 38, No. 17, 1988, pp. 12675-12678.
34. S. Q. Wang, H. Q. Ye and S. Yip, "First Principles Studies on the Pressure Dependences of the Stress Strain Relationship and Elastic Stability of Semiconductors," Physics: Condensed Matter, Vol. 18, No. 2, 2006, pp. 395- 409.
35. C. Narayana, V. J. Nesamony, and A. L. Ruoff, "Phase Transformation of BeS and Equation-Of-State Studies to 96 GPa," Physical Review B, Vol. 56, No. 22, 1997, pp. 14338-14343.
36. R. Asokamani and R. Rita, "Electronic Structure and Physical Properties of ABC_2 (A = Zn, B = Si, Ge, Sn, and C = P, As) Ternary Penictide Semiconductors," Physica Status Solidi B, Vol. 226, No. 2, 2001, pp. 375-384.
37. W. Kim, M. Kim, Y. Chang, J. Shin and J. Bae, "Fatigue Crack Growth Behavior of NR and HNBR Based Vulcanizates with Potential Application to Track Pad for Heavy Weight Vehicles," Macromolecular Research, Vol. 11, No. 2, 2003, pp. 73-79.
38. G. D. Raj, "Solid State Physics," 1st Edition, Anmol Publication Pvt. Ltd, New Delhi, 2004.
39. V. Kumar and B. S. R. Sastry, "Thermal Expansion Coefficient of Binary Semiconductors," Crystal Research and Technology, Vol. 36, No. 6, 2001, pp. 565-569.
40. J. P. Connerade, P. Kengkan, P. A. Lakshmi and R. Semaoune, "Scaling Laws for Atomic Compressibility," Journal of Physics B: Atomic and Molecular Optical Physics, Vol. 33, No. 22, 2000, pp. L847-L854.
41. S. M. Sze, "Semiconductor Devices: Physics and Technology," Dar Al-Hikma, Mosul, 1990.
42. J. Z. Jiang, H. Lindelov, L. Gerward, K. Ståhl, J. M. Recio, P. Mori-Sanchez, S. Carlson, M. Mezouar, E. Dooryhee, A. Fitch and D. J. Frost, "Compressibility and Thermal Expansion of Cubic Silicon Nitride," Physical Review B, Vol. 65, No. 16, 2002, pp. 161202-161205.
43. Y. Sun, S. E. Thompson and T. Nishida, "Physics of Strain Effects in Semiconductors and Metal-Oxide-Semiconductor Field-Effect Transistors," Journal of Applied Physics, Vol. 101, No. 10, 2007, pp. 104503-104522.

44. M. P. D'Evelyn and T. Taniguchi, "Elastic Properties of Translucent Polycrystalline Cubic Boron Nitride as Characterized by the Dynamic Resonance Method," GE Research & Development Center, General Electric Company, 1998.
45. M. Mattesini and S. F. Matar, "First Principle Characterization of New Ternary Heterodiamond BC_2N Phases," Computational Materials Science, Vol. 20, No. 2, 2001, pp. 107-119.
46. A. M. Pendás, A. Costales, M. A. Blanco, J. M. Recio and V. Luanã, "Local Compressibilities in Crystals," Physical Review B, Vol. 62, No. 21, 2000, pp. 13970-13978.
47. A. M. Pendás, "Stress, Virial, and Pressure in the Theory of Atoms in Molecules," Journal of Chemical Physics, Vol. 117, No. 3, 2002, pp. 965-979.
48. B. T. Wang, P. Zhang, H. L. Shi, B. Sun and W. D. Li, "Mechanical and Chemical Bonding Properties of Ground State BeH_2," The European Physical Journal B, Vol. 74, No. 3, 2010, pp. 303-308.
49. G. E. Fernandez, S. A. Serebrinsky, J. L. Gervasoni and J. P. Abriata, "Calculation of the Pressure Dependence of the Bulk Modulus Using a Jelliumm Model," International Journal of Hydrogen Energy, Vol. 29, 2004, pp. 93-95.
50. S. Barzilai, I. Halevy and O. Yeheskel, "Bulk Modulus of Sc_2O_3: Ab Initio Calculations and Experimental Results," Journal of Applied Physics, Vol. 110, No. 4, 2011, p. 043532.
51. B. Yu, D. Chen, Q. Tang, C. Wang and D. Shi, "Structural, Electronic, Elastic and Thermal Properties of Mg_2Si," Journal of Physics and Chemstry of Sollids, Vol. 71, No. 5, 2010, pp. 758-763.
52. L. C. M. Miranda and D. ter Haar, "Plasma Effects in Sound Amplification in Piezo-Electric Semiconductors," Revista Brasileira de Física, Vol. 2, No. 2, 1972, pp. 77- 86.
53. J. A. Sanjurjo, E. López-Cruz, P. Vogl and M. Cardona, "Dependence on Volume of the Phonon Frequencies and the IR Effective Charges of Several III-V Semiconductors," Physical Review B, Vol. 28, No. 8, 1983, pp. 4579- 4584.
54. E. Burstein, S. Perkowitz and M. H. Brodsk, "The Dielectric Properties of the Cubic IV-VI Compound Semiconductors," Journal De Physique, Vol. 29, No. 11-12, 1968, pp. 78-83.
55. K. Parliński, "First-Principles Calculations of Vibrational and Thermodynamical Properties of Solids," Materials Science-Poland, Vol. 23, No. 2, 2005, pp. 357-363.
56. T. Iitaka and T. Ebisuzaki, "First-Principles Calculation of Elastic Properties of Solid Argon at High Pressures," Riken Review, No. 48, 2001, pp. 12-15.
57. J. Xie, S. de Gironcoli, S. Baroni and M. Scheffler, "First-Principles Calculation of the Thermal Properties of Silver," Physical Review B, Vol. 59, No. 2, 1999, pp. 965-969.
58. C. Malinowska-Adamska, P. Sloma and J. Temaszewski, "Physical Properties of Ni fcc Lattice in Terms of the Self-Consistent Phonon Theory," Journal of Physics: Condensed Matter, Vol. 18, No. 2, 2006, pp. 751-758.
59. J. Polit, E. M. Sheregii, J. Cebulski, B. V. Robouch, A. Marcelli, M. Cestelli Guidi, M. Piccinini, A. Kisiel, P. Zajdel, E. Burattini and A. Mycielski, "Phonon and Vibrational Spectra of Hydrogenated CdTe," Journal of Applied Physics, Vol. 100, No. 1, 2006, p. 013521.

Index